AF522610

BOOLEAN ALGEBRA

BOOLEAN ALGEBRA

By

A.K. Sharma

DISCOVERY PUBLISHING HOUSE
NEW DELHI-110002

Published by:
Namit Wasan

DISCOVERY PUBLISHING HOUSE PVT. LTD.
4383/4B, Ansari Road, Darya Ganj
New Delhi-110 002 (India)
Phone : +91-11-23279245; 23253475; 43596065
E-mail : discoverybooksindia@gmail.com
discoverypublishinghouse@gmail.com
namitwasan9@gmail.com
web : www.discoverypublishinggroup.com

***Edition:* 2020**

ISBN: 978-81-8356-230-0

Boolean Algebra

Printed at:
Infinity Imaging Systems
Delhi

Preface

This book is "Boolean Algebra" is intended to supply the typical freshman and related disciplines with a first exposure to the mathematical topics, essential their study of computer science. The author feel great pleasure in bringing out this book. This book has been written keeping in mind all point at which student feel difficulty. Effort have been made to explain such points in depth, so that students can follow the subject early.

Suggestion for improvement from all corners will be thankfully acknowledge.

Author

CONTENTS

1

BOOLEAN ALGEBRA

INTRODUCTION

In order to define a Boolean Algebra, we need the additional concept of compensation.

Definition: *Complemented Lattice: Let $[L; \wedge, \vee]$ be a lattice which contains a least element, denoted by 0, and a greater element, denoted by 1 (such a lattice is called bounded). $[L; \wedge, \vee]$ is called a complemented lattice if and only if for every element $a \in L$, there exists an element $\bar{a}$ in L such that $a \wedge \bar{a} = 0$ and $a \vee \bar{a} = 1$. The element $\bar{a}$ is called a complement of the element a.*

Example 1:

Let M = P (A), where A= {a, b, c}. Then $[M; \cup, \cap]$ is a bounded lattice with 0 = f and 1 = A. Then, to find if it exists, the complement, $\bar{B}$, of, say B = {a, b} $\in$ M, we want $\bar{B}$ such that

$$\{a, b\} \cap \bar{B} = f$$

and

$$\{a, b\} \cup \bar{B} = A.$$

Here, $\bar{B}$= {c}, and since it can be shown that each element of M has a complement, $[M; \cup, \cap]$ is a complemented lattice. Note that if A is any set and M = P (A), then $[M; \cup, \cap]$ is a complemented lattice where the complement of $B \in M$ is $\bar{B} = B^c = A - B$.

We observe that the complement of each element of M is unique. Is this is always the case? The answer is no. Consider the following.

Example 2:

Let L = {1, 2, 3, 5, 30} and consider the lattice $[L, \vee, \wedge]$ (under "divides"). The least element of L is 1 and the greatest element is 30. Let *us compute the complement of the element a = 2. We want to determine* $\bar{a}$ *such that* $2 \wedge \bar{a} = 1$ *and* $2 \vee \bar{a} = 30$*. Certainly,* $\bar{a} = 3$ *works, but so does* $\bar{a} = 4$*, so the complement of a = 2 in this lattice is not unique. However,* $[L; \vee, \wedge]$ *is still a complemented lattice since each element does have at least one complement.*

The following theorem gives us an insight into when uniqueness of complements occurs.

FINITE BOOLEAN ALGEBRAS AS N-TUPLES OF 0'S AND 1'S

From the previous section we know that all finite Boolean algebras are of order 2^n, where n is the number of atoms in the algebra. We can therefore completely describe every finite Boolean algebra by the algebra of power sets. Is there a more convenient, or at least an alternate way, of defining finite Boolean algebras? We could produce new groups by taking artesian products of previously known groups. We imitate this process for Boolean algebras.

The simplest non-trivial Boolean algebra is the Boolean algebra on the set $B_2 = \{0, 1\}$. The ordering on B_2 is the natural one, $0 \le 0$, $0 \le 1$, $1 \le 1$. If we treat 0 and 1 as the truth values "false" and "true" respectively, we see that the Boolean operations $\vee$ (join) and $\wedge$ (meet) are nothing more than the logical connectives $\vee$ (or) and $\wedge$ (and). The Boolean operations, $-$, (complementation) is the logical ~ (negation). In fact, this is why the symbols $-$, $\vee$, and $\wedge$ were chosen as the names of the Boolean operations. The operation tables for $[B_2; -, \vee, \wedge]$ are simply those of "or", "and" and "not" which we repeat here:

$\vee$	0	1
0	0	1
1	1	1

$\wedge$	0	1
0	0	0
1	0	1

u	$\bar{u}$
0	1
1	0

Its corollaries, all Boolean algebras of order 2 are isomorphic to this one.

We know that if we form $B_2 \times B_2 = B_2^2$ we obtain the set {(0, 0), (0, 1), (1, 0), (1, 1)}, a set of order 4. We define operations on B_2^2 the natural way, namely, components wise, so that $(0, 1) \vee (1, 1) = (0 \vee 1), (1 \vee 1) = (1, 1)$, $(0, 1) \wedge (1, 1) = (0 \wedge 1), (1 \wedge 1) = \overline{(0,1)}, (\bar{0}, \bar{1}) = (1, 0)$. We claim that B_2^2 is a Boolean algebra under the componentwise operations. Hence, $[B_2^2; -, \vee, \wedge]$ is Boolean algebra of order 4. Since all Boolean algebras of order 4 are isomorphic to each other, we have found a simple way of describing all Boolean algebras of order 4.

It is quite clear that we can describe any Boolean algebra of order 8 by considering $B_2 \times B_2 \times B_2 = B_2^2$ and, in general, any Boolean algebra of order 2n — that is, all finite Boolean algebras by $B_2^n = B_2 \times B_2 \times ... \times B_2$ (n number of times). The Hasse diagrams of the $[B_2^n ; -, \vee, \wedge]$ for n = 1, 2, 3, 4, are given in Figure 9.4.5. Just think of the strings as n-tuples.

Boolean algebra is an algebra of logic. George Boole in 1854 invented a systematic way of manipulating logic symbols which became known as Boolean algebra. Now-a-days Boolean algebra has became an indispensable tool to computer scientists because of it direct applicability to switching theory and the logical design of digital computers.

LAW OF BOOLEAN ALGEBRA

The Boolean Algebra consists of a non–empty set B with two binary operations + and (.), one unary operation (') *Not*, and two distinct elements 0 and 1, denoted by (B, + , . , ', 0, 1) which follow the following postulates:

Postulate I. (Commutative Laws)

Both the binary operations are commutative, i.e.,

(i) $a + b = b + a, \forall a, b \in B$

(ii) $a . b = b . a, \forall a, b \in B$

Postulate II. (Associative Laws)

Both the operations are associative, i.e.,

(i) $\forall a, b, c \in B,$ $a + (b + c) = (a + b) + c$

(ii) $\forall a, b, c \in B,$ $a. (b . c) = (a . b) .c$

Postulate III. (Distributive Laws)

Each operation distributes over the other i.e.,

(i) $\forall a, b, c \in B,$ $a. (b + c) = a . b + a . c$

(ii) $\forall a, b, c \in B,$ $a + (b.c) = (a + b) . (a + c)$

Postulate IV. (Existence of Identity Elements)

There exist distinct identity elements 0 and 1 belonging to B such that

(i) $a + 0 = 0 + a = a, \forall a \in B$

(ii) $a . 1 = 1 . a = a, \forall a \in B$

Postulate V. (Existence of Inverse or complement)

$\forall a \in B$, there exists an inverse element a' of B such that

(i) $a + a' = a' + a = 1$, where $1 \in B$ is the identity for '+'

(ii) $a \cdot a' = a' \cdot a = 0$, where $0 \in B$ is the identity element for (.).

SIMILARITY IN LOGIC, THEORY OF SETS AND BOOLEAN ALGEBRA

It can be easily shown that

(i) Algebra of statements under the binary relation '$\wedge$' and '$\vee$' is a Boolean algebra.

(ii) Algebra of sets under the binary operations $\cap$ and $\cup$ is a Boolean algebra. The following table shows the similarity in logic, theory of sets and Boolean Algebra.

	Logic	Sets	Boolean Algebra
Statements	p, q, r, ..	A, B, C, ...	Elements of B denoted by a, b, ...
Operator AND	$p \wedge q$	$A \cap B$	$a \cdot b$
Operator OR	$p \vee q$	$A \cup B$	$a + b$
Operator NOT	$\sim p$	A'	a'
Implication	$p \Rightarrow q$ or $\sim p \vee q$	mapping one-one	$a' + b$
Equivalence	$p \Leftrightarrow q$	mapping	$a' \cdot b' + a \cdot b$
Tautology	T	Universal Set	1
Fallacy	F	ϕ	0

Theorem:

For any $a \in B$, prove that (i) $a + a = a$; (ii) $a \cdot a = a$

Proof:

(i) Now

$a = a + 0$ (by postulate I)

$= a + a \cdot a'$ (by postulate V)

$= (a + a) \cdot (a + a')$ (by postulate III)

$= (a + a) \cdot 1$ (by postulate V)

$= a + a$ (by postulate IV)

Hence, $a = a + a$.

(ii) To provea $= a \cdot a$,

consider	$a = a \, . \, 1$	(Postulate IV)
	$= a \, . \, (a + a')$	(Postulate V)
	$= a \, . \, a + a \, . \, a'$	(Postulate III)
	$= a \, . \, a + 0$	(Postulate V)
	$= a \, . \, a$	(Postulate IV)

Hence, $a = a \, . \, a.$

DUALITY

The dual of any statement in Boolean Algebra (B, +, .) is the statement which is derived by interchanging the operations + and . and their corresponding identity elements 1 and 0, in the original statement, for example, the dual of $0. \, a + b \, . \, 1 = b$ is $(1 + a) \, . \, (b + 0) = b$ and the dual of $a \, . \, (b + c) = a \, . \, b + a \, . \, c$ is $a + (b \, . \, c) = (a + b) \, . \, (a + c)$

Theorem 1:

In a Boolean Algebra B, prove the following statements and state their duals if any

(i) $a + 1 = 1, \quad \forall a \in B$

(ii) $a \, . \, 0 = 0, \quad \forall a \in B$

(iii) $a + b = a + a'b \quad \forall a,b \in B$

(iv) $a' + a \, . \, b = a' + b \quad \forall a,b \in B$

Proof:

(i) $a + 1 = 1$

Now, $(a + 1) = (a + 1) \, . \, 1$

$= 1 \, . \, (a + 1)$

$=(a + a') \, . \, (a + 1)$

$= a + (a' \, . \, 1)$

$= a + a' = 1$

$\therefore \quad a + 1 = 1 \, .$

Its dual is $a \, . \, 0 = 0$

(ii) $a \, . \, 0 = 0$

Here $a. \, 0 = (a \, . \, 0) + 0$

$= 0 + (a \, . \, 0)$

$= (a \, . \, a') + (a' + 0)$

$= a \,.\, (a' + 0)$

$= a \,.\, a' = 0$

Hence $a \,.\, 0 = 0$

Its dual is $a + 1 = 1$

(iii) $a + b = a + a' \,.\, b$

L.H.S. $= a + b = (a + b) \,.\, 1$

$= (a + b) \,.\, (a + a')$

$= (a + b) \,.\, a + (a + b) \,.\, a'$

$= a. \,(a + b) + a' \,.\, (a + b)$

$= (a \,.\, a + a \,.\, b) + (a' \,.\, a + a' \,.\, b)$

$= a + (a \,.\, b) + 0 + a' \,.\, b$

$= a + a' \,.\, b$

Hence $a + b = a + a' \,.\, b$

Its dual is $a \,.\, b = a \,.\, (a' + b)$

(iv) $a' + a \,.\, b = a' + b$

R.H.S. $a' + b = a' \,.\, 1 + b \,.\, 1$

$= a' \,.\, (a + 1) + b \,.\, (a + a')$

$= a' \,.\, a + a' \,.\, 1 + b \,.\, a + b \,.\, a'$

$= 0 + a' \,.\, 1 + a \,.\, b + a' \,.\, b$

$= a' \,.\, 1 + a' \,.\, b + a \,.\, b$

$= a' \,(1 + b) + a \,.\, b$

$= a' \,.\, 1 + a \,.\, b$

$= a' + a \,.\, b$

Hence $a' + b = a' + a \,.\, b$

Its dual is $a' \,.\, b = a', (a + b)$

Theorem 2:

Prove that any complement a' for a $\in$ B is unique

Proof:

Let us suppose that a_1' and a_2' are two complements of $a \in B$. Then

$$a + a_1' = 1, \qquad a + a_2' = 1$$

and $$a \,.\, a_1' = 0, \qquad a \,.\, a_2' = 0.$$

Now $a_1' = a_1' \,.\, 1$

$= a_1' (a + a_2')$

$= (a_1' \,.\, a) + a_1' \,.\, a_2'$

$= a \,.\, a_1' + a_1' \,.\, a_2'$

$= 0 + (a_1' \,.\, a_2')$

$= (a_1' \,.\, a_2') + 0 = a_1' \,.\, a_2'$ ...(1)

Also, $a_2' = a_2' \,.\, 1$

$= a_1' (a + a_1')$

$= (a_2' + a) + (a_2' \,.\, a_1')$

$= (a \,.\, a_2') + (a_1' \,.\, a_2')$

$= 0 + (a_1' \,.\, a_2')$

$= (a_1' \,.\, a_2') + 0 = a_1' \,.\, a_2'$ ...(2)

Now, (1) and (2) $\Rightarrow a_1' = a_2'$

Hence the complement of any element in B is unique.

Theorem 3:

(Absorption Laws) Prove that

(i) $a + (a \,.\, b) = a \;\forall\; a, b, \in B$

(ii) $a \,.\, (a + b) = a \;\forall\; a, b, \in B$

Proof:

(i) $a + (a \,.\, b) = a$

L.H.S. $= a + (a \,.\, b) = (a \,.\, b) + a$

$= a \,.\, b + a \,.\, 1$

$= a \,.\, (b + 1)$

$= a \,.\, 1 = a$

$\therefore a + (a \,.\, b) = a$

(ii) $a \,.\, (a + b) = a$

L.H.S. $= (a + b) \,.\, a$

$= (a + b) \,.\, (a + 0)$

$= a + (b + 0)$

$= a + 0 = a$

$\therefore a \,.\, (a + b) = a$

Theorem 4:

(De-Morgan's Laws) : *Prove that*

(i) $(a + b)' = a' . b'$, *i.e.*, $(a + b) . (a' . b') = 0$ *and* $(a + b) + (a' . b') = 1, \forall a, b \in B$

(ii) $(a . b)' = a' + b'$, *i.e.*, $(a . b) . (a' + b') = 0$ *and* $(a + b) + (a' + b') = 1$.

Proof:

(i) We shall show that a' . b' is the complement of (a + b), i.e., a' . b' = (a + b)'

Consider (a + b) + (a' b')

$$= [(a + b) + a'] . [(a + b) + b']$$
$$= [a' + (a + b)] . [a + (b + b')]$$
$$= [(a' + a) + b] . (a + 1)$$
$$= 1 + b = b + 1 = 1.$$

Also, $(a + b) . (a' . b')$

$$= (a' . b') . (a + b)$$
$$= [(a' . b') . a] + [(a' . b') . b]$$
$$= [a . (a' . b')] + [a' . (b' . b)]$$
$$= [(a . a') . b'] + [a' . (b' . b)]$$
$$= (0 . b') + (a' . 0) = 0 + 0 = 0$$

$\Rightarrow$ a' . b' is the complement of (a + b).

But we know that every element of B has a unique complement.

Hence the result follows.

(ii) Proceeding exactly in (i), we can easily prove that a' + b' is the complement of a .b.

ALGEBRA OF BOOLEAN SYSTEM

I. Associative Law: For any a, b, c $\in$ B, the following are true:

(i) a + (b + c) = (a + b) + c

(ii) a . (b . c) = (a . b) . c

II. Commutative Law: For all a, b $\in$ B, we have

(i) a + b = b + a

(ii) a .b = b . a

III. Identity Law: For all a ∈ B, the following relations are true

$$a + 0 = a$$

$$a + 1 = 1$$

$$1 . 1 = 1$$

$$a . 0 = 0$$

IV. Complement Law: For all a ∈ B there exists a' ∈ B, where a, is the complement of a, such that

$$a + a' = 1$$

$$(a')' = a$$

$$a . a' = 0$$

$$0' = 1$$

$$1' = 0$$

V. De-Morgan's Law: For all a, b ∈ B there exists a', b' ∈ B such that

(i) (a + b)' = a'. b'

(ii) (a . b)' = a' + b'

(iii) a. (b + c) = (ab') . (ac')

(iv) a (b + c)' = (ab') + (ac')

VI. Idempotent Law: For all a, b, c ∈ B

(i) a + (b . c) = (a + b) . (a + c)

(ii) a . (b + c) = (a . b) + (a . c)

SYSTEMS OF LINEAR EQUATIONS

The method of solving systems of equitation by matrices that we will look at is based on procedures involving equations that we are familiar with from previous mathematics courses. The main idea is to reduce a given system of equations to another simpler system which has the same solutions.

Definition: *Solution Set: Given a system of equations involving real variables x_1, ... x_n the solution set of the system is the set of n-tuples in* R^n, *(a_1, ..., a_n) such that the substitutions $x_1 = a_1$, .., $x_n = a_n$ make all the equations true.*

If the variables are from a set S, then the solution set will be a subset of S^n.

Definition: Equivalent Systems of Equations. *Two systems of linear equations are called equivalent if they have the same set of solutions.*

Example:

The previous definition tells us that if we know that the system

$$\begin{cases} 4x_1 + 2x_2 + x_3 = 1 \\ 2x_1 + x_2 + x_3 = 4 \\ 2x_1 + 2x_2 + x_3 = 3 \end{cases}$$

is equivalent to the system

$$\begin{cases} x_1 + 0x_2 + 0x_3 = -1 \\ 0x_1 + x_2 + 0x_3 = -1 \\ 0x_1 + 0x_2 + 1x_3 = 7 \end{cases}$$

then both systems have the solution set {(–1, –1, 7)}: i.e.

$x_1 = -1$, $x_2 = -1$, $x_3 = 7$.

MATRIC INVERSION

Tthe inverse of an n × n matrix. We noted that not all matrices have inverse, but when the inverse of a matrix exists, it is unique. This enables us to defined the inverse of an n × n matrix A as the unique matrix B such that AB = BA = I, where I is the n × n identity matrix. In order to obtain some practical experience, we developed a formula which allowed us to determine the inverse of invertible 2 × 2 matrices. We will now use the Guass-Jordan procedure for solving systems of linear equations to compute the inverse; when they exist, of n × n matrices, n ≥ 2. The following procedure for a 3 × 3 matrix c an be generalized for n × n matrices, n ≥ 2.

Example:

Given the matrix

$$A = \begin{bmatrix} 1 & 1 & 2 \\ 2 & 1 & 4 \\ 3 & 5 & 1 \end{bmatrix}$$

we want to find the matrix

$$B = \begin{bmatrix} x_{11} & x_{12} & x_{13} \\ x_{21} & x_{22} & x_{23} \\ x_{31} & x_{32} & x_{33} \end{bmatrix},$$

if it exists, such that (a) AB = I and (b) BA = I. We will concentrate on finding a matrix which satisfies Equation a and then verify that the matrix B obtained satisfies Equation b.

$$\begin{bmatrix} 1 & 1 & 2 \\ 2 & 1 & 4 \\ 3 & 5 & 1 \end{bmatrix} \begin{bmatrix} x_{11} & x_{12} & x_{13} \\ x_{21} & x_{22} & x_{23} \\ x_{31} & x_{32} & x_{33} \end{bmatrix} = \begin{bmatrix} 1 & 0 & 0 \\ 0 & 1 & 0 \\ 0 & 0 & 1 \end{bmatrix}$$

$$\Leftrightarrow \begin{bmatrix} 1x_{11}+1x_{21}+2x_{31} & 1x_{12}+1x_{22}+2x_{32} & 1x_{13}+1x_{23}+2x_{33} \\ 2x_{11}+1x_{21}+4x_{31} & 2x_{12}+1x_{22}+4x_{32} & 2x_{13}+1x_{23}+4x_{33} \\ 3x_{11}+5x_{21}+1x_{31} & 3x_{12}+4x_{22}+1x_{32} & 3x_{13}+5x_{23}+1x_{33} \end{bmatrix} \quad \text{...(1)}$$

$$= \begin{bmatrix} 1 & 0 & 0 \\ 0 & 1 & 0 \\ 0 & 0 & 1 \end{bmatrix}$$

By definition of equality of matrices, this gives us three systems of equations to solve. The matrix form of one of the (1) system is:

$$\begin{bmatrix} 1 & 1 & 2 & 1 \\ 2 & 1 & 4 & 0 \\ 3 & 5 & 1 & 0 \end{bmatrix} \quad \text{...(2)}$$

Using the Gauss-Jordan technique of (2) we have:

$$\begin{bmatrix} 1 & 1 & 2 & 1 \\ 2 & 1 & 4 & 0 \\ 3 & 5 & 1 & 0 \end{bmatrix} \xrightarrow[-3R_1+R_3]{-2R_1+R_2 \text{ and}} \begin{bmatrix} 1 & 1 & 2 & 1 \\ 0 & -1 & 0 & -2 \\ 0 & 2 & -5 & -3 \end{bmatrix}$$

$$\xrightarrow[R_2+R_3]{R_2+R_1 \text{ and}} \begin{bmatrix} 1 & 0 & 2 & -1 \\ 0 & -1 & 0 & -2 \\ 0 & 0 & -5 & -7 \end{bmatrix}$$

$$\xrightarrow{2/5\,R_3+R_1} \begin{bmatrix} 1 & 0 & 0 & -19/5 \\ 0 & -1 & 0 & -2 \\ 0 & 0 & -5 & -7 \end{bmatrix}$$

$$\xrightarrow[-1R_3]{-1R_2 \text{ and}} \begin{bmatrix} 1 & 0 & 0 & -19/5 \\ 0 & 1 & 0 & 2 \\ 0 & 0 & 1 & 7/5 \end{bmatrix}$$

So $x_{11} = -19/4$, $x_{21} = 2$, and $x_{31} = 7/5$, which gives us the first column of the matrix B. The matrix form of the system to obtain x_{12}, x_{22} and x_{32}, the second column of B, is:

$$\begin{bmatrix} 1 & 1 & 2 & 0 \\ 2 & 1 & 4 & 1 \\ 3 & 5 & 1 & 0 \end{bmatrix}, \qquad \text{...(3)}$$

which reduces to

$$\begin{bmatrix} 1 & 0 & 0 & 9/5 \\ 0 & 1 & 0 & -1 \\ 0 & 0 & 1 & -2/5 \end{bmatrix} \qquad \text{...(4)}$$

The critical idea to note here is that the coefficient matrix in (3) is the same as the matrix in (4), hence the sequence of row operations that we used to reduce the matrix in 12.2b can be used to reduce the matrix in (4). To determine the third column of B, we reduce

$$\begin{bmatrix} 1 & 1 & 2 & 0 \\ 2 & 1 & 4 & 0 \\ 3 & 5 & 1 & 1 \end{bmatrix}$$

to obtain $x_{13} = 2/5$, $x_{23} = 0$ and $x_{33} = -1/5$. Here again it is important to note that the sequence of row operation used to "solve" this system is exactly the same as those we used in the first system. Why not save ourselves a considerable amount of time and effort and solve all three systems simultaneously? This we can effect by augmenting the coefficient matrix by the identity matrix I. We then have

$$\begin{bmatrix} 1 & 1 & 2 & 1 & 0 & 0 \\ 2 & 1 & 4 & 0 & 1 & 0 \\ 3 & 5 & 1 & 0 & 0 & 1 \end{bmatrix}$$

$$\xrightarrow[\text{row opreations as above]}]{\text{[The same sequence of}} \begin{bmatrix} 1 & 0 & 0 & -19/5 & 9/5 & 2/5 \\ 0 & 1 & 0 & 2 & -1 & 0 \\ 0 & 0 & 1 & 7/5 & -2/5 & -1/5 \end{bmatrix}$$

So that

$$B = \begin{bmatrix} -19/5 & 9/5 & 2/5 \\ 2 & -1 & 0 \\ 7/5 & -2/5 & -1/5 \end{bmatrix}$$

(The reader should verify that BA = I so that $A^{-1} = B$.)

As the following theorem indicates, the verification that BA = I is not necessary. The proof of the theorem is beyond the scope of this text. The interested reader can find it in most linear algebra texts.

Theorem 1:

Let A be an n × n matrix. If a matrix B can be found such that AB = I, then AB = I, so that B = A^{-1}. In fact, to find A^{-1}, we need only find a matrix B that satisfies one of the two conditions AB = I or BA = 1.

This chapter that not all n × n matrices have inverse. How do we determine whether a matrix has an inverse using this method? The answer is quite simple: the technique we developed to computer inverse is a matrix approach to solving several systems of equations simultaneously.

Example:

The reader can verify that if

$$A = \begin{bmatrix} 1 & 2 & 1 \\ -1 & -2 & -1 \\ 0 & 1 & 2 \end{bmatrix}$$

then
$$\begin{bmatrix} 1 & 2 & 1 & 1 & 0 & 0 \\ -1 & -2 & -1 & 0 & 1 & 0 \\ 0 & 5 & 8 & 0 & 0 & 1 \end{bmatrix}$$

reduces to
$$\begin{bmatrix} 1 & 2 & 1 & 1 & 0 & 0 \\ -1 & 0 & 0 & 1 & 1 & 0 \\ 0 & 5 & 8 & 0 & 0 & 1 \end{bmatrix}$$

Although, this matrix can be further row reduced, it is not necessary to do so since in equation form we have:

$$\text{(i)} \begin{cases} x_{11} + 2x_{21} + x_{31} = 1 \\ 0x_{11} + 0x_{21} + 0x_{31} = 1 \\ 0x_{11} + 5x_{31} + 8x_{31} = 0 \end{cases}$$

$$\text{(ii)} \begin{cases} x_{12} + 2x_{22} + x_{32} = 0 \\ 9x_{12} + 0x_{22} + 0x_{32} = 1 \\ 0x_{12} + 5x_{22} + 8x_{32} = 0 \end{cases}$$

$$\text{(iii)} \begin{cases} x_{13} + 2x_{23} + x_{33} = 0 \\ 0x_{13} + 0x_{23} + 0x_{33} = 0 \\ 0x_{13} + 5x_{23} + 8x_{33} = 1 \end{cases}$$

Clearly, there is no solution to System i (or to System ii), therefore A^{-1} does not exist. From this discussion it should be obvious to the reader that the zero row of the coefficient matrix together with the non-zero entry in the fourth column of that row in matrix 12.2e tell us that A^{-1} does not exists.

AN INTRODUCTION TO VECTOR SPACES AND THE DIAGONALIZATION PROCESS

When we encountered various types of matrices it became apparent that a particular kind of matrix, the diagonal matrix, was much easier to use in computations. For example, if $A = \begin{bmatrix} 2 & 1 \\ 2 & 3 \end{bmatrix}$, then A^5 can be found. Its computation is tedious, however. If

$$A = \begin{bmatrix} 1 & 0 \\ 0 & 4 \end{bmatrix},$$

$$\text{then } A = \begin{bmatrix} 1^5 & 0 \\ 0 & 4^5 \end{bmatrix} = \begin{bmatrix} 1 & 0 \\ 0 & 1024 \end{bmatrix}$$

In a variety of applications it is beneficial to be able to *diagonalize* a matrix. In this section we will investigate what this means and consider a few applications. In order to understand when the diagonalization process can be performed, it is necessary to develop several of the underlying concepts of *linear algebra.*

By now, you realize that mathematicians tend to generalize. Once we have found a "good thing," something that is useful, we apply it to as many different concepts as possible. In doing so, we frequently find that the "different concepts" are not really different but only look different. Four sentences in four different languages might look dissimilar, but when they are translated into a common language, they might very well express the exact same idea.

Early in the development of mathematics, the concept of a vector red to a variety of applications in physics and engineering. We can certainly picture vectors, or "arrows," in the xy-plane and even in the three-dimensional space. Does it make sense to talk about vectors in four-dimensional space, in ten dimensional space, or in any other mathematical situation? If so, what is the essence of a vector? Is it its shape or the properties, or rules, it follows? The shape in two or three-space is just a picture, or geometric interpretation, of a vector. The essence is the rules, or properties, we wish vectors to follow so we can manipulate thcm algebraically. What follows is a definition of what is called a vector space, "space of vectors". It is a list of all the essential properties of vectors, and it is the basic definition of what is called linear algebra.

UNIQUE FEATURES OF BOOLEAN ALGEBRA

There are some unique features of Boolean Algebra which distinguish it from other branches of algebra. Some of them are :

(i) Boolean algebra has only a finite set of elements, which are 1 and 0. These are usually called true and false. Ordinary algebra, on the other hand, deals with a set of an infinite number of elements. Thus all the variables and constants are allowed to have only two possible values 0 and 1.

(ii) Boolean algebra does not have operations equivalent to subtraction and division.

(iii) In Boolean algebra the cancellation law does not hold,

i.e., $a + b = a + c$

$\Rightarrow$ $b = c$

and $a.b = a.c \Rightarrow b = c.$

(iv) In ordinary algebra, there is no equivalent of the unary operation of bar or (') over a variable known as complementing.

BASIC OPERATIONS

The basic operations used in Boolean Algebra are logical addition, logical multiplication and complementation. These basic operations are called *logic operations*. Boolean 1 and 0 look like the binary numbers but they are not. They do not represent any numerical values. They only have logical significance.

Boolean algebra uses only three operations on its variables. These operations are :

(i) The *OR addition* represented by a + (plus) sign.

(ii) The *AND multiplication* represented by a × (cross), or a . (dot) sign.

(iii) The *NOT operation* represented by a bar or dash over a variable.

The Boolean algebra tells us that for binary variables (0 and 1), the following holds :

OR	AND	NOT
$0 + 0 = 0$	$0 \cdot 0 = 0$	$0'$ or $\overline{0} = 1$
$0 + 1 = 1$	$0 \cdot 1 = 0$	$1'$ or $\overline{1} = 0$
$1 + 0 = 1$	$1 \cdot 0 = 0$	
$1 + 1 = 1$	$1 \cdot 1 = 1$	

De Morgan's Theorem

De Morgan's theorems (or rules) are very useful in simplifying complicated logical expressions and can be stated as under:

Theorem 1:

The complement of the sum of two (or more) variables is equal to the product of the complements of the variables, i.e.,

$$(A + B)' = A' \cdot B'.$$

Theorem 2:

The complement of the product of two (or more) variables in equal to the sum of complements of the variables, i.e.,

$$(A \cdot B)' = A' + B'.$$

Proof:

These theorems can be proved by substituting the two permitted values of the variables, i.e., 0 and 1 on both sides of the identity. In each case the left hand side will equal the right hand side, thus proving the theorem.

1. To prove $(A + B)' = A' \cdot B'$.

Case 1 : When $A = 0, B = 0$

$(A + B)' = (0 + 0)' = 0' = 1$

$A' \cdot B' = 0' \cdot 0' = 1 \cdot 1 = 1$

Case 2 : When $A = 0, B = 1$

$(A + B)' = (0 + 1)' = 1' = 0$

$A' \cdot B' = 0' \cdot 1' = 1 \cdot 0 = 0.$

Case 3 : When $A = 1, B = 0$

$(A + B)' = (1 + 0)' = 1' = 0$

$A' \cdot B' = 1' \cdot 0' = 0 \cdot 1 = 0.$

Case 4 : When $A = 1, B = 1$

$(A + B)' = (1 + 1)' = 1' = 0$

$A' \cdot B' = 1' \cdot 1' = 0 \cdot 0 = 0.$

Since in each case, the left hand side equals the right hand side, the theorem stands proved. Thus, for all possible combinations $(A + B)' = A' \cdot B'$.

2. To prove $(A \cdot B)' = A' \cdot B'$

Case 1 : When $A = 0, B = 0$

$(A \cdot B)' = (0 \cdot 0)' = 0' = 1$

$A' + B' = 0' + 0' = 1 + 1 = 1$

Case 2 : When A = 0, B = 1

$(A \cdot B)' = (0 \cdot 1)' = 0' = 1$

$A' + B' = 0' + 1' = 1 + 0 = 1$

Case 3 : When A = 1, B = 0

$(A \cdot B)' = (1 \cdot 0)' = 0' = 1$

$A' + B' = 1' + 0' = 0 + 1 = 1$

Case 4 : When A = 1, B = 1

$(A \cdot B)' = (1 \cdot 1)' = 1' = 0$

$A' + B' = 1' + 1' = 0 + 0 = 0.$

Since in each case left hand side is equal to the right hand side, this theorem is also proved, i.e., for all combinations

$$(A \cdot B)' = A' + B'.$$

Demorganisation, i.e., Reduction of Simple Boolean Expressions Using De Morgen's Theorem

It may be seen from De Morgan's theorem that the procedure needed for bringing out an expression from under a NOT sign is as follows :

Step 1 : Remove the overall NOT sign, i.e., complement the given expression.

Step 2 : Change all OR's to AND's and ANDs to ORs.

Step 3 : Complement all individual variables.

To make the rule clear, consider the following examples :

Example 1:

$$(A + BC)' = A + BC \qquad \text{...By step 1}$$
$$= A(B + C) \qquad \text{...By step 2}$$
$$= A'(B' + C') \qquad \text{...By step 3}$$

Example 2:

$$[(A + B) \cdot (C + D)]' = (A + B) \cdot (C + D) \qquad \text{... By step 1}$$
$$= (A \cdot B) + (C \cdot D) \qquad \text{...By step 2}$$
$$= A' \cdot B' + C' \cdot D' \qquad \text{...By step 3}$$

In order to bring an expression under the NOT sign, the opposite procedure would be followed. To illustrate it, let us consider the example.

$$A' + B' + C' = (A')' + (B')' + (C')' \quad \text{...By step 3}$$
$$= A + B + C$$
$$= A \cdot B \cdot C \quad \text{...By step 2}$$
$$= A' \cdot B' \cdot C' \quad \text{...By step 1}$$

BOOLEAN FUNCTIONS

We have seen that the AND, OR and NOT operations performed on Boolean variables, result in Boolean variables. This gives the 'closure' property. As in ordinary algebra, we have the concept of a function of Boolean variables or an expression containing Boolean variables. For example consider the equation

$$X = A \cdot B + C' \cdot D + E \cdot F$$

Here the variables X is a *function* of A, B, C, D, E and F. This written as X = f(A, B, C, D, E, F) and the right hand of the equation is called an *expression*. Each of the symbols A, B, C, D etc. are referred to as *literals*.

The value of a Boolean function is obtained by substituting 0 and 1 for variables in the expression. It is clear that different Boolean expressions may determine the same Boolean function. For example, $A \cdot (B + C)$ and $(A \cdot B) + (A \cdot C)$ determine the same Boolean function. Two Boolean expressions are said to be *equivalent* if they assume the same value for every assignment of values to the variables.

A Boolean function of degree 2 is a function with four elements, namely pairs of elements from B {0, 1} to B, a set with two elements. Hence, there are 16 different Boolean functions of degree 2.

Precedence of Operators

In order to evaluate Boolean expressions it is necessary to define the procedence of operators. The precedence is as follows :

(i) Scan the expression from left to right.

(ii) First evaluate expressions enclosed in parentheses.

(iii) Perform all the complement (NOT) operations.

(iv) Perform all (·) (AND) operations next in the order in which they appear.

(v) Perform all '+' (OR) operations last.

Venn Diagram

To fix our ideas of Boolean Algebra, a pictorial model known as the Venn diagram is useful. This diagram consists of a rectangle inside which are drawn a number of circles, one circle for each variable. The region inside the circle belongs to the variable and the region outside belongs to its complement. Fig. 1.1 shows a Venn diagram for two variables A and B. Region inside the circle named A, represents A and that outside A'. With two intersecting circles A and B we have 4 regions corresponding to A · B, A · B'. A' · B, A' · B' respectively as shown in the figure.

Venn diagrams may be used to illustrate postulates and theorems of Boolean algebra. Fig. 1.2 illustrates the identities a + b · c = (a + b) · (a + c). In Fig. 1.2 (a), the area covered by the slanted lines corresponds to the expression a + (b · c). In Fig. 1.2 (b) area (a + b) is marked by horizontal lines and area (a + c) by vertical lines. The area intersected by the horizontal and the vertical lines correspond to the expression (a + b) · (b + c) and this is identical to the area in Fig. 1.2 (a). Venn diagrams thus provide an aid to remember or prove the theorems of Boolean algebra.

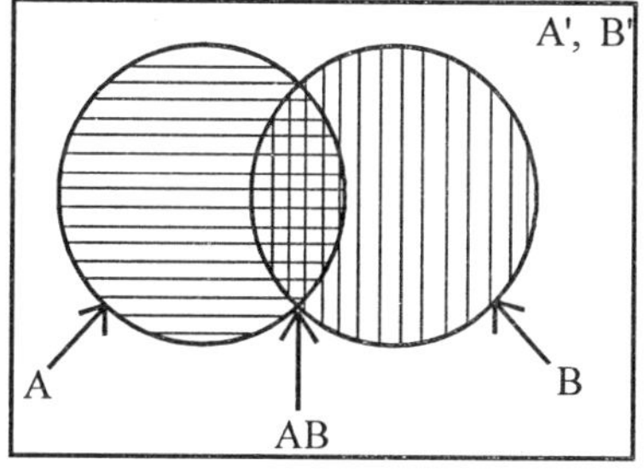

Fig. 1.1

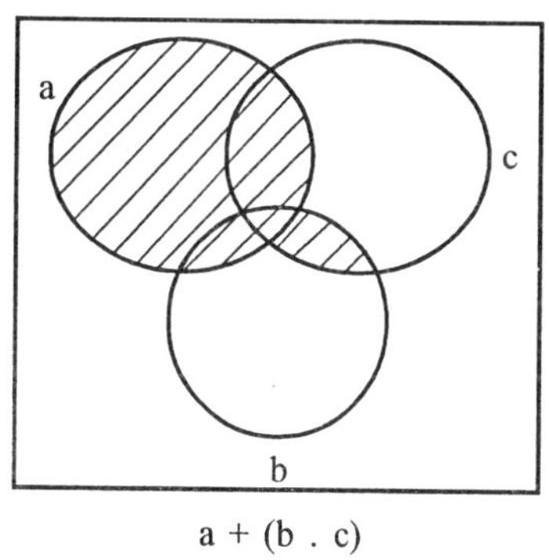

a + (b . c)

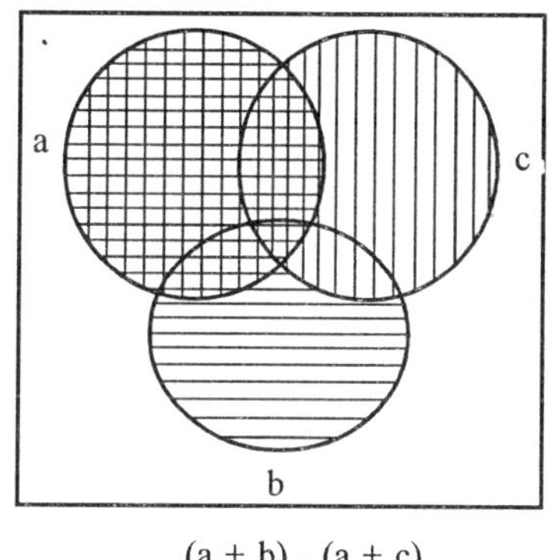

(a + b) . (a + c)

Fig. 1.2

Truth Table

Consider a Boolean function z = f(a, b, c). A table which lists the value of the dependent variable z for each set of values of the independent variables a, b and c is known as a *Truth Table.*

We will now obtain a truth table for the function z = a' + b · c. To obtain the table following steps will be followed :

(i) Form a table with one column for each of the independent variables and one column for the dependent variable. Use a vertical double line to separate the dependent and the independent variables.

(ii) Assuming that there are n independent variables in the function, start with 00...0 (n bits) as values of the independent variables and form successive rows (each containing n bits) in the natural binary counting sequence till the count becomes 2^n. In our example there are three variables and hence eight rows starting from 000 and ending with 111.

(iii) Substitute the values of the independent variables in each row and enter the evaluated value of the function in the same row to the right of the double vertical line.

Following this procedure the truth table for the function z = a' + b . c is as follows :

a	b	c	z
0	0	0	1
0	0	1	1
0	1	0	1
0	1	1	1
1	0	0	0
1	0	1	0
1	1	0	0
1	1	1	1

It is thus easy to obtain the Truth Table.

SUM OF PRODUCTS AND PRODUCT OF SUMS FORM

Let we have a set of variables $a_1, a_2, ..., a_n$. A *Boolean expression* E or E $(a_1, a_2, ..., a_n)$ in these variables is any variable or any expression built up from the variables using the Boolean operations +, ·, and '. For example

E_1 = (a + b' c)' + (a b c' + a' b)' and

E_2 = (a b' c' + b)' + a' c)'

are Boolean expressions in a, b and c.

A *literal* is a variable or complemented variable, such as a, a', b, b', and so on. A *fundamental product* is a literal or a product of two or more literals in which two literals involve the same variable. Thus

a c', a b' c, a, b', a'bc

are fundamental products, but a b a' c and a b c b are not. Any product of literals can be reduced to either 0 or a fundamental product. For example. a b a' c = 0 since a a' = 0 (complement law), and abcb = abc since bb = b (idempotant law).

A fundamental product P_1 is said to be *contained in* or *included in* another product P_2 if the literals of P_1 are also literals of P_2. For example a'c is contained in a' b c, but a' c is not contained in a b' c since a' is not a literal of a b' c. If P_1 is contained in P_2, i.e., $P_2 = P_1 \cdot Q$, then, by the absorption law,

$$P_1 + P_2 = P_1 + P_1 \cdot Q = P_1.$$

A Boolean expression E is said to be a *sum-of-products* expression if E is a fundamental product or the sum of two or more fundamental products none of which is contained in another.

Example 1:

(i) $ac' + a'bc' + ab'c$

(ii) $xyz + xz + x'yz'$

(iii) $ab + a'bc' + b'c' + d$

A Boolean expression E is said to be in a product-of-sums if it consists of several sum terms logically multiplied. The variables may or may not be complemented.

Example 2:

(i) $(a + b)(a' + b)$,

(ii) $(a + b' + c)(a + c)$,

(iii) $(x + y')(z' + w)x$

Method of Finding Sum-of-Products Forms

Following method in four steps may be used to transform any Boolean expression E into an equivalent sum-of-products expression with the help of laws of Boolean algebra :

(i) Use De-Morgan's laws and involution to move the complement operation into any parenthesis until the complement operation only applies to variables. Then E will consist only of sums and products of literals.

(ii) Use the distributive operation to next transform E into a sum of products.

(iii) Use the commutative, idempotent and complement laws to transform each product in E into 0 or a fundamental product.

(iv) Use the absorption and identity laws to finally transform E into a sum-of-products expression.

SUBALGEBRAS AND ISOMORPHIC BOOLEAN ALGEBRAS

Subalgebras : Let S be a nonempty subset of a Boolean algebra B. If S itself be a Boolean algebra (with respect to the operations of B) then it is called *subalgebra* of B. It is clear that S is a subalgebra of B if and only if S is closed under the three operations of B, i.e., '+', '·' and '.

Example: *Let we have a Boolean Algebra B = {1, 2, 5, 7, 10, 14, 35, 70} then its subset S = {1, 2, 35, 70} is a subalgebra of B.*

Isomorphic Boolean Algebras

Two Boolean algebras B_1 and B_2 are said to be *isomorphic* if there is a one-to-one correspondence $f : B_1 \rightarrow B_2$ which preserves the three operations, i.e.,

$$f(x + y) = f(x) + f(y),$$

$$f(x \cdot y) = f(x) \cdot f(y)$$

and $f(x') = f(x)'$

for any elements, x, y in B_1.

BOOLEAN ALGEBRAS AS LATTICES

By the definition of Boolean algebra, we know that every Boolean algebra B satisfies the associative, commutative and absorption laws. Hence it is a lattice where + and · are the join and meet operations respectively. With respect to this lattice,

$$a + 1 = 1$$

$$\Rightarrow \quad a \leq 1 \text{ and } a \cdot 0 = 0\backslash$$

$$\Rightarrow \quad 0 \leq a,$$

for any element $a \in B$. Thus B is a bounded lattice. Furthermore, we know that B is distributive and complemented also. Thus, we have

A Boolean algebra B is a bounded, distributive and complemented lattice.

Since a Boolean algebra is a lattice, it has a natural partial ordering. We define $a \leq b$ when the conditions $a + b = b$ and $a \cdot b = a$ hold. In Boolean algebra, we have equivalent conditions.

$a + b = b$; $a \cdot b = a$; $a' + b = 1$; $a + b' = 0$.

Thus, whenever any of the above four condition is known to be true in a Boolean algebra, we can write a $\leq$ b.

LOGIC GATES

Computer science involves designing devices to produce appropriate outputs from given inputs. In digital electronics, a gate is a logic circuit which has two or more inputs and one output. The output waveforms of a logic gate depend upon the input waveforms and the input-output characteristic of the circuit described in terms of mathematical logic. These gates are represented by symbols, the lines entering the symbol from left are inputs, and the line on the right is the output. Placing a small circle on an input or output complements the signal on that line.

THE OR GATE

The OR circuit has two or more inputs and would give an output if a pulse is applied to any one of the inputs. If the n inputs to the OR logic circuit be designated by A, B, ..., N and the output by Y, then the OR operation may be defined as "Y equals A or B or ... or N". The Boolean expression for OR operation is written as

$$Y = A + B + \ldots + N,$$

where the + symbol indicates the OR concept. It is to be noted that each of the symbols A, B, ..., N may assume one or two possible values, either 0 or 1.

Fig. 1.3(a) shows a two output OR gate using two ideal diodes D_1 and D_2. Here A and B represent the two inputs and Y the output. R represents the output load inverter. Fig. 1.3(b) gives the symbolic representation of the OR gate.

There are four possibilities in which the input voltages may appear :

(i) When both A = 0V, and B = 0V, i.e., there is no voltage either at A or B, none of the diode conducts and output Y = 0V.

(ii) When A = 0V and B = 1V, diode D_2 is forward biased and hence conducts. The circuit current flows via R dropping 1V across it. In this way output Y achieves a potential of 1 volt. Thus Y = 1V.

(iii) When A = 1V and B = 0V, D_1 is forward biased and hence conducts causing output Y = 1V.

(iv) When both A = 1V and B = 1V, both diodes are forward biased and conduct but the drop across R is 1V, i.e., output Y = 1V because voltages of A and B are in parallel.

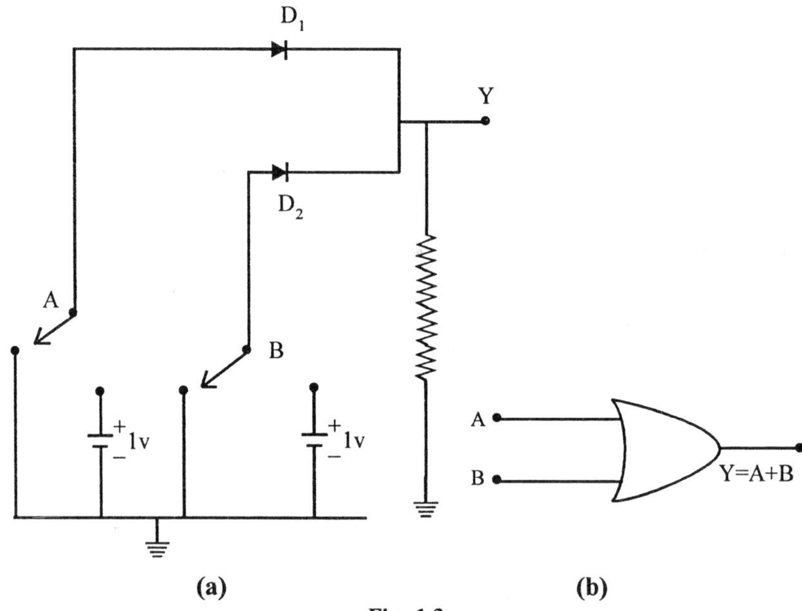

Fig. 1.3

All the possible conditions in a two input OR circuit can be represented in a tabular fashion known as truth table.

Truth Table OR Operation

Input		Output
A	**B**	**Y = A + B**
0	0	0
0	1	1
1	0	1
1	1	1

There exists a parallelism between the logic OR gate and the connection of electrical switches. Fig 1.4 shows an equivalent OR switch.

There exists a circuit between terminals 1 and 2 if switch A or switch B or both are closed.

Thus *OR gate is equivalent to a parallel circuit* in its logic function. It is also called *inclusive OR gate* because it includes the case when both inputs are true.

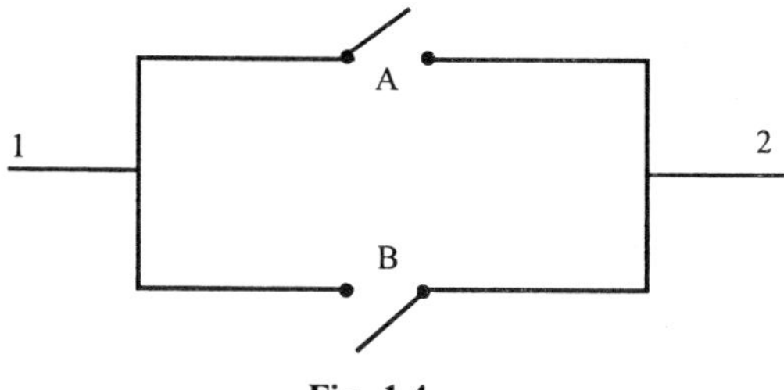

Fig. 1.4

LOGIC CIRCUITS

For various decision making functions a combination of AND, OR and NOT logic gates is employed. The logic circuits so formed are known as *combinational logic circuits.* There are many types of combinational logic circuits which perform specific functions such as encoders, decoders, multiplexers, etc. While these combinational logic circuits employ a variety of circuit configurations, we will discuss here only two configurations, which occur frequently in combinational logic circuits. These circuits are :

(a) Sum-of-Products (SOP) Circuits : Fig. 1.5 shows a circuit in which the output of two AND gates constitutes the input of two AND gates constitutes the input of an OR gate. The final output is thus (AB + CD).

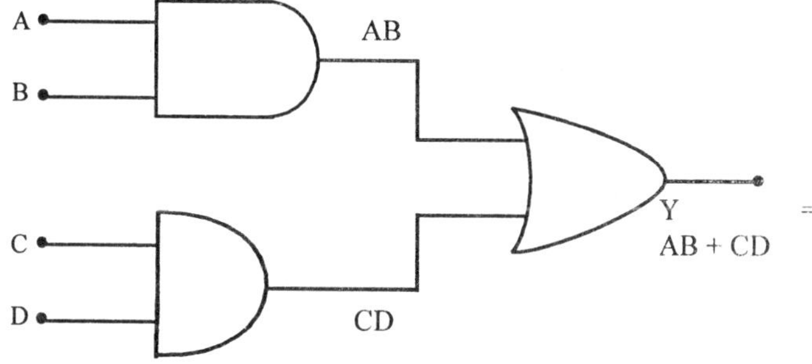

Fig. 1.5

Clearly, AB and CD are the two product terms and the output of the OR gates a sum term (AB + CD), which is in the form sum-of-products. The distinguishing feature of this circuit is that, inputs are coupled to AND gates and the output is obtained from an OR gate.

The sum-of-products circuit given in Fig. 1.5 is in the form of a two-level network, which means that the signal passes from input to output through two gates. There may, however, be SOP circuits which employ more than two levels.

(b) Product-of-Sums

Circuit : The other basic decision making circuit is referred to as a product-of-sums circuit, in its simplest form, is shown in Fig. 1.6.

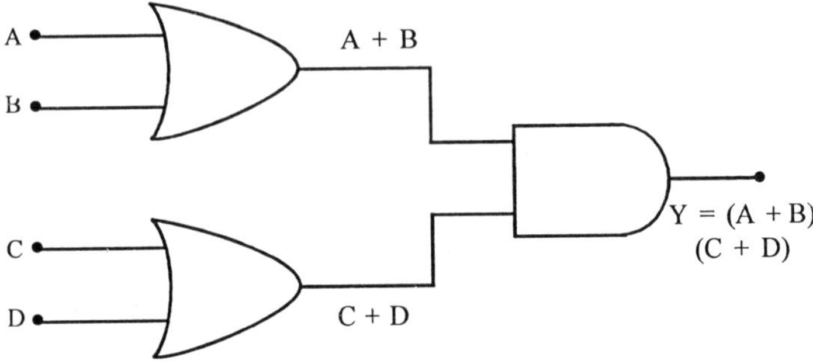

Fig. 1.6

In this type of circuit inputs feed OR gates and the output is derived from an AND gate. The expression for the output of this circuit is

$$Y = (A + B) \cdot (C + D)$$

Like the SOP circuit this is also a two level circuit, in which both the sum terms have two variables. However, a sum term may have a single variable or a sum of several variable. The POS circuit of Fig. 1.6 generates a product of two sum terms, but these may be POS expressions having several sum terms.

TRANSLATING ALGEBRA TO LOGIC

All logic diagrams are so drawn that the input is on the left hand side and the output is on the right hand side. To draw a logic diagram which corresponds to a given Boolean expression, we should begin at the output and develop the logic circuit as we work backwards from the output to the input.

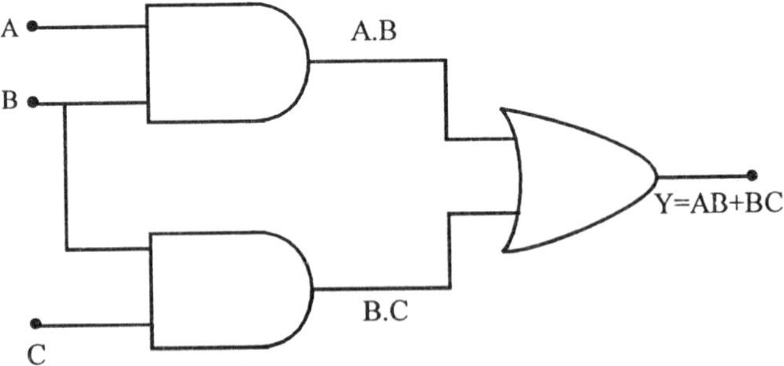

Fig. 1.7

Let we want to develop a logic circuit which corresponds to the following Boolean expression :

$$Y = A \cdot B + B \cdot C$$

The expression is a sum-of-products form, which points to an OR gate at the output. The OR gate has two inputs $A \cdot B$ and $B \cdot C$. These inputs point to an AND function which provide inputs $A \cdot B$ and $B \cdot C$. Therefore, the following logic circuit will correspond to the given Boolean expression.

TRANSLATING LOGIC CIRCUIT TO ALGEBRA

In developing a Boolean expression for a given logic circuit, we start at the input and develop an expression for the output of each logic gate advancing from left to right, until the output has been reached. For example, consider the logic diagram given in Fig. 1.8. First we will translate its function into a Boolean expression and then we will attempt to simplify the Boolean expression and recreate a logic circuit which will perform the same function. Steps for translating the given logic circuit to Boolean expression are as follows :

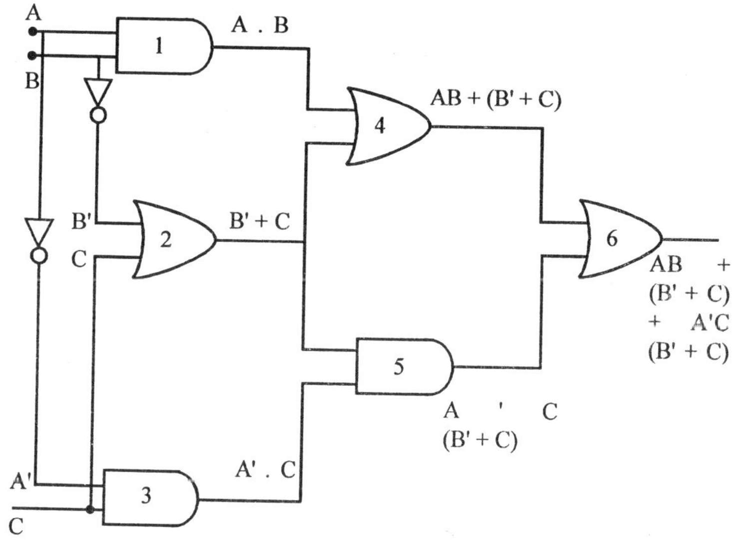

Fig. 1.8

(i) Inputs A and B feed AND gate 1, so its output is $A \cdot B$.

(ii) Inputs B' and C feed OR gate 2, so its output is $B' + C$.

(iii) Inputs A' and C feed AND gate 3, so its output is $A' \cdot C$.

(iv) Outputs of gates 1 and 2 feed OR gate 4, so its output is $A \cdot B + (B' + C)$.

(v) Outputs of gates 2 and 3 feed AND gate 5, so its output is $(A' \cdot C)(B' + C)$.

(vi) Outputs of gates 4 and 5 feed OR gate 6, so the final output is

$$A \cdot B + (B' + C) + A' \cdot C(B' + C).$$

This Boolean expression is capable of simplification as below :

$$Y = A \cdot B + (B' + C) + A' \cdot C(B + C)$$

$$= A \cdot B + (B' + C)(1 + A'C)$$

$$= A \cdot B + B' + C$$

$$= A + B' + C$$

This expression can be implemented by a 3-input OR gate as shown in Fig. 1.9.

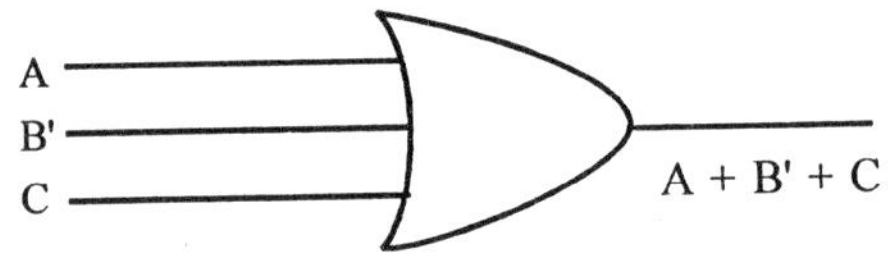

Fig. 1.9

TRUTH TABLE FROM LOGIC CIRCUIT

For analyzing the performance of a logic circuit, it is useful to develop a truth table from which the output can be evaluated for all possible combinations of input values. We will analyze the performance of the logic circuit given in Fig. 1.10.

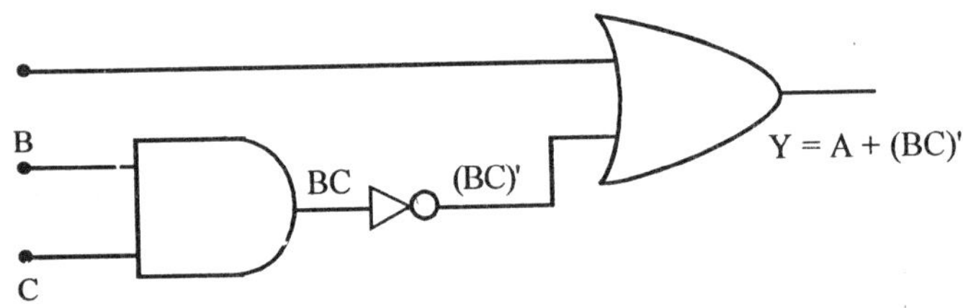

Fig. 1.10

(i) As a first step develop an output equation for the logic circuit following the guidelines given in Fig. 9.16.

The output equation for the logic circuit of Fig. 9.23 is

$$Y = A + (BC)'$$

(ii) The next step is to determine the number of variables from which the number of input combinations should be determined.

In the given circuit the number of variables is 3 and, therefore, the number of input combination is 8.

(iii) Next draw up a truth table as below :

Truth Table

Inputs			Outputs		
A	B	C	BC	(BC)'	A + (BC)'
0	0	0	0	1	0 + 1 = 1
0	0	1	0	1	0 + 1 = 1
0	1	0	0	1	0 + 1 = 1
0	1	1	1	0	0 + 0 = 0
1	0	0	0	1	1 + 1 = 1
1	0	1	0	1	1 + 1 = 1
1	1	0	0	1	1 + 1 = 1
1	1	1	1	0	1 + 0 = 1

LOGIC CIRCUIT FROM TRUTH TABLE

Starting from a truth table, it is possible to design logic circuits in the sum-of-products and product-of-sums forms. When we design logic circuit, we know the number of inputs for which the circuit is to be designed and, according to the design creteria, we also know the output states for the various input combinations. Let us consider the following truth table for a 2-input logic circuit :

Truth Table

Input		Output	Product	Sum
A	B	X	Terms	Terms
0	0	1	A' · B'	
0	1	1	A' · B	
1	0	0		A' + B
1	1	1	A · B.	

This table also gives the product terms for which output is 1 and the sum terms for which the output is 0.

Sum-of-Products Expression

For the sum-of-products expression, we only consider the product terms for the rows for which the output is 1. Output is 1 for rows 1, 2 and 4. The output of row 1 will be 1 only when both the inputs A and B are complemented. The output of row 2 will be 1 only when A is complemented and B is not complemented. For row 4 neither input is to be complemented. The product of terms for the three output states will be $A' \cdot B'$, $A' \cdot B$ and $A \cdot B$. The logical sum of these three product terms will give the required SOP expression, i.e.,

$$\begin{aligned} Y &= A' \cdot B' + A' \cdot B + A \cdot B \\ &= A'(B' + B) + A \cdot B \\ &= A' + B \end{aligned}$$

This leads to the logic circuit given in Fig. 1.11.

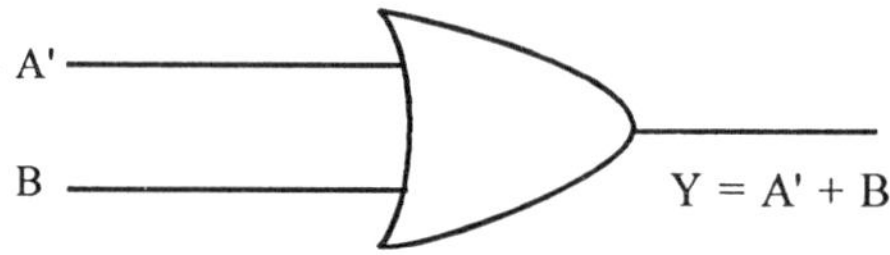

Fig. 1.11

Product-of-Sums Expression

For the product-of-sums expression, we need to consider the rows for which the output is zero. The output is 0 only for row 3. The sum term for row should be $A' + B$, i.e., input A should be complemented and input B should remain as it is. If we had two or more sum terms, we would have taken the logical product of these sum terms to give the required expression. Since there is only one sum term, the expression will be given by

$$Y = A' + B.$$

This expression is the same as for the sum-of-products expression.

Example:

Design a logic circuit for the following truth table :

Truth Table

A	Input B	C	Output X	Product Terms	Sum Terms
0	0	0	1	$A' \cdot B' \cdot C'$	
0	0	1	1	$A' \cdot B' \cdot C$	
0	1	0	1	$A' . B . C'$	$A + B' + C'$
0	1	1	0		
1	0	0	1	$A \cdot B' \cdot C'$	
1	0	1	1	$A \cdot B' \cdot C$	
1	1	0	0		$A' + B' + C$
1	1	1	0		$A' + B' + C'$

Solution:

The sum-of-products expression will be as follows :

$$Y = A' \cdot B' \cdot C' + A' \cdot B' \cdot C + A' \cdot B \cdot C' + A \cdot B' \cdot C' + A \cdot B' \cdot C$$
$$= A' \cdot B'(C' + C) + A \cdot B' (C' + C) + A' \cdot B \cdot C'$$
$$= A' \cdot B' + A \cdot B' + A' \cdot B \cdot C'$$
$$= B' + A' \cdot B \cdot C' = B' + A' \cdot C'.$$

The logic circuit required to implement this function will be as shown in Fig. 1.12.

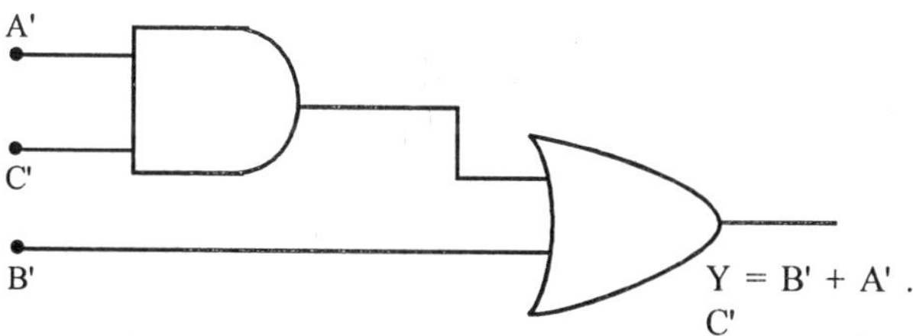

Fig 1.12

The product-of-sums expression will be the product of the three sum terms as follows :

$$Y = (A + B' + C') (A' + B' + C) (A' + B' + C')$$
$$= (A + B' + C') (A' + B')$$
$$= B'(A + A') + B'(1 + C') + A' \cdot C'$$
$$= B' + A' \cdot C'.$$

This expression is the same as for the sum-of-products solution.

KARNAUGH MAP TECHNIQUES

Before a logic equation can be translated into a logic circuit, it is necessary to reduce it to the simplest possible form. For reducing logic expressions we apply Boolean laws. The application of these laws is sometimes inconvenient when the number of application of these laws does not always lead to the simplest possible expression, which increases the chances of error. There is a better and faster system for reducing the Boolean expressions which is known as Karnaugh map technique. The technique provides a graphical method for this purpose.

Minterms and Maxterms

Before dealing with the graphical representation of a Boolean expression, we will define some of the terms used in mapping techniques.

Minterm

A *minterms of* n variables is a product of n literals in which each variable appears exactly once in either true or complemented form, but not both. For example, the list of all the minterms of the two variables a and b are

ab, a'b, ab', a'b'

Similarly, the list of all the minterms of three variables a, b and c are

abc, abc', ab'c, a'bc, ab'c', a'b'c a'bc'. a'b'c'.

In a similar way, n variables can be combined to form 2^n minterms.

Maxterms

A maxterm of n variables is a sum of n literals in which each variable appears exactly once in either true or complemented form, but not both. For example, all maxterms of two variables a and b are

a + b, a' + b, a + b', a' + b',

and all maxterms of three variables a, b and c are

a + b + c, a' + b + c, a + b + c', a + b' + c,

a + b' + c', a' + b + c', a' + b' + c, a' + b' + c'.

In a similar manners, n variable can be combined to form 2^n maxterms.

When a Boolean function *f* is written as a sum of minterms, it is called as a *minterm expansion* or the *disjunctive normal form* of the Boolean function. It is also called *canonical sum of products* or *standard sum of products*. The following are minterm expansion of Boolean functions :

f(a, b) = ab + a'b

f(a, b, c) = abc + ab'c' + a'bc + a'b'c.

The canonical sum-of-products form can be implemented simply by an interconnection of gates. Each minterm is realised using single AND gate. The outputs of all the minterms are connected to the inputs of a single OR gate whose output is the desired function. Clearly, if there are n minterms in the canonical sum of products, there will be n + 1 gates used in its realisation. For example f(a, b, c) given above can be realised as shown in Fig. 1.13.

When a Boolean function f is written as a product of maxterms, it is called a *maxterm expansion* or the *conjuctive normal form* of the Boolean function. It is also referred to as *canonical product of sums or standard product of sums.*

The following are maxterm expansions of Boolean functions

f(a, b) = (a + b) (a' + b)

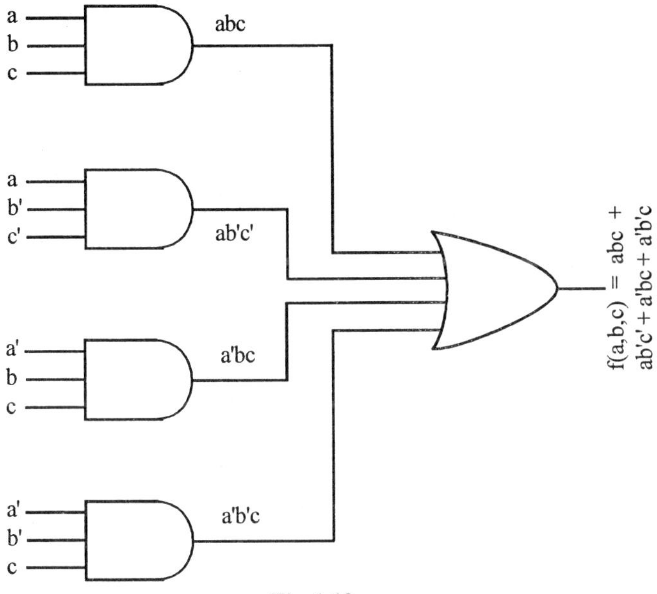

Fig 1.13

f(a, b, c) = (a + b + c) (a + b' + c) (a' + b + c') (a' + b' + c).

The gate circuit or realisation of f(a, b, c) is shown in Fig. 1.14.

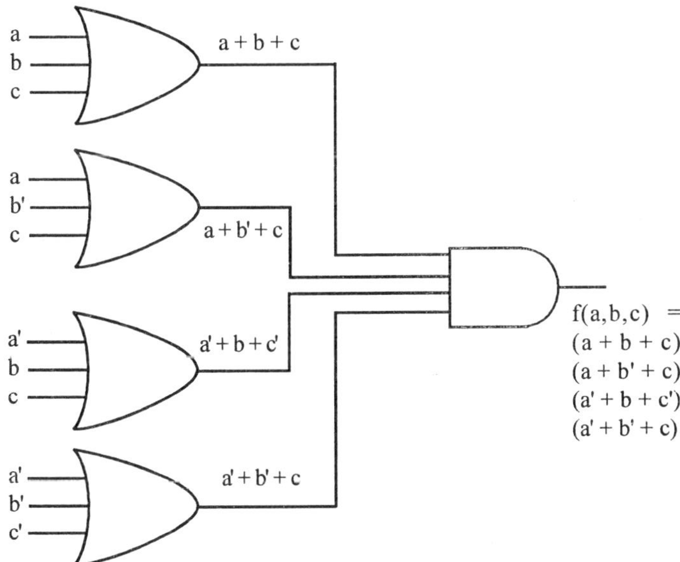

Fig. 1.14

Boolean functions expressed as a sum of minterms or product of maxterms are said to be in *canonical form.*

Karnaugh Maps

A Karnaugh map (k-map) is a diagram made up of a number of squares. If the Boolean expression contains n variables, the map will have 2^n squares. Each square represents a minterm and 1s are written in the corresponding squares for the minterms present in the expression and 0s are written in those squares which correspond to the minterms not present in the expression. As soon as the map is filled with 1s and 0s, the canonical sum-of-products expression for the output can be obtained by OR ing together those squares that contain 1. It is a two dimensional representation of a truth table. It can display in a visual form all the information contained in a truth table.

2-Variable Karnaugh Maps

As n = 2 here, the map will have $2^2 = 4$ squares. The values of one variable, say A, are listed above the top horizontal line and the values of other variable (B, say), are listed on the left side. Four possible mintems with two variables A and B

AB, AB', A'B, A'B'

are represented by the four squares in the map as labelled in Fig. 1.15 and alternative representation. Squares are said to be *adjacent* if the minterms that they represent differ in exactly one literal. For example, the square representing A'B is adjacent to the squares representing AB and A'B'.

	A'	A
B'	A'B'	AB'
B	A'B	AB

B \ A	A' 0	A 1
B' 0	A'B'	AB'
B 1	A'B	AB

Fig. 1.15

The expression can be simplified by properly combining those squares in the K-map which contain 1s. The process of combining 1s is called *looping*. Whenever there are 1s in two adjacent squares in the K-map, the minterms represented by these squares can be looped and it eliminates the variable that appear in complemented and uncomplemented form.

3-Variable Karnaugh Map

A Karnaugh map in three variables is a rectangle divided into $2^3 = 8$ squares. To accommodate the eight product terms, we can have two squares in a row and four in each column, or four in each row and two in each column, which will in either case accommodate all the eight product terms. We will use the latter alternative as given in Fig. 1.16.

A \ BC	B'C' 0 0	B'C 0 1	BC 1 1	BC' 1 0
A' 0	A'B'C' 0 0 0 1	A'B'C 0 0 1 1	A'BC 0 1 1 0	A'BC' 0 1 0 1
A 1	AB'C' 1 0 0 0	AB'C 1 0 1 1	ABC 1 1 1 1	ABC' 0 1 0 0

Fig. 1.16

In this map the rows are labelled A' and A, as a result of which when we move down from the top row to the row below in any column, only one of the variables that is A' changes to A. The columns are also so labelled that only one variable is different between adjacent squares horizontally or vertically (not diagonally).

Input	**Minterm**			
A	**B**	**C**	**Fundamental Product**	**Output**
0	0	0	A'B'C'	1
0	0	1	A'B'C	1
0	1	0	A'BC'	1
0	1	1	A'BC	0
1	0	0	AB'C'	0
1	0	1	AB'C	1
1	1	0	ABC'	0
1	1	1	ABC	1

The fundamental products of a 3-variable system are given in following table and are entered in this map. Product terms in each square represent the product of three variables, which correspond to the row and column of their positions. Also each product term finds a place in one of the squares and that each entry has an address in binary form. The addresses correspond to the binary representation of the product terms. For example, the address of the term AB'C is 101 and the address of A'BC' is 010.

It is to be noted that the address of every term differs from those of the adjacent terms in the same row and column by only one digit, which means that there is a change in only one of the variables in the product terms. Also, addresses of terms at the opposite extremities of the same row differ by only one digit. They can therefore be regarded to be adjacent as indicated below :

(i) A'B'C' and A'BC' are adjacent.

(ii) AB'C' and ABC' are adjacent.

This map gives all the information contained in the truth table. In this map the following minterms are adjacent, which can be combined to form groups for the purpose of reduction of the Boolean function.

(i) Minterms A'B'C and AB'C are adjacent as only variable A' changes to A. Variables B'C do not change and so this group yields the product B'C.

(ii) Minterms AB'C and ABC are adjacent and form a group yielding the product AC, as only variable B' changes to B. The first and second groups overlap, but that is permissible.

(iii) Minterms A'B'C' and A'BC' be considered to be adjacent, although they are the extreme ends of a rows, since only variable B' has changed to B. This group yields a product A'C'.

(iv) Although minterms A'B'C' and A'B'C are adjacent but they have not been grouped, as grouping them does not cover any minterm which has not already been covered.

According to the Karnaugh map technique, the SOP expression is obtained by ORing all the reduced terms covering all the logic 1s on the map. Therefore, the reduced function is

$$Y = B'C + AC + A'C'.$$

Four Variable Karnaugh Map

A Karnaugh map in four variables is a square divided into $2^4 = 16$ squares. The squares represent the 16 possible minterms in four variables. The definition of adjacent squares must be extended so that not only are left most and right most column adjacent as in the 3-variables map but also the first and last rows are adjacent. Karnaugh map for four variables A, B, C and D. Note that A' B' C' D' square in the first row is adjacent to the square A' B' C D' in the last row, since they differ only in the variable C.

Essential and Nonessential Prime Implicants

Essential prime implicants consist of those groups of minterms which contain a minterm that does not form a past of any other group of minterms constituting a prime implicant. Group containing minterms 0, 1, 2 and 3 has two minterms 0 and 1, which do not form a part of any other group and it is therefore an essential prime implicant. For the same reason, the enclosure containing minterms 2, 3, 6 and 7 is an essential prime implicant. The group of minterms 13 and 15 constitutes an essential prime implicant, as minterm 13 cannot form a part of any other group, whereas the group of minterms 7 and 15 is a nonessential prime implicant, as both these minterms have already been covered by other essential prime implicants. A useful test to judge whether a prime implicant is essential, is to see whether it can be removed without leaving any minterm unenclosed.

Sum-of-Products Reduction

Karnaugh maps will be found very useful in simplifying logic expressions before translating them into logic circuits. The starting point for designing a logic circuit is either a truth table or a Boolean equation. For a sum-of-products solutions, we have to develop product terms for which we may proceed as follows :

(i) If the starting point is a Boolean expression, examine if each term of the equation has all the variables. If not expand the expression.

(ii) If the starting point is a truth table, develop all the product terms.

(iii) Construct an SOP expression containing those product terms which correspond to logic 1 state.

(iv) Now draw a Karnaugh map having 2^n squares, where n is the number of variables and enter 1 in those squares in the map, which correspond to the product terms.

(v) Now form groups of 1s.

(vi) Now take all the non-variables corresponding to the enclosures and write the SOP expression by O Ring the reduced variable terms.

MINIMAL BOOLEAN EXPRESSIONS

Minimal Sum-of-Products

Let us have a Boolean sum-of-products expression E. Let E_L be the number of literals and E_s be the number of summands in E. For example, suppose

$$E = abc' + a'b'd + ab'c'd + a'bcd$$

Then $E_L = 3 + 3 + 4 + 4 = 14$ and $E_s = 4$.

Suppose E and F are equivalent Boolean sum-of-products expressions. Then E is said to be *simpler* than F if :

(i) $E_L < F_L$ and $E_s \leq F_L$, or

(ii) $E_L \leq F_L$ and $E_s < F_L$.

E is known to be *minimal* if there is no equivalent sum-of-products expression which is simpler than E. There can be more than one equivalent minimal sum-of-products expressions.

Prime Implicants

A fundamental product P is called a *prime implicant* of a Boolean expression E if

$$P + E = E$$

and no other fundamental product contained in P has this property. For example, let

$$E = ab' + abc' + a'bc'$$

We have

$$E = ab'(c + c') + abc' + a'bc'$$

$$= ab'c + ab'c' + abc' + a'bc'$$

Now, $\quad ac' = ac'(b + b) = abc' + ab'c'$

Since the summands of ac' are among those of E, we have

$$ac' + E = E.$$

Similarly, we can show that

$$a + E \neq E \text{ and } c' + E \neq E.$$

Thus ac' is a prime implicant of E.

Consensus of Fundamental Products

Let P_1 and P_2 be fundamental products such that exaclty one variable, say a_i, appears uncomplemented in one of P_1 and P_2 and complemented in the other. Then the *consensus* of P_1 and P_2 is the product (without repetition) of the literals of P_1 and the literals of P_2 after a_i and a_i' are deleted.

COMPLETE SUM-OF-PRODUCTS FORMS

A Boolean expression $E = E(a_1, a_2, ..., a_n)$ is said to be a *complete sum-of-products* expression if E is a sum-of-products expression where each product involves all the n variables. Such a fundamental product P which involves all the variables is called a *minterm,* and there is a maximum of 2^n

such products for n variables.

Following method may be used to transform a sum-of-products form into a complete sum-of-products form :

(i) Find a product P in E which does not involve the variable a_i, and then multiply P by $(a_i + a'_i)$, deleting any repeated products. [$\because a_i + a_i' = 1$ and $P + P = P$].

(ii) Repeated step (i) until every product P in E is a minterm.

THE DIAGONALIZATION PROCESS

We now have the background to understand the main ideas behind the diagonalization process.

Definition : Eigen value, Eigenvector. *Let A be an n × n matrix over* R., λ is an eigenvalue of A if for some non-zero column vector $X \in R^n$ we have AC = λX. X is called an eigenvector corresponding to the eigenvalue λ.

Example:

Find the eigenvalues and corresponding eigenvectors of the matrix.

$$A = \begin{bmatrix} 2 & 1 \\ 2 & 3 \end{bmatrix}$$

We want to find $X = \begin{bmatrix} x_1 \\ x_2 \end{bmatrix}$

and λ, such that **AX** = λ**X**.

$$\Leftrightarrow \begin{bmatrix} 2 & 1 \\ 2 & 3 \end{bmatrix}\begin{bmatrix} x_1 \\ x_2 \end{bmatrix} = \lambda\begin{bmatrix} x_1 \\ x_2 \end{bmatrix}$$

$$\hat{U}` \begin{bmatrix} 2 & 1 \\ 2 & 3 \end{bmatrix}\begin{bmatrix} x_1 \\ x_2 \end{bmatrix} - \lambda\begin{bmatrix} x_1 \\ x_2 \end{bmatrix} = \begin{bmatrix} 0 \\ 0 \end{bmatrix}$$

$$\hat{U} \left(\begin{bmatrix} 2 & 1 \\ 2 & 3 \end{bmatrix} - \lambda\begin{bmatrix} 1 & 0 \\ 0 & 1 \end{bmatrix}\right)\begin{bmatrix} x_1 \\ x_2 \end{bmatrix} = \begin{bmatrix} 0 \\ 0 \end{bmatrix}$$

$$\hat{U} \begin{bmatrix} 2-\lambda & 1 \\ 2 & 3-\lambda \end{bmatrix}\begin{bmatrix} x_1 \\ x_2 \end{bmatrix} = \begin{bmatrix} 0 \\ 0 \end{bmatrix} \quad \text{...(1)}$$

The last matrix equation will have non-zero solutions if and only if

$$\det \begin{bmatrix} 2-\lambda & 1 \\ 2 & 3-\lambda \end{bmatrix} = 0$$

or $(2 - \lambda)(3 - \lambda) - 2 = 0$, which becomes $\lambda^2 - 5\lambda + 4 = 0$; so $\lambda = 1$ or $\lambda = 4$. We now have to find the eigenvectors x associated with each eigenvalue.

Case 1: If $\lambda = 1$, then Equation (1) becomes:

$$\begin{bmatrix} 2-1 & 1 \\ 2 & 3-1 \end{bmatrix}\begin{bmatrix} x_1 \\ x_2 \end{bmatrix} = \begin{bmatrix} 0 \\ 0 \end{bmatrix}$$

$$\begin{bmatrix} 1 & 1 \\ 2 & 2 \end{bmatrix}\begin{bmatrix} x_1 \\ x_2 \end{bmatrix} = \begin{bmatrix} 0 \\ 0 \end{bmatrix}$$

which reduces to the single equation, $x_1 + x_2 = 0$. Frora this, $x_2 = -x_1$. This means the solution set of this equation is (in column notation).

$$E_1 = \left\{\begin{bmatrix} x_1 \\ -x_1 \end{bmatrix} x_1 \in \mathbf{R}\right\}$$

So any column vector of the form

$$\begin{bmatrix} x_1 \\ -x_1 \end{bmatrix}$$

where x_1 is any non-zero real number, is an eigenvector associated with the eigenvalue $\lambda = 1$. The reader should verify that, for example,

$$\begin{bmatrix} 2 & 1 \\ 2 & 3 \end{bmatrix}\begin{bmatrix} 2/3 \\ -2/3 \end{bmatrix} = 1\begin{bmatrix} 2/3 \\ -2/3 \end{bmatrix},$$

so that $\begin{bmatrix} 2/3 \\ -2/3 \end{bmatrix}$ is an eigenvector of A with associated eigenvalue 1.

Case 2. If $\lambda = 4$ then equation 12.4.1 becomes:

$$\begin{bmatrix} 2-4 & 1 \\ 2 & 3-4 \end{bmatrix}\begin{bmatrix} x_1 \\ x_2 \end{bmatrix} = \begin{bmatrix} 0 \\ 0 \end{bmatrix},$$

which reduces to the single equation $-2x_1 + x_2 = 0$, so that $x_2 = 2x_1$. The solution set of the equation is

$$E_2 = \left\{\begin{bmatrix} x_1 \\ 2x_1 \end{bmatrix} \middle|\ x_1 \in \mathbf{R}\right\}.$$

Therefore, all eigenvectors of A associated with the eigenvalue $\lambda = 4$ are of the form $\begin{bmatrix} x_1 \\ 2x_1 \end{bmatrix}$, where x, can be any non-zero number.

SAME APPLICATIONS

A large and varied number of applications involve computations of powers of matrices. These applications can be found in science, the social sciences, economics, the analysis of relationships with groups, engineering, and indeed, any area where mathematics is used and, therefore, where programs are to be developed. We will consider a few diverse examples here. A good introductory text for additional examples is the one by William.

The aid you understanding of the following examples, we develop a helpful technique to compute Am for m ≥ 1. If A can be diagonalized, then there is a matrix P such that $P^{-1}AP = D$, a diagonal matrix.

$$\Leftrightarrow A = PDP^{-1}$$

$$\Leftrightarrow A^m = PDmP^{-1}$$

by Exercise 8a of Section 5.4. Of course, the last equation can also be derived, by the simple observation that

$$\begin{aligned} A^m &= (PDP^{-1})^m \\ &= (PDP^{-1})\,(PDP^{-1}) \ldots (PDP^{-1}) \text{ (m factors)} \\ &= PDmP^{-1}. \end{aligned}$$

SOLVED EXAMPLES

Example 1:

Indicate the procedure of operations in the Boolean expression

$$A \cdot A' + B \cdot (A + C)' + (A \cdot C).$$

Solution:

$A \cdot A' + B\ (A + C)' + (A \cdot C)$

$(A + C')$ First scan

$A' \cdot C'$ Second scan

$0 + B \cdot A' \cdot C' + A \cdot C$ Third scan.

Example 2:

Reduce the following Boolean expression E into sum-of-products form :

$$E = ((ab)'c)'\ ((a' + c)\ (b' + c'))'.$$

Solution:

(i) Using De-Morgan's laws and involution, we get

$$E = (a(b')' + c')\ ((a' + c)' + (b' + c')')$$

$= (ab + c')(ac' + bc)$

Now, E consists only of sums and products of literals.

(ii) Using the distributive laws, we get

$E = abac' + abbc + ac'c' + bcc'$

(iii) Using the commutative, idempotent and complement laws, we get

$E = abc' + abc + ac' + 0$

Now, each term in E is a fundamental product or zero.

(iv) The product ac' is contained in abc'; hence by the absorption law

$ac' + (ac' \cdot b) = ac'$

Thus we may delete abc' from the sum. Also, by the identity law for 0, we may delete 0 from the sum. Then we have

$E = abc + ac'$

Now E is represented by a sum-of-products expression.

Example 3:

Express $E(a, b, c) = a(b'c)'$ in its complete sum-of-products form.

Solution:

We have

$E = a(b'c)' = a(b + c') = ab + ac'$

Now E is represented by a sum-of-products expression.

Since $c + c' = 1$, $b + b' = 1$ hence given expression becomes

$E = ab(c + c') + ac'(b + b')$

$= abc + abc' + abc' + ab'c'$

$= abc + abc' + ab'c'$

This is complete sum-of-products form.

Example 4:

Simplify the following expression using laws of Boolean Algebra.

$$AB + ABC + A'B + AB'C$$

Solution:

We have $AB + ABC + A'B + AB'C$

$= AB + A'B + ABC + AB'C$

$= B(A + A') + AC(B + B')$

$= B \cdot 1 + AC \cdot 1$ $\quad [\because A + A' = B + B' = 1]$

$= B + AC$

Example 5:

Demogranise the expression

$[(A' + B + C') (A' + B + C)]$.

Solution:

$[(A' + B + C') (A' + B + C)]$

$= (A' + B' + C') (A' + B + C)$ ...By step 1

$= A' B C' + A' B C$ ...By step 2

$= (A')' B' (C')' + (A')' B' C'$

$= A B' C + A B' C'$.

Example 6:

Find the consensus Q of P_1 and P_2 where

(i) $P_1 = a'bc$ and $P_2 = a'bd$

(ii) $P_1 = abc'd$ and $P_2 = ab'e$

(iii) $P_1 = ab'$ and $P_2 = b$

(iv) $P_1 = a'bc$ and $P_2 = abc'$.

Solution:

(i) In P_1 and P_2 no variable appears uncomplemented in one of the products and complemented in the other. Hence P_1 and P_2 have no consensus.

(ii) Here b appears in P_1 and its complement b' appears in P_2, hence delete b and b' and then multiply the literals of P_1 and P_2 (without repetition) to get Q = ac'de.

(iii) Here b appears in P_2 and its complement b' in P_1. Hence deleting b and b', we get Q = a.

(iv) Each a and c appear complemented in one of the products and uncomplemented in the other. Hence P_1 and P_2 have no consensus.

Example 7:

Recursion. Consider the Fibonacci sequence.

$F_0 = 1, F_1 = 1$

$F_2 = F_{k-1} + F_{k-2}$

Solution:

In order to formulate the problem in matrix form, we introduced the "dummy equation" $F_{k-1} = F_{k-1}$, so that now we have

$$F_k = F_{k-1} + F_{k-2} \qquad F_{k-1} = F_{k-1},$$

Hence, in matrix form,

$$\begin{bmatrix} F_k \\ F_{k-1} \end{bmatrix} = \begin{bmatrix} 1 & 1 \\ 1 & 0 \end{bmatrix} \begin{bmatrix} F_{k-1} \\ F_{k-2} \end{bmatrix}$$

$$= A \begin{bmatrix} F_{k-1} \\ F_{k-2} \end{bmatrix} \text{ if } A = \begin{bmatrix} 1 & 1 \\ 1 & 0 \end{bmatrix}$$

$$= A^2 \begin{bmatrix} F_{k-2} \\ F_{k-3} \end{bmatrix} \qquad \text{Why?}$$

$$= A^3 \begin{bmatrix} F_{k-3} \\ F_{k-4} \end{bmatrix}$$

$$\vdots$$

$$= A^{k-1} \begin{bmatrix} F_1 \\ F_0 \end{bmatrix}$$

$$= A^{k-1} \begin{bmatrix} 1 \\ 1 \end{bmatrix} \qquad \text{Why?}$$

Next, by diagonalizing A and using the fact that $A^m = PDP^mP^{-1}$, we can show that

$$F_k = \frac{1}{\sqrt{5}} \left[\left(\frac{1+\sqrt{5}}{2} \right)^k - \left(\frac{1-\sqrt{5}}{2} \right)^k \right].$$

Comments:

(1) An equation of the form $F_k = aF_{k-1} + bF_{k-2}$, where a and b are given constants, is sometimes referred to as a *homogeneous second-order* (to determine a term, we need to know the two predecessor terms) *linear difference equation.* The conditions $F_0 = c_0$ and $F_1 = c_1$ are called *initial conditions* and are given constants. Those of you who are familiar with differential equations may recognize that the above language parallels that in differential equations. Difference (recurrence) equations move forward discretely—that is in a finite number of finite steps—while a differential equation moves continuously—that is, takes an infinite number of infinitesimal steps.

(2) A recurrence relationship of the form $F_k = aF_{k-1} + b$, where a and b are constants, is called a first-order difference equation. In order to write out the sequence, we need to know that (one) initial condition and a and be. Equations of this type can be solved similarly to the method outlined in Example 12.5.1. by introducing the superfluous equation $1 = 0 * F_{k-1} + 1$ to obtain in matrix form:

$$\begin{bmatrix} F_1 \\ 1 \end{bmatrix} = \begin{bmatrix} a & b \\ 0 & 1 \end{bmatrix} \begin{bmatrix} F_{k-1} \\ 1 \end{bmatrix}$$

$$= \begin{bmatrix} a & b \\ 0 & 1 \end{bmatrix}^k \begin{bmatrix} F_0 \\ 1 \end{bmatrix} \quad \text{Why?}$$

Example 8:

Graph Theory. Consider the graph in Figure 12.5.1. From the procedures outlined in Section 6.4., the adjacency matrix of this graph is

$$A = \begin{bmatrix} 1 & 1 & 0 \\ 1 & 0 & 1 \\ 0 & 1 & 1 \end{bmatrix}$$

Recall that A^k is the adjacency matrix of the relation r^k, where r is the relation r = {(a, a), (a, b), (b, a), (b, c), (c, b), (c, c)} of the above graph. Also recall that in computing A^k, we used Boolean arithmetic. What happens if we use “regular” arithmetic? Then, for example,

$$A^2 = \begin{bmatrix} 2 & 1 & 1 \\ 1 & 2 & 1 \\ 1 & 1 & 2 \end{bmatrix}$$

How can we interpret this? We note that $A_{33} = 2$ and that there are two paths of length two from node 3 (that is, c) to node 3 (c). Also, $A_{13} = 1$, and there is one path of length 2 from node a to node c.

Example 9:

Prove the associative laws for B

(i) $a + (b + c) = (a + b) + c, \ \forall\, a, b, c \in B.$

(ii) $a \,.\, (b \,.\, c) = (a \,.\, b) \,.\, c, \ \forall\, a, b, c \in B.$

Solution:

Let A = a + (b + c), B = (a + b) + c.

Hence A . a = [a + (b + c)] . a

$a \cdot [a + (b + c)]$

$= a \cdot [a + (b + c)]$

$= a$. (by Absorption Law)

and $B \cdot a = [(a + b) + c] \cdot a$

$= (a + b) \cdot a + c \cdot a = a \cdot (a + b) + c \cdot a$

$= a + c \cdot a = a + a \cdot c = a \cdot 1 + a \cdot c$

$= a (1 + c) = a \cdot 1 = a.$

Thus $A \cdot a = B \cdot a$ ...(1)

Again $A \cdot a' = [a + (b + c)] \cdot a'$

$= a \cdot a' + (b + c) \cdot a'$

$= 0 + (b + c) \cdot a' = (b + c) \cdot a'$

and $B \cdot a' = [(a + b) + c] \cdot a'$

$= (a + b) \cdot a' + c \cdot a'$

$= [a \cdot a' + b \cdot a'] + c \cdot a'$

$= [0 + b \cdot a'] + c \cdot a' = b \cdot a' + c \cdot a'$

$= (b + c) \cdot a'$ ...(2)

Thus $A \cdot a' = B \cdot a'$ ...(2)

From (1) and (2), we get $A = B$

Hence $a + (b + c) = (a + b) + c$.

(ii) Let $X = (a \cdot b) \cdot c$ and $Y = a \cdot (b \cdot c)$

Now, $a + X = a + [(a \cdot b) \cdot c] = (a + a \cdot b) \cdot (a + c)$

$= a \cdot (a + c) = a \cdot a + a \cdot c$

$= a + a \cdot c = a$

Similarly $a + Y = a$

Again $a' + X = a' + [(a \cdot b) \cdot c]$

$= (a' + a \cdot b) \cdot (a' + c)$

$= [(a' + a) \cdot (a' + b)] \cdot (a' + c)$

$= [1 \cdot (a' + b)] \cdot (a' + c)$

$= (a' + b) \cdot (a' + c) = a' + b \cdot c$

$= 1 \cdot (a' + b \cdot c) = (a' + a) \cdot (a' + b \cdot c)$

$= a' + [a \cdot (b \cdot c)] = a' + Y$

Now $(a + X) \cdot (a' + X) = (a + Y) (a' + Y)$

$\Rightarrow \quad (X + a) . (X + a') = (Y + a) (Y + a')$

$\Rightarrow \quad X + a . a' = Y + a . a'$

$\Rightarrow \quad X + 0 = Y + 0$

$\Rightarrow \quad X = Y$

Hence, $(a . b) . c = a . (b . c)$.

Example 10:

Show that the set B with elements 0 and 1 and operations + and '.' defined below is Boolean Algebra.

+	0	1
0	0	1
1	1	1

.	0	1
0	0	0
1	0	1

Solution:

From the above tables, we conclude

I. Commutative Laws

(i) $0 + 1 = 1 = 1 + 0$, (ii) $0.1 = 0 = 1.0$

II. Associative Laws

(i) $1 + (0 + 1) = 1+1 = (1 + 0) + 1$,

(ii) $1.(0.1) = 1.0 = 0 = 0.1 = (1.0) . 1$

III. Distributive Laws

(i) $1 . (0 + 1) = 1.1 = 1 = 0 + 1 = 1. 0 + 1 . 1$

(ii) $1 + (0.1) = 1 + 0 = 1 = 1.1 = (1 + 0) . (1 + 1)$.

IV. Existence of Identity Elements

Since $0 + 0 = 0$, $1 + 0 = 1$, 0 is the identity element with respect to +, and since $1.0 = 0$, $1.1 = 1$, 1 is the identity element with respect to (.).

V. Existence of Complements: In the set B = [0, 1], the complement of 0 is 1 and that of 1 is 0. Therefore, corresponding to each element 0 and 1 of B, we have

$0 + 0', = 0 + 1, 1 + 1' = 1 + 0 = 1,$

1 being the identity element corresponding to b . 0., ad

$0.0' = 0.1 = 0, 1.1' = 1.0 = 0.$

0 being the identify element corresponding to b.0 +.

Thus the algebra given by the table of operations, and the elements mentioned above, is a Boolean Algebra.

Example 11:

In a Boolean Algebra B prove that $(a')' = a$, $\forall a \in B$, where a' is the complement of a.

Solution:

Since a' is the complement of a,

$$\left.\begin{array}{ll} \therefore & a + a' = a' + a = 1 \\ \text{and} & a \,.\, a' = a' \,.\, a = 0 \end{array}\right\} \qquad \text{...(1)}$$

Let (a')' be the complement of (a') so that

$$\left.\begin{array}{l} (a') + (a')' = (a') + a' = 1 \\ (a') \,.\, (a')' = (a')' \,.\, (a') = 1 \end{array}\right\} \qquad \text{...(2)}$$

Now (1) and (2) imply that a and (a')' are the two complements of the element $a' \in B$. But in Boolean algebra every element of B has a unique complement.

Hence $(a')' = a$.

Example 12:

Establish the following results in a Boolean Algebra B

(i) $a \,.\, (a + b) = a$,

(ii) $(a \,.b) \,.\, c = a \,.\, b \,.\, c$.

Solution:

(i) $a \,.\, (a + b) = a$

Now $a = a \,.\, 1$

$= a \,.\, (1 + b)$ $\qquad (\because b + 1 = 1)$

$= a \,.\, 1 + a \,.\, b$ $\qquad$ (Postulate III)

$= a + a \,.\, b$ $\qquad$ (Postulate IV)

$= a \,.\, a + a \,.\, b$ $\qquad (\because a \,.\, a = a)$

$= a \,.\, (a + b)$ $\qquad$ (Postulate II)

$\therefore a \,.\, (a + b) = a.$

(ii) $(a \,.\, b) \,.\, c = a \,.\, b \,.\, c.$

Consider $(a \,.\, b) \,.\, c - [a \,.\, b \,.\, (a + a')] \,.\, c$ $\qquad [\because a + a' = 1 \text{ and } a \,.\, 1 = 0]$

$= (a \,.\, b \,.\, a + a \,.\, b \,.\, a') \,.\, c$ $\qquad$ (Postulate III)

$= (a \,.\, a \,.\, b + a \,.\, a' \,.\, b) \,.\, c$ $\qquad$ (Postulate I)

$= (a \,.\, b + 0 \,.\, b) \,.\, c$ $\qquad (\because a \,.\, a = a \text{ and } a \,.\, a' = c)$

$$= (a . b + 0 . b) . c \qquad (\because a . 0 = 0)$$

$$= c . (a . b + 0) \qquad \text{(Postulate I)}$$

$$= c . a . b + c . 0 \qquad \text{(Postulate III)}$$

$$= c . a . b + 0 \qquad (\because a . 0 = 0)$$

$$= c . a . b \qquad \text{(Postulate IV)}$$

$$= a . b . c. \qquad \text{(Postulate I)}$$

Example 13:

Using Boolean Algebra, Show that $p . q. r + p . q . r' + p . q' . r + p' . q . r = p . q + q . r + r . p.$

Solution:

L.H.S. $p . q . r + p . q . r' + p . q' . r + p' . q . r$

$$= (p . q . r + p . q . r') + (p . q' . r + p' . q . r)$$
$$= p . q . (r + r') + p . q' . r + p' . q . r$$
$$= (p . q) . 1 + p . q' . r + p' . q . r$$
$$= (p . q + p . q' . r) + p' . q . r$$
$$= (p . q) + p . q' . r + p' . q . r$$
$$= p . (q + q' . r) + p' . q . r$$
$$= p . (q + r) + p' . q . r$$
$$= p . q + p . r + p' . q . r$$
$$= p . r + (p . q + p' . q . r)$$
$$= p . r + q . (p + p' . r)$$
$$= p . r + q . (p + r)$$
$$= p . r + q . p + q . r$$
$$= q . p + q . r + p . r$$
$$= p . q + q . r + r . p = \text{R.H.S.}$$

Example 14:

Using Boolean Algebra determine the validity of the following argument: "The triangle is isosceles if and only if two of its sides are equal. No two sides of the triangle are equal. Therefore, triangle is not isosceles".

Solution:

Let p stands for "The triangle is isosceles" and q for "Two sides of the triangle are equal". Then the above argument can be written as $(p \Leftrightarrow q) \wedge \sim q \Rightarrow \sim p$. In the language of Boolean Algebra, this may be written a

$$[a \,.\, b + a' \,.\, b'] + a'$$
$$= b' \,.\, (a \,.\, b + a' \,.\, b']' + a'$$
$$= [b' \,.\, (a \,.\, b) + b' \,.\, (a' \,.\, b')]' + a'$$
$$= [(b' \,.\, b) \,.\, a\, b' \,.\, (a' \,.\, b')]' + a'$$
$$= [0 + b' \,.\, (a' + b')]' + a'$$
$$= [b' \,.\, (a' \,.\, b')]' + a' = b + (a + b) + a'$$
$$= b + (a + a') = b + 1$$
$$= 1, \text{ which is a tautology}$$

Thus the given argument is a valid argument.

Example 15:

Test the validity of the following argument based on Boolean Algebra.

If my son stands first in the class, I will given him a watch. Either he stood first or I was out of station. I did not give my son a watch this time. Therefore, I was out of station.

Solution:

Let a : My son stands first in the class.

b : I will give him a watch

c : I was out of station.

Thus the premises are a

$\Rightarrow$ b, $a \vee c$, $\sim c$, conclusion : c.

Using Boolean Algebra, we have

$$(a' + b) \,.\, (a + c) \,.\, b'$$
$$= (a' + b) \,.\, b' \,.\, (a + c)$$
$$= (a'\, b' + bb')\, (a + c)$$
$$= (a'\, b' + 0)\, (a + c)$$
$$= (a'\, b')\, (a + c)$$
$$= a'\, a\, b' + a'\, b'\, c$$
$$= 0 + a'\, b'\, c$$
$$= a'\, b'\, c = 1 \text{ since all the premises are true.}$$

Hence $a' = 1$, $b' = 1$, $c = 1$. But we are concerned with only c, so we conclude $c = 1 \Rightarrow$ I was out of station.

Example 16:

Consider the following Boolean expression :

$Y = (A \cdot B + B' \cdot C)'$ *draw a corresponding logic circuit.*

Solution:

The given expression is a NOTed function which points to an inverter at the output, the input of which is A · B + B' · C. This is a sum-of-products function, which points to an OR gate before the inverter. These are three input signals. The inputs to the OR gate are in a product form, which points to AND functions. The input B' is scalized by using an inverter.

Example 17:

Draw a logic circuit corresponding to following Boolean expression :

$$Y = (A + C') \cdot (A' + B) + (B' + C)\ A' \cdot C'$$

Solution:

The given expression points to an OR gate at the output. It also suggests that the OR gate is fed by AND gates, which are proceeded by OR gates and an AND gate. The entire circuit is given by Fig. 1.17.

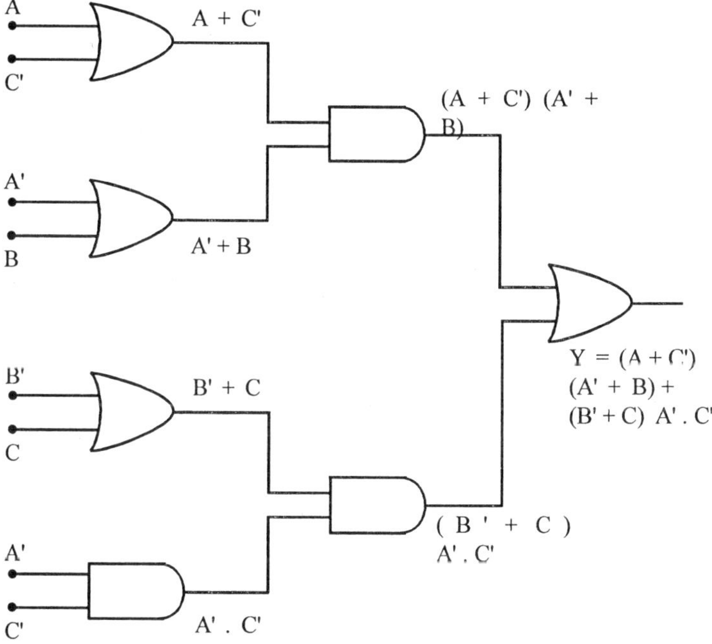

Fig. 1.17

Note : In many cases considerable simplicity and saving in cost can be achieved by reducing an expression to its simplest form before translating it into a logic circuit. Boolean expression of Example 15 can be reduced to the following form :

$$Y = B(A + C') + A' \cdot C'$$

This leads to a much simpler logic circuit as shown in Fig. 1.18 which requires only four gates.

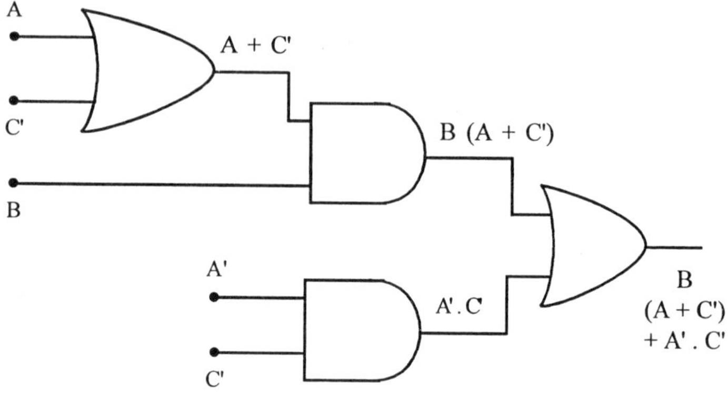

Fig. 1.18

Example 18:

Draw Karnaugh maps and simplify the expressions

(a) AB' + A'B; (b) AB' + A'B'.

Solution:

Since, K-map of the contain any adjacent squares of minterms containing 1, hence the expression can not be simplified further.

In the above figure two adjacent squares A'B' and AB' containing a have been grouped together. They have been encircled. These two terms can be looped that eliminates the A variable since it appears both in complemented and uncomplemented forms. This can be verified as follows :

$$AB' + A'B' = (A + A')B'$$
$$= I \cdot B' = B'.$$

Example 19:

Find Karnaugh map and simplify the expression

AB' + A'B + A'B'.

Solution:

The required K-map is shown in Fig. 1.19.

There are two pairs of 1s and they can be combined as shown in Fig. 1.19 in first column and first row has been enclosed twice, as it is permissible to use the same 1 more than once.

Looping of horizontal 1 squares gives the result B' and vertical 1 squares gives A'. Hence the given expression reduces to the simplified form as A' + B'. This can be verified as follows :

$$AB' + A'B + A'B' = AB' + A'(B + B')$$
$$= AB' + A'$$
$$= A' + B'.$$

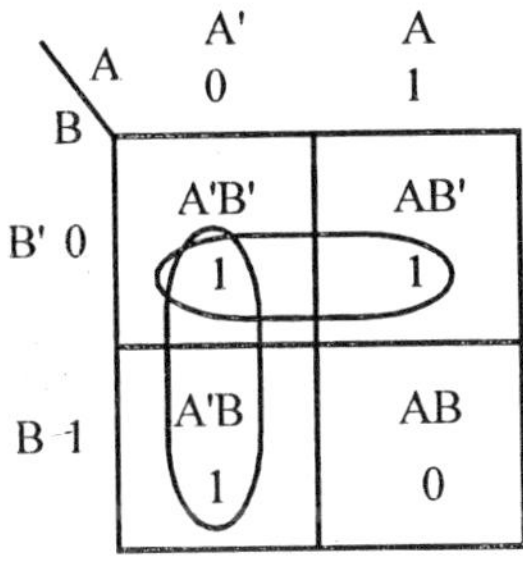

Fig. 1.19

Example 20:

Is it possible to reduce the following function with the help of a Karnaugh map?

$$Y = A'B + AB.$$

Solution:

In entered in the squares corresponding to the product terms. As these 1s are in squares located diagonally, there is a change in both the variables. Therefore, they cannot be grouped together, as only adjacent 1s located horizontally or vertically can be grouped. The given Boolean function is not therefore capable of any reduction.

Example 21:

Reduce the following Boolean function :

$$Y = A'B'C + A'BC + ABC + ABC'$$

with the help of Karnaugh map.

Solution:

In Fig. 1.20 1s have been entered in squares, which correspond to the four product terms.

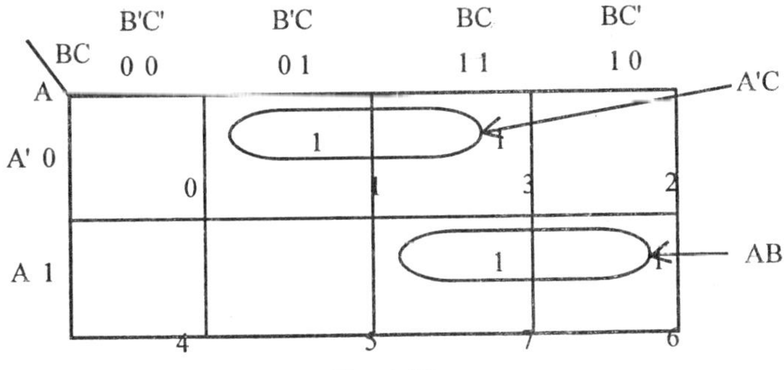

Fig. 1.20

From the pair comprising of minterms of squares 1 and 3, we can drop the variable B' which has changed and retain variables A' and C which have not changed. This gives the reduced product A'C. Similarly, from the other pair the unchanged product AB is retained. Although minterms of squares 3 and 7 are adjacent, and can form a pair, they have not been grouped, as these minterms have already been covered by other groups. Therefore, the reduced function is as follows.

$$Y = A'C + AB.$$

Example 22:

Reduce the following Boolean function

$$Y = A'B'C' + A'BC' + ABC' + AB'C.$$

Solution:

In Fig. 1.21, 1s have been entered in squares which correspond to the product terms.

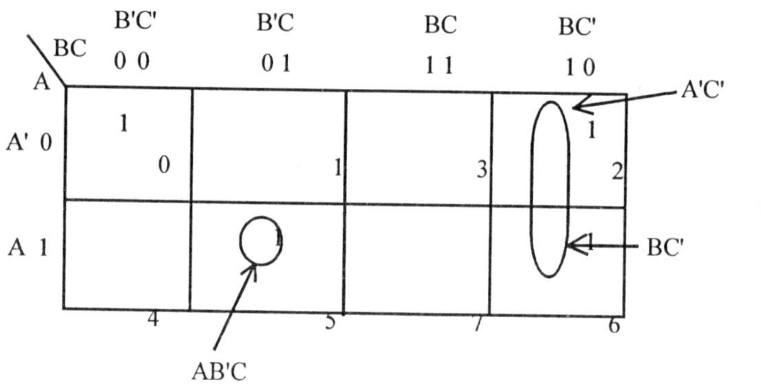

Fig. 1.21

As minterm of the square 5 is isolated, it can not form a group with any other minterm. Minterms of squares 2 and 6 on the extreme right form a convenient group and, after dropping the variable A, we are left with B and C', which have not changed.

Squares on the extreme left and right of a row (or column) can be considered to be adjacent, as between them only one variable is different.

Thus, minterms of squares 0 and 2 can form a pair. Form this pair, after dropping the variable B, we are left with A'C'. The reduced function is now as follows :

$$Y = A'C' + BC' + AB'C.$$

Example 23:

Use a Karnaugh map to simplify the following Boolean function :

$$Y = A'B'C' + A'B'C + A'BC + A'BC' + AB'C + ABC.$$

Solution:

In have been entered in squares which correspond to the product terms.

Notice from this diagram that the minterms have been covered by two quartelets, one of which yields A', and the other C. The reduced Boolean function is as follows :

$$Y = A' + C.$$

Example 24:

Plot the following Boolean function on a Karnaugh map and reduce it :

$$Y = A'B'C + A'B'C + A'BC + ABC' + AB'C + ABC.$$

Solution:

The function has been plotted in Fig. 1.22.

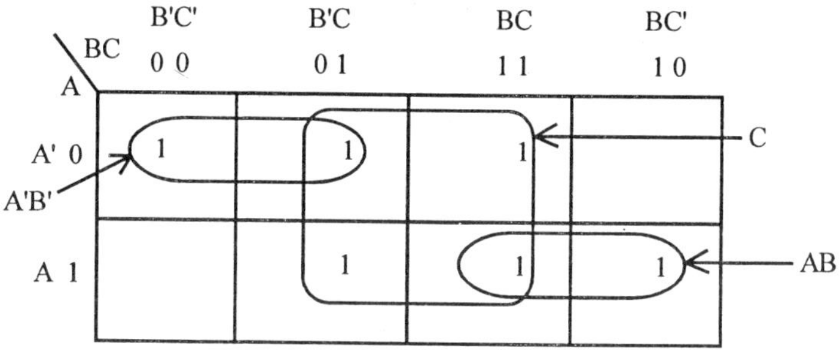

Fig. 1.22

Minterms of the squares 1, 3, 5 and 7 form a quarter. The variables in this are A and B and, after dropping them, we are left with only one variable that is C.

Minterms of the squares 0 and 1 form one pair, which yields A'B' and the other pair comprising of minterms of squares 6 and 7 yields AB, after dropping the variable C.

Note that two 1s which are part of a quartet, have been covered more than once, but that is permissible. The reduced function is as follows :

$$Y = A'B' + AB + C.$$

Example 25:

Draw a Karnaugh map to represent the following Boolean function :

$$Y = AC' + A'C + B.$$

Solution:

The first product term has two nonvariables A and C'. Obviously the variable is B. Therefore we should look for two squares for which the variable is B and nonvariable are A and C'. These squares are therefore squares 4 and 6. The other product term also points to two squares, for which the nonvariables are A' and C, and variable is B. Two squares which satisfy this condition are 1 and 3. The last term is a single nonvariable B, which points to a quartet for which the variables are A and C. The squares which satisfy this condition are 3, 2, 6 and 7. The squares have been marked accordingly in Fig. 1.23 and also grouped to show the reduced terms.

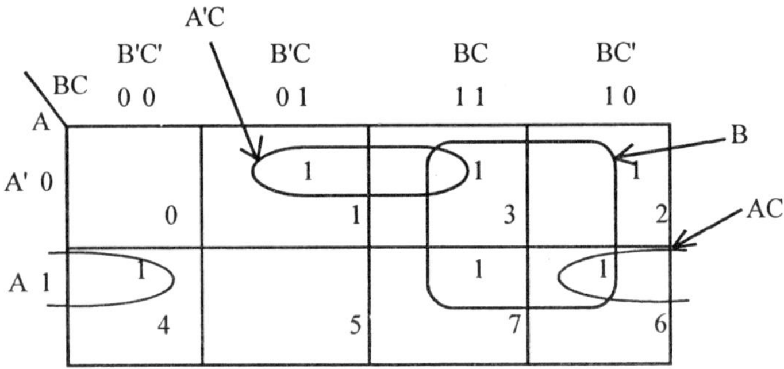

Fig. 1.23

Example 26:

Write the Boolean function represented by the Karnaugh map of Fig. 9.55 and simplify it.

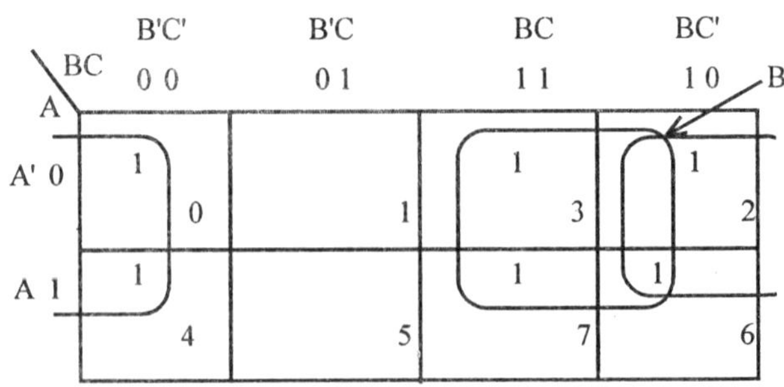

Fig. 1.24

Solution:

The Boolean function which is represented by Fig. 9.55 is as follows :

$$Y = A'B'C' + A'BC + A'BC' + AB'C' + ABC + ABC'.$$

In order to simplify the expression, the minterms can be enclosed by two quartets as shown in the diagram. Notice that two 1s in the left column can be grouped with two 1s in the right column to form a quartets. This quartet has two variables A and B, which can be dropped and one nonvariable C' which can be retained. The nonvariable yielded by the other quartet is B. The simplified function is as follows :

$$Y = B + C'.$$

Example 27:

Write the Boolean function represented by the map and simplify it.

Solution:

Minterms of the squares 0, 1 have been grouped with those of 8, 9 as they are at opposite ends of rows and form a quartet. This group yields nonvariables B'C'.

Similarly, the remaining four minterms are at opposite ends of columns and term the other quartet. This group yields variables BD'. The reduced function is $Y = B'C' + BD'$.

Example 28:

Draw a Karnaugh map to represent the following Boolean function

$$Y = A'B'C'D + ACD + BD' + AB + BC.$$

Solution:

The first term represents a four variable term, from which it is clear that its minterm will be in square 1. It has therefore been assigned to square 1. The term ACD represents a pair. Squares 11 and 15 satisfy the requirement of its position.

There are three quartets, each of which yields a product of two variables. The term BD' points to squares 4, 6, 12 and 14. The nonvariables in the other quartets are AB, for which the proper squares are 12, 13, 14 and 15.

The last quartet yields the product BC. Its proper position is in squares 6, 7, 14 and 15. The map is marked accordingly.

Example 29:

Design a logic circuit to satisfy the truth table given below :

Input				Product Term	Output
A	B	C	D		
0	0	0	0	A'B'C'D'	1
0	0	0	1	A'B'C'D	1
0	0	1	0	A'B'CD'	1
0	0	1	1	A'B'CD	1
0	1	0	0	A'BC'D'	1
0	1	0	1	A'BCD'	1
0	1	1	1	A'BCD'	0
0	1	1	1	A'BCD'	1
1	0	0	0	AB'C'D	0
1	0	0	1	AB'C'D	1
1	0	1	0	AB'CD'	0
1	0	1	1	AB'CD	0
1	1	0	0	ABC'D'	1
1	1	0	1	ABC'D	1
1	1	1	0	ABCD'	1
1	1	1	1	ABCD	0

Solution:

The SOP equation corresponding to the product terms for which the output is 1 will be as follows.

$$Y = A'B'C'D' + A'B'C'D + A'B'CD' + A'B'CD + A'BC'D' + A'BC'D + A'BCD' + AB'C'D + ABC'D' + ABC'D + ABCD'.$$

As it is much easier to simplify this function with the help of Karnaugh map, 1s have been entered in those squares of a Karnaugh map (Fig. 1.25), which correspond to the product terms of the logic function.

These 1s have been enclosed in groups as shown in the map. The nonvariables are indicated against each enclosure in the map.

The reduced logic function is as follows and it can be implemented by the logic circuit given in Fig. 1.25.

$$Y = A'B' + BD' + C'D.$$

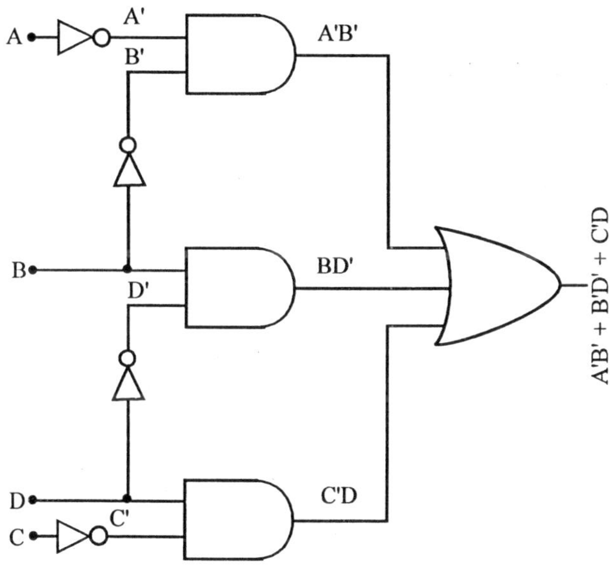

Fig. 1.25

Example 30:

Draw a Karnaugh map to represent the following Boolean function :

$Y = A + B'D'$.

Solution:

Since the first term is a single variable, it points to an octet in which the only nonvariable is A.

The squares which satisfy this condition are from 8 to 15 as shown in the map. The second term has two nonvariables. Which points to a quartet. The squares in which B' and D remain unchanged are 0, 2, 8 and 10, as shown in the map.

EXERCISES

1. Define Boolean Algebra and establish the following results :

   (i) if $a + c = b + c$ and $a \cdot c = b \cdot c$, then $a = b$, $\forall$ a, b, c $\in$ B.

   (ii) if $b \cdot a = c \cdot a$ and $b \cdot a' = c \cdot a'$, then $b = c$, when a, b, c $\in$ B.

2. Prove the following for all a, b, c $\in$ B.

   (i) $a + (a' \cdot b) = a + b$

   (ii) $a \cdot (a' + b) = a \cdot b$

(iii) $(a + b) \cdot (a' + c) = (a \cdot c) + a' \cdot b$.

3. If a, b $\in$ B, where B is a Boolean Algebra, prove that

   $a \cdot b + a \cdot b' + a' \cdot b + a' \cdot b' = 1$.

4. If a, b, c $\in$ B, where B is a Boolean algebra, prove that

   $(a + b) \cdot (b + c) \cdot (c + a) = a \cdot b + b \cdot c + c \cdot a$.

5. Use the technique of Boolean algebra to comment on the following statements

   (i) If a girl gets high marks in the examination she is either industrious or intelligent. The girl has got high marks in the examination. She is either industrious or intelligent.

   (ii) The lecturer is good if and only if he imparts knowledge to his students.

6. Test the validity of the statement : A democracy can survive only if the electorate is well informed or no candidate for public office is dishonest. The electorate is well informed only if education is free. If all candidates for public office are dishonest, then God exists. Therefore a democracy can survive only if education is free or God exists.

   [**Hint :** Argument is $(a \Rightarrow b \wedge c) \wedge (b \Rightarrow d) \wedge (c \Rightarrow c) \Rightarrow (a \Rightarrow d \vee c)$ in Boolean algebra $[(a' + b + c)(b' + d)(c' + c)' + (a' + d + c) = 1]$.

7. Prove the following identities

   (i) $a \cdot (a' + b) = a \cdot b$.

   (ii) $x \cup (x' \cap y) = x \cup y$

   (iii) $x \cap (x \cup y) = x \cup (x \cap y)$

   (iv) $x + y = (x' \cdot y')'$

   (v) $x \cdot y' = 0 \Leftrightarrow x \cdot y = x$

8. Prove that

   (i) $(a \cup b) \cap a' \cap b' = 0$

   (ii) $(a \cap b \cap c) \cup (a' \cap b' \cap c') = 1$.

9. Simplify the following using Boolean algebra :

   (a) $Y = AB + (AB)' + AB'C(AB + C)$

   (b) $Y = A[B + C(AB + AC)']$

   (c) $Y = (A + B + AB)(A + C)$

   (d) $Y = AB + ABC + A'B + AB'C$.

10. Simplify the following using Boolean algebra
    (a) Y = (AB' + C) (A + B')C
    (b) Y = AB + A'BC' + BC
    (c) Y = AB' + C + (A' + B)C'
11. Simplify the following Boolean expressions :
    (a) A = XY' + X'Y' + X'Y + XY
    (b) A = X(X + Y) (X + XY)
    (c) A = XYZ + XY'Z + XYZ' + XY'Z'
12. Simplify the following
    (a) A = (XY' + Y) (XY + Y')
    (b) A = X'Y(Y + Z) + YZ(Y' + X')
    (c) A = X'Y(X' + Z) + XY'(Y' + Z)
    (d) A = X'YZ' + XYZ' + XY'Z' + X'Y'Z'
14. Draw logic diagram to represent the following Boolean expressions.
    (a) a(ab)' + abc
    (b) (a + b)c + d.
15. Simplify the following Boolean expressions :
    (a) Y = [(A' + B) (A' + C) (B + C)]'
    (b) Y = [{(A · B')' · C'}' · D]'
    (c) Y = (A · B · C)' ((A · B)' + C)'
    (d) Y = [(A' · B) (B' · C) (C' · D')]'
    (e) Y = (A' · B · C)' + (B' · C' · A)'
    (f) Y = [((A' · B' · C)' + A · B)' (C · B')]'
16. Without simplifying convert the following expressions to AND/OR/INVERT logic :
    (a) Y = (A + B') (C' + D)
    (b) Y = (AB + C') (A'C + B)
    (c) Y = A(B + C') (A' + C)
    (d) Y = AD(B' + D)
17. Enter the following functions on a Karnaygh map :
    (a) Y = A + B'
    (b) Y = A'B + AB'

18. Find the minterms for the following expressions :

    (a) A + B + C'

    (b) AB + A' + C

19. Implement the following function with NAND/NOR logic :

    Y = AD(B' + D).

20. Prove that

    (A + B) (A' + C) (B + C) = (A + B) (A' + C).

21. Find the minterms for the following expressions :

    (a) AB + BC'

    (b) A'C + B'C + AC

22. Enter the following functions an a Karnaygh map :

    (a) Y = A'BCD + AB'CD + ABC'D + ABCD'

    (b) Y = A'BC'D + AB'CD' + A'BCD' + A'B'CD

23. Simplify the following functions using Karnangh maps :

    (a) Y = ABC'D' + ABC'D + ABCD + ABCD' + A'B'CD'

    (b) Y = A'BC'D' + A'BC'D + ABC'D + ABCD + ABCD' + AB'C'D

2

POSETS AND LATTICES

INTRODUCTION TO POSETS

Definition 1: *A set L on which a partial ordering relation (reflexive, antisymmetric, and transitive) r is defined is called a partially ordered set, or poset, for short.*

We recall a few examples of posets:

(1) L = R and r is the relation $\leq$.

(2) L = P (A) where A = {a, b} and r is the relation $\subseteq$.

(3) L = {1, 2, 3, 6} and r is the relation | (divide). We remind the reader that the pair (a, b) as an element of the relation r can be expressed as (a, b) $\in$ r, or arb, depending on convenience and readability.

The posets we will concentrate on in this chapter will be those which have maxima and minima. These partial orderings resemble that of $\leq$ on R, so the symbol $\leq$ is used to replace the symbol r in the definition of a partially ordered set. Hence, the definition of a poset becomes:

Definition 2: *Poset. A set on which a partial ordering, $\leq$, is defined is called a partially ordered set, or, in brief, a poset. Here $\leq$ is a partial ordering on L if and only if for all a, b, c $\in$ L.*

(1) $a \leq a$ (reflexivity),

(2) $a \leq b$ and $b \leq a \Rightarrow a = b$ (antisymmetry), and

(3) $a \leq b$ and $b \leq c \Rightarrow a \leq c$ (transitivity)

We now proceed to introduce maximum and minimum concepts. To do this, we will first define these concepts for two elements of the poset L, and then define the concepts over the whole poset L.

Definitions 3: Lower Bound, Upper Bound. *Let a, b $\in$ L, a poset. Then c $\in$ L is a lower bound of a and b if $c \leq a$ and $c \leq b$, d $\in$ L is an upper bound of a and b if $a \leq d$ and $b \leq d$.*

Definitions 4: Greatest Lower Bound, Least Upper Bound. *Let L be a poset and ≤ be the partial ordering on L. Let a, b ∈ L, then:*

(1) *g ∈ L is called the greatest lower bound (glb) of a and b if and only if g ≤ a and g ≤ b, and further, if there exists an element g' ∈ L such that if g' ≤ a and g' ≤ b, then g' ≤ g. This latter phrase says if g' is also a lower bound, then g is "greater" than g', so g is the greatest lower bound.*

(2) *An element l ∈ L is called the least upper bound (lub) of a and b if and only if a ≤ l and b ≤ l, and further, if there exists an element l' ∈ L such that a ≤ l' and b ≤ l', then l ≤ l'; that is, l is the least of the upper bounds.*

Definitions 5: Greatest Element, Least Element. *An element m is a poset L is called the greatest (maximum) element if, for all a ∈ L, a ≤ m. An elements s ∈ L is called the least (minimum) element if for all a ∈ L, s ≤ a. The greatest and least elements, when they exist, are frequently denote dby 1 and 0 respectively.*

Example 1:

Let L = {1, 3, 5, 7, 15, 21, 35, 103} and let ≤ be the relation | (divides) on L. Then L is a poset. To determine the lub of 3 and 7, we look for all l ∈ L, such that 3|l and 7 | l. certainly, both l = 21 and l = 105 work. Next, since 21|105, then 21 = lub of 3 and 7. Similarly, the lub of 3 and 5 is 15. The greatest element of L is 105 since a |105 for all a ∈ L. To find the glb of 15 and 35 , we firs consider all elements g of L such that g | 15 and g | 35. Certainly, both g = 4 and g = 1 satisfy these conditions. But since 1 | 5, then g = 5 is the glb of 15 and 35. The least element of L is 1 since 1 | a for all a ∈ L.

Hence fourth, for any positive integer n. D_n will denote the set of all positive integers which are divisors of n. For example, the set L of Example 13.1.1. is D_{105}.

Example 2:

If M = P (A), where A = {a, b, c}, with the relation ⊆ on M, then M is a poset. The gth of the elements {a, b} and {a, c} is g = {a}. For any other element g' of M which is subset of {a, b} and {a, c} (there is only one; what is it?} g' ⊆ g. The least element of M is φ and the greatest element of M is {a, b, c}.

With a little practice, it is quite easy to find the least upper bounds and greatest lower bounds of all possible pairs in M directly from the graph of M.

The previous examples and definitions indicate the lub and glb are define in terms of the partial ordering of the given poset. It is not yet clear whether all posets are such that every pair of elements has both a lub and a glb. Indeed, this is not the case.

Here, we shall study more about partially ordered sets (Posets) which finally lead the concepts of lattices. There we introduced ordered sets and partially ordered sets.

PARTIALLY ORDERED SETS OR POSETS

A relation R on a set A is called a *partial order* if R is reflexive, antisymmetric, and transitive. The set A together with the partial order R is called a *partially ordered set,* or simply a poset, and we denote this poset by (A, R).

Example:

(i) The relation of divisibility (a R b if and only if a|b) is a partial order on Z^+.

(ii) Let A be a collection of subsets of a set S. The relation $\subseteq$ of set inclusion is a partial order on A, so (A, $\subseteq$) is a poset.

(iii) If Z^+ be the set of positive integers, then the usual relation $\leq$ (less than or equal to) is a partial order on Z^+, as is $\geq$ (greater than or equal to)

(iv) The relation $<$ on Z^+ is not a partial order, since it is not reflexive.

The most familiar partial orders are the relation $\leq$ and $\geq$ on Z and R. Generally, if these are the partial orders on a set, then symbollically R is replaced by $\leq$ or $\geq$.

Dual of Poset

Let R be a partial order on a set A and let R^{-1} be the inverse relation of R. Then R^{-1} is also a partial order. The poset (A, R^{-1}) is called the *dual* of the poset (A, R), and the partial order R^{-1} is called the *dual* of the partial ordcr R.

Whenever (A, $\leq$) is a poset, we will always use the symbol $\geq$ for the partial order $\leq^{-1}$, and thus (A, $\geq$) will be the dual poset. Similarly, the dual of poset (A, $\geq$) will be denoted by (A, $\leq$).

Quasi-order : Let $>$ is a relation on a set S satisfying the following two properties.

(i) For any $a \in A$, we have $a \not> a$. (Irreflexive).

(ii) If a > b, and b > c, than a > c. (Transitive).

Then > is called a *quasi-order* on S.

Similarly, if ≤ is a partial order on a set S and we define a < b such that a ≤ b but a ≠ b, then < is a quasi-order on S. Conversely, if < is a quasi-order on a set S and we define a ≤ b to mean a < b or a = b, then ≤ is a partial order on S.

Comparability : Let (A, ≤) is a poset, the elements a and b of are said to be **comparable** if

$$a \leq b \text{ or } b \leq a.$$

In a partially ordered set every pair of elements need not be comparable. For example, (a R b if and only if a|b) is a poset on Z^+. In this poset the elements 3 and 7 are not comparable, since 3 ≠ 7 and 7 ≠ 3. Thus the word partial in partially ordered set means that some elements may not be comparable.

Linearly Ordered Sets : If every pair of elements in a poset A is comparable, we say that A is a *linearly ordered* set and the partial order is called a *linear order*. A is also called a *chain*. Although an ordered set A may not be linearly ordered, it is still possible that a subset S of A to be linearly ordered.

Every subset of a linearly ordered set A must also be linearly ordered.

Product and Lexicographical Order : If S and T be two ordered sets then we may define the order relation on the Cartesian product S × T as follows :

(i) Product Order : Let S and T are posets. Then the following is an order relation on the product set S × T, called the *product order :*

$(a, b) \leq (a', b')$ if $a \leq a'$ and $b \leq b'$

(ii) Lexicographical Order : Let S and T are linearly ordered sets. Then the following is an order relation on the product set S × T, called the *lexicographical* or *dictionary order :*

$(a, b) < (a', b')$ if $a < b$ or if $a = a'$ and $b < b'$

This order can be extended to $S_1 \times S_2 \times \dots \times S_n$ as follows :

$(a_1, a_2, \dots, a_n) < (a'_1, a'_2, \dots, a'_n)$ and $a_k < a'_k$.

Lexicographical order is linear.

Theorem : *If (A, ≤) and (B, ≤) are posets, then (A × B, ≤) is also a poset, with partial order ≤ defined by*

$$(a, b) \leq (a', b') \text{ if } a \leq a' \text{ in } A \text{ and } b \leq b' \text{ in } B.$$

If $(a, b) \in A \times B$, then $(a, b) \leq (a, b)$ since $a \leq a$ in A and $b \leq b$ in B, so $\leq$ satisfies the reflexive property in $A \times B$.

Let $(a, b) \leq (a', b')$ and $(a', b') \leq (a, b)$, where a and $a' \in A$ and b and $b' \in B$. Then

$$a \leq a' \text{ and } a' \leq a \text{ in } A$$

and $$b \leq b' \text{ and } b' \leq b \text{ in } B.$$

Since A and B are posets, the anti-symmetry of the partial orders on A and B implies that

$$a = a' \text{ and } b = b'$$

Hence $\leq$ satisfies the anti-symmetry property in $A \times B$.

Again, suppose that

$$(a, b) \leq (a', b') \text{ and } (a', b') \leq (a'', b''),$$

where $a, a', a'' \in A$ and $b', b', b'' \in B$.

Then $a \leq a'$ and $a' \leq a''$,

so $a \leq a''$, by the transitive property of the partial order on A. Similarly

$$b \leq b' \text{ and } b' \leq b'',$$

so $b \leq b''$ by the transitive property of the partial order on B. Hence

$$(a, b) \leq (a'', b'').$$

Obviously, the transitive property holds for the partial order on $A \times B$.

Hence, $A \times B$ is a poset.

Representation and Hasse Diagram

A partial order $\leq$ on a set X can be represented by a diagram known as *Hasse diagram* of $(X, \leq)$.

The Hasse diagram of a finite partially ordered set S is the directed graph whose vertices are the elements of S and there is a directed edge from a to b whenever $a < b$ in S. In the diagram we place b higher than a and draw a line between them. It is understood that movement upwards indicates succession. In the diagram thus formed, there is a directed path from vertex x to vertex y if and only if $x < y$. There can be no directed cycles in the diagram of S since the order relation is anti-symmetric. The Hasse diagram of a poset S is a picture of S; hence it is very useful in describing types of elements in S.

Theorem 1:

Let A be a finite non-empty poset with partial order $\leq$. Then A has atleast one maximal element and at least one minimal element.

Let a be any element of A. If a is not maximal, we can find an element $a_1 \in A$ such that $a < a_1$. If a_1 is not maximal, we can find an element $a_2 \in A$ such that $a_1 < a_2$. This argument can not be continued indefinitely, since A is a finite set. Thus, we set

$$a < a_1 < a_2 < \ldots < a_{r-1} < a_r$$

which cannot be extended. Hence, we cannot have $a_r < b$ for any $b \in A$. Therefore, a_k is a maximal element of $(A, \leq)$.

By the same argument we can prove that the dual poset $(A, \geq)$ has a maximal element. So $(A, \leq)$ has a minimal element.

Theorem 2:

A poset has at most one greatest element and at most one least element.

Let a and b be greatest elements of a poset A. Then, since b is a greatest element, we have $a \leq b$. Similarly, since a is a greatest element, we have $b \leq a$. Hence $a = b$ by the anti-symmetry property. Thus, if the poset has a greatest element, it only has one such element. Since this fact is true for all posets, the dual poset $(A, \geq)$ has at most one greatest element, so $(A, \leq)$ also has at most one least element.

LATTICE

A poset $(L, \leq)$ is called a *lattice* if every 2-element subset of L has path, a least upper bound and a greatest lower bound, i.e., if l.u.b. (a, b) and g. l. b. (a, b) exist for every a and b in L. In this case we denote

$$a \vee b = \text{l. u. b. } (a, b) = y \text{ (read as a join b)}$$

$$a \wedge g = \text{g. l. b. } (a, b) = a \text{ (read as a meet b)}$$

Properties of Lattices

Theorem 1:

If L be a lattice, then for every a and b in L

(i) $a \vee b =$ *if and only if* $a \leq b$

(ii) $a \wedge b = a$ *if and only if* $a \leq b$

(iii) $a \wedge b = a$ *if and only if* $a \vee b = b$

(i) Let $a \vee b = b$. Since $a \leq a \vee b = b$, we get $a \leq b$. Conversely, if $a \leq b$, then, since $b \leq b$, b is an upper bound of a and b. So, by definition of l. u. b., we have $a \vee b \leq b$. Again, since $a \vee b$ is an upper bound , $b \leq a \vee b$, so $a \vee b = b$.

(ii) Similar to (i)

(iii) Combining (i) and (ii), we get (iii).

Theorem 2:

If L be any lattice, then for any a, b, c $\in$ L,

1. *Idempotent Properties*

   (a) $a \vee a = a$

   (b) $a \wedge a = a$

2. *Commutative Properties*

   (a) $a \vee b = b \vee a$

   (b) $a \wedge b = b \wedge a$

3. *Associative Properties*

   (a) $a \vee (b \vee c) = (a \vee b) \vee c$

   (b) $a \wedge (b \wedge c) = (a \wedge b) \wedge c$

4. *Absorption Properties*

   (a) $a \vee (a \wedge b) = a$

   (b) $a \wedge (a \vee b) = a$

1. (a) $a \vee a$ = l. u. b. {a, a} = l. u. b. {a} = a

   (b) $a \wedge a$ = g. l. b. {a, a} = g. l. b. {a} = a

2. (a) $a \vee b$ = l. u. b. {a, b} = l. u. b. {b, a} = $b \vee a$

   (b) $a \wedge b$ = g. l. b. {a, b} = g. l. b. {b, a} = $b \wedge a$

3. (a) We have $(a \vee b) \vee c$ = l. u. b. {$a \vee b$, c}.

   So $a \vee b \leq (a \vee b) \vee c$ and $c \leq (a \vee b) \vee c$.

   Also, $a \leq a \vee b$ and $b \leq a \vee b$.

   Hence, $a \leq (a \vee b) \vee c$ and $b \leq (a \vee b) \vee c$ by transitivity.

   Thus $(a \vee b) \vee c$ is an upper bound of a and b.

   By definition of l. u. b. of b and c, viz. $b \vee c$.

   $$a, b \vee c < (a \vee b) \vee c$$

   So, it is an upper bound of a and $b \vee c$.

   Thus, by definition of $\vee$, we have

   $$a \vee (b \vee c) \leq (a \vee b) \vee c$$

   Similarly $(a \vee b) \vee c \leq a \vee (b \vee c)$.

By antisymmetry of $\leq$, we have

$$(a \vee b) \vee c = a \vee (b \vee c)$$

(b) Same as part (a)

4. (a) Since $a \wedge b \leq a$ and $a \leq a$, a is an upper bound of $a \wedge b$ and a. So $a \vee (a \wedge b) \leq a$. By definition of l. u. b., we have $a \leq a \vee (a \wedge b)$.

Hence by the property of antisymmetry of $\leq$, we have

$$a \vee (a \wedge b) = a$$

(b) Same as part (a)

Theorem 3:

In any lattice, L, prove the distributive inequalities.

(i) $a \wedge (b \vee c) \geq (a \wedge b) \vee (a \wedge c)$

(ii) $a \vee (b \wedge c) \leq (a \vee b) \wedge (a \vee c)$

hold for any $a, b, c \in L$.

(i) $a \wedge b \leq a$

$a \wedge b \leq b \leq b \vee c$

$\Rightarrow$ $a \wedge b$ is lower bound of $\{a, b \vee c\}$.

$\Rightarrow$ $a \wedge b \leq a\ (b \vee c)$ ...(1)

Again, $a \wedge c \leq a$

$a \wedge c \leq c \leq b \vee c$

$\Rightarrow$ $a \wedge c \leq a \wedge (b \vee c)$ ...(2)

From (1) and (2), it is clear that $a \wedge (b \vee c)$ is an upper bound of $\{a \wedge b, a \wedge c\}$.

$$\Rightarrow \quad (a \wedge b) \vee (a \wedge c) \leq a \wedge (b \vee c)$$

Hence $a \wedge (b \vee c) \geq (a \wedge b) \vee (a \wedge c)$

(ii) The proof is similar to (i)

Remark : If in any lattice the operations of union and intersection are distributive over each other then the above inequalities became equalities.

PRINCIPLE OF DUALITY

We know that if $\leq$ is a partial order any set, then the inverse relation $\geq$ is also a partial order. Also, l. u. b. of a and b with respect to $\leq$ is the same as the g. l. b. with respect to $\geq$ and vice-versa. This observation leads to the principal of duality for lattices, according to which if we interchange $\vee$ with $\wedge$ and $\leq$ with $\geq$ in a true statement about lattices, we get another true statement. The corresponding statements are called dual of each other.

Isomorphic Lattices : If f : L $\rightarrow$ L' is an isomorphism from the poset (L, $\leq$) to the poset (L', $\leq'$) then L is a lattice if and only if L' is a lattice. In fact, if a and b are elements of L, then f (a $\wedge$ b) = f (a) $\wedge$ f (b) and f (a $\vee$ b) = f (a) $\vee$ f (b).

If two lattices are isomorphic, we say that they are *isomorphic lattices.*

Lattice as Algebraic System

An algebraic system is a set together with a few rules for combining elements of the set to form other elements of the set.

We define lattices as partially ordered sets in which for each a, b $\in$ A

l. u. b. (a, b) = a $\vee$ b ; g. l. b. (a, b) = a $\wedge$ b exists in the set.

By definitions, we have in a lattice (A, $\leq$), with every two elements a, b of A there is associated an element a $\vee$ b of A and there is associated an element a $\wedge$ b of A.

Thus, we can interpret the operations $\vee$ and $\wedge$ as binary operations on A.

Any set L together with two binary operations satisfying the commutative, associative and absorption laws is a lattice in the sense that a partial order relation can be defined in terms of either of these operations with respect to which it becomes a lattice with l. u. b. and g. l. b. with respect to this order coinciding with the given binary operations. Thus any lattice can be said to be an algebraic system.

Sublattices

A non-empty subset L' of a lattice L is called a sublattice of L if a, b $\in$ L' $\Rightarrow$ a $\vee$ b, a $\wedge$ b $\in$ L', i.e., the algebra (L', $\vee$, $\wedge$) is a sublattice of (L, $\wedge$, $\vee$) if L' is closed under both operations $\wedge$ and $\vee$.

A sub-lattice itself is a lattice. However, any subset of L which a lattice need not be a sublattice.

SOME SPECIAL LATTICES

Bounded Lattices

A lattice L is said to have a *lower bound* O if for any element x in L we have O $\leq$ X. Again, L is said to have an *upper bound* I if for any X in L we have X $\leq$ I. If a lattice L has both a lower bound O and an upper bound I then it is said *bounded lattice.*

In bounded lattices we have

$a \vee I = I$, $a \wedge I = a$, $a \vee 0 = a$,

$a \wedge O = O$, for any element a in L.

Example:

(i) Lattice P (U) of all subsets of any universal set U is a bounded lattice with U as an upper bound and the empty set ϕ as a lower bound.

(ii) The non-negative integers with the usual ordering,

$$O < 1 < 2 < 3 < 4 < \ldots$$

have O as a lower bound but have no upper bound. Hence they do not constitute a bounded lattice.

Theorem:

Let L = $\{a_1, a_2, \ldots, a_n\}$ be a finite lattice. Then L is bounded.

The greatest element of L is $a_1 \vee a_2 \vee \ldots \vee$ an and its least element is $a_1 \vee a_2 \wedge \ldots \wedge a_n$. There are also the upper and lower bounds for L, respectively. Thus, every finite lattice L is bounded.

Distributive Lattices

A lattice L is said to be *distributive* if for any elements a, b, c in L we have the following distributive laws :

(i) $a \wedge (b \vee c) = (a \wedge b) \vee (a \wedge c)$

(ii) $a \vee (b \wedge c) = (a \vee b) \wedge (a \vee c)$

otherwise L is said to be *nondistributive.*

Example:

For a set S, the lattice P (S) is distributive, since union and intersection (the join and meet respectively) each satisfy the distributive property.

Complements : Let L be a bounded lattice with upper bound I and lower bound O, and let $a \in L$. An element $a' \in L$ is called a *complement* of a if

$$a \vee a' = I \text{ and } a \wedge a' = O,$$

obviously, $O' = I$ and $I' = O$,

Complements need not exist and need not be unique.

Theorem:

Let L be a bounded distributive lattice. If a complement exists, it is unique.

Let a' and a" be complements of the element $a \in L$. Then

$$a \vee a' = I,\ a \vee a'' = I$$

$$a \wedge a' = O,\ a \wedge a'' = O$$

Using the distributive laws, we obtain

$$\begin{aligned} a' = a' \vee O &= a' \vee (a \wedge a'') \\ &= (a' \vee a) \wedge (a' \vee a'') \\ &= I \wedge (a' \vee a'') = a' \vee a'' \qquad ...(1) \end{aligned}$$

Also,

$$\begin{aligned} a'' = a'' \vee O &= a'' \vee (a \wedge a') \\ &= (a'' \vee a) \wedge (a'' \vee a') \\ &= I \wedge (a' \vee a'') = a' \vee a'' \qquad ...(2) \end{aligned}$$

From (1) and (2) it is clear that a' = a".

Complemented Lattice

A lattice L is called *complemented* if it is bounded and if every element in L has a complement.

SOLVED EXAMPLES

Example 1:

Show that the set N of natural numbers under divisibility forms a poset.

Solution:

We have

(i) $n|n$ for all $n \in N \Rightarrow |$is reflexive

(ii) $n|n$ and $m|n \Rightarrow n = m$, |is anti-symmetric

(iii) $n|m$ and $m|p \Rightarrow n|p$, | is transitive

Hence, | is a partial ordering on N and (N, |) is a poset.

Example 2:

Show that the relation $\leq$ is a partial ordering on the set of integers.

Solution:

We have

(i) $a \leq a$ for every $a \Rightarrow \leq$ is reflexive

(ii) $a \leq b$ and $b \leq a \Rightarrow a = b$, $\leq$ is anti-symmetric.

(iii) a ≤ b and b ≤ c ⇒ a ≤ c, ≤ is transitive.

Hence, ≤ is a partial ordering on the set of integers.

Example 3:

Let P(S) be the power set; i.e., set of all subsets of a given set S. Show that the inclusion relation ⊆ is a partial ordering on the power set P(S).

Solution:

We have

(i) A ⊆ A for all A ⊆ S ⇒ ⊆ is reflexive.

(ii) A ⊆ B and B ⊆ A ⇒ A = B, ⊆ is anti-symmetric.

(iii) A ⊆ B and B ⊆ C ⇒ A ⊆ C, ⊆ is transitive.

Hence, ⊆ is a partial ordering on P(S) and (P(S), ⊆) is a poset.

Example 4:

Determine whether

(a) (2, 5) > (3, 7)

(b) (1, 7) < (1, 10)

(c) (1, 2, 3, 5) < (1, 2, 7, 9) in the poset (Z × Z, ⪯) where ≤ is lexicographic ordering constructed from the relation ≤ on Z.

Solution:

(a) Since 2 < 3, it follows that (2, 5) < (3, 7).

(b) Since 1 = 1 and 7 < 10, if follows that (1, 7) < (1, 10).

(c) Since 1 = 1, 2 = 2 but 3 < 7, it follows that (1, 2, 3, 5) < (1, 2, 7, 9).

Example 5:

Let S = {a, b, c} and A = P(S) be a poset with partial order ⊆ (set inclusion). Find the least and the greatest elements.

Solution:

The given poset is :

$$A = \{\phi, \{a\}, \{b\}, \{c\}, \{a, b\}, \{a, b\}, \{b, c\}, S\}$$

Clearly, the empty set ϕ is a least element of A, and the set S is a greatest element of A.

Example 6:

Find the least and greatest element in the poset (Z^+, |), if they exist.

Solution:

The least element of the poset $(Z^+, |)$ is 1. There is no greatest element since there is no integer which is divisible by all positive integers.

Example 7:

Determine the greatest, least, minimal and maximal element of the posets represented by following Hasse diagrams.

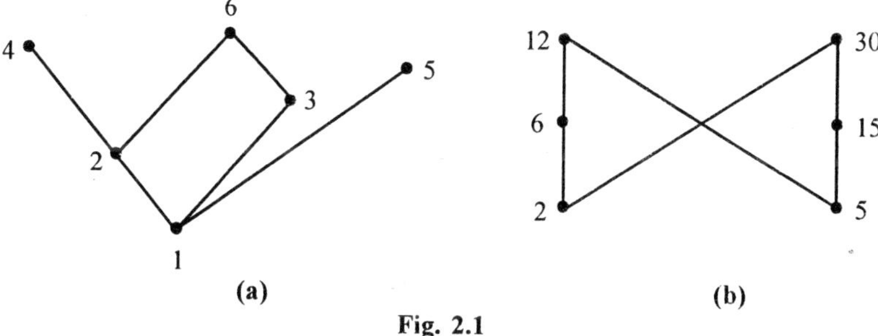

Fig. 2.1

Solution:

(a) The least element of the poset represented by Hasse diagram (a) is 1 and is the only minimal element. There is no greatest element and 4, 6 and 5 are maximal elements.

(b) The poset represented by Hasse diagram (b) has neither greatest element nor least element. Two minimal elements are 2, 5 and two maximals are 12 and 30.

Upper and Lower Bound : Let A be a poset and B be a subset of A. An element $a \in A$ is called an *upper bound* of B if $b \leq a$ for all $b \in B$. An element $a \in A$ is called a *lower bound* of B if $a \leq b$ for all $b \in B$.

An element $a \in A$ is called a *least upper bound (l.u.b.)* of B if a is an upper bound of B and $a \leq a'$, whenever a' is an upper bound of B.

Thus a = l.u.b. (B) if $b \leq a$ for all $b \in B$, and if whenever $a' \in A$ is also an upper bound of B, then $a \leq a'$. It is also termed as *supremum* of A and we write it as sup (A) instead of l.u.b (A).

Similarly, an element $a \in A$ is called a *greatest lower bound (g.l.b.)* of B if a is a lower bound of B and $a' \leq a$, whenever a' is a lower bound of B.

Thus a = g.l.b. (B) if $a \leq b$ for all $b \in B$, and if whenever $a' \in A$ is also a lower bound of B, then $a' \leq a$. We also say it *infimum* of A and write inf (A) instead of g.l.b. (A).

Remarks:

(i) A subset B of a poset may or may not have upper or lower bounds.

(ii) There can be more than one upper bound and lower bound of a set.

(iii) An upper or lower bound may or may not belong to B itself.

(iv) The greatest element is always the l.u.b. but the converse is not true. a = l.u.b. (B) is the greatest element iff a $\in$ B.

(v) The least element is always the g.l.b. but the converse is not true. b = g.l.b. (B) is the least element iff b $\in$ B.

Consistent Enumeration : Let A be a finite partially ordered set. If there be a function f : A $\to$ B so that if a < b then f(a) < f(b) then such a function is called a *consistent enumeration* of A.

Isomorphism : Let (A, $\leq$) and (B, $\leq$) be two posets. A one-to-one (injective) mapping

f : A $\to$ B is called an *isomorphism* or a *similarity mapping* from A into B if f preserves the order relation, i.e., if a $\leq$ a' then f(a) $\leq$ f(a') for all a and a' $\in$ A.

If two ordered sets A and B are *isomorphic* or *similar* then we write as A $\simeq$ B.

Hasse diagram of both the posets are identical except for the labelling of vertices.

Remarks **:** Let (A, $\leq$) and (A', $\leq$') be isomorphic posets under the isomorphism f : A $\to$ A'.

(i) If a is a maximal (minimal) element of (A, $\leq$), then f(a) is a maximal (minimal) element of (A', $\leq$').

(ii) If a is the greatest (least element) of (A, $\leq$), then f(a) is the greatest (least) element of (A', $\leq$').

(iii) If a is an upper bound (lower bound, l.u.b., g.l.b.) of a subset B of A, then f(a) is an upper bound (lower bound, l.u.b., g.l.b.) for the subset f(B) of A'.

(iv) If every subset of (A, $\leq$) has a l.u.b. (g.l.b.), then every subset of (A', $\leq$') has a l.u.b. (g.l.b.).

Example 8:

Consider the poset A = (1, 2, 3, 4, 6, 9, 12, 18, 36, |}. Find the g.l.b. and l.u.b. of the sets {6, 18} and {4, 6, 9}.

Solution:

If 6 and 18 are divisible by an integer then that integer will be a lower bound {6, 18}. Only such integers are 1 and 6. Since 1|6, 6 is the greatest lower bound of {6, 18}. This will also be the g.l.b.

An integer is an upper bound of {6, 18} if and only if it is divisible by 6 and 18 which is 18. Hence l.u.b. {6, 18} = 18.

The only lower bound of {4, 6, 9} is 1. Hence g.l.b. {4, 6, 9} = 1. The only upper bound of {4, 6, 9} is 36. Hence l.u.b. {4, 6, 9} = 36.

Example 9:

Consider $S = \{x \in R : 1 < x < 2\}$ *with* $\leq$ *as the partial order. Find*

(i) All the upper and lower bounds of S.

(ii) g.l.b. and l.u.b. of S.

Solution:

(i) Every real number ≥ 2 is an upper bound of S and every real number ≤ 1 is a lower bound of S.

(ii) 1 is a g.l.b. and 2 is the l.u.b. of S.

Example 10:

Let $A = \{1, 2, 4, 8\}$ *with divisibility and* $B = \{0, 1, 2, 3\}$ *with usual* $\leq$ *are partial order. Consider the function* $f : A \to B$ *defined by* $f(1) = 0$, $f(4) = 2$, $f(8) = 3$. *Show that f is an isomorphism from A onto B.*

Solution:

Here |A| = |B| = 4. It is evident that the given f is one - one onto. Again for all x, y $\in A$.

$x|y \Leftrightarrow f(x) \leq f(y)$

Hence the given function is an isomorphism from (A, |) onto (Bm $\leq$).

Example 11:

Let S be a set and let $L = P(S)$ *with partial order* $\subseteq$. *Show that L is a lattice.*

Solution:

Let A and B belong to poset (L, $\subseteq$). The $A \vee B$ is the set $A \cup B$. To see this, we have $A \subseteq A \cup B$, $B \subseteq A \cup B$, and if $A \subseteq C$ and $B \subseteq C$, then $A \cup B \subseteq C$. Hence $A \vee B$ is $A \cup B$. Similarly, we can show that the element $A \wedge B$ in (L, $\subseteq$) is the set $A \cap B$. Thus, L is a lattice.

Example 12:

Let L be a lattice and let a and b be elements of L such that $a \leq b$. The interval [a, b] is defined as the set of all $x \in L$ such that $a \leq x \leq b$. Prove that [a, b] is a sublattice of L.

Solution:

Let x, y be in {a, b}. Then $x, y \in L \Rightarrow x \vee y, y \wedge x \in L$ as L is a lattice. Now a is a lower bound of x and y. Then $a \leq x \leq x \vee y \leq b$, so $x \vee y \in$ [a, b] and similarly $a \leq x \wedge y \leq x \leq b$ so $x \wedge y \in$ {a, b). Hence {a, b} is a sublattice.

Example 13:

Show that in a complemented, distributive lattice, the following are equivalent:

(i) $a \leq b$, (ii) $a \wedge b' = o$, (iii) $a' \vee b = 1$, (iv) $b' \leq a'$

Solution:

(i) $a \leq b \Rightarrow a \vee b = b$

$\Rightarrow (a \wedge b) \wedge b' = 0$ $[\because b \wedge b' = o]$

$\Rightarrow (a \wedge b') \vee (b \wedge b') \Rightarrow 0$ [by distribution law]

$\Rightarrow a \wedge b' = 0$

Hence (i) $\Rightarrow$ (ii)

Again, $a \wedge b' = 0 \Rightarrow (a \wedge b')' = I$

$\Rightarrow a' \vee (b')' = I$

$\Rightarrow a' \vee b = 1$

Hence (ii) $\Rightarrow$ (iii)

Now, $a' \wedge b = I \Rightarrow (a' \;\; b') \wedge b' = b'$ $[I \wedge b' = b']$

$\Rightarrow (a' \wedge b') \vee (b \wedge b') = b'$ [by distribution law]

$\Rightarrow a' \wedge b' = b'$ $[\because b \wedge b' = o]$

$\Rightarrow b' \leq a'$

Hence (iii) $\Rightarrow$ (iv)

Now, $b' \leq a' \Rightarrow a' \wedge b' = b'$

$\Rightarrow (a' \wedge b') = b$ [By De Morgan's law]

$\Rightarrow a \vee b = b$

$\Rightarrow a \leq b$

Hence (iv) ⇒ (i); Thus (i), (ii), (iii) and (iv) are equivalent.

Example 14:

Consider the lattice L in Fig. 2.2

(a) Which of the following are sublattices of L ?

$L_1 = \{0, a, b, I\}$, $L_2 = \{0, a, e, I\}$

$L_3 = \{a, c, d, I\}$

(b) Find complements, if exist, for elements, a, b and c.

(c) Is L distributive ?

(d) Is L a complemented Lattice.

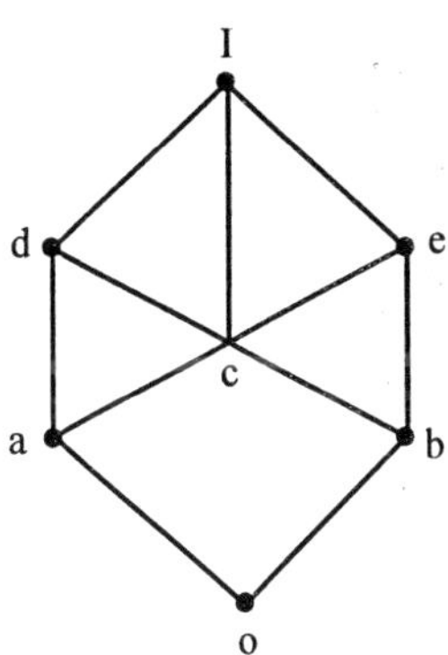

Fig. 2.2

Solution:

(a) A sub-set of a lattice is a sub-lattice if it is closed under $\wedge$ and $\vee$. L_1 is not a sublattice since $a \vee b = c$, which does not belong to L_1. L_2 is a sub-lattice since $a \vee e = I$ and $a \wedge e = 0$ belonging to L_2. L_3 is a sublattice since $c \vee d = I$ and $c \wedge d = a$, $a \vee c = I$ and $a \wedge c = d$, $a \vee d = I$, $a \wedge d = e$.

(b) We have $a \wedge e = 0$ and $a \vee e = I$, so a and e are compliments. Also, $b \wedge d = 0$ and $b \vee d = I$, so b and d are complements. But c has no complement.

(c) L is not distributive, since $L' = \{0, a, d, e, I\}$ is a sublattice which forms pentagon.

(d) L is not complemented since c has no complement.

EXERCISES

1. Which of the following are partial order?

   (a) The relation $R = \{(a, b) \in Z \times Z : |a - b| \leq 1\}$ on Z.

   (b) The relation $R = \{(a, b) \in Z \times Z : a - b \leq 0\}$

2. Which of the following are posets?

   (i) $(Z, >)$, (ii) $(Z, =)$, (iii) $(Z, \geq)$.

3. Let $A = \{a_1, a_2\}$. Describe all partial order relations on A.

4. Let $A = \{1, 2, 3, 4\}$ and consider the relation.

   $R = \{(1, 1), (1, 2), (1, 3), (2, 2), (3, 2),$
   $(3, 3), (4, 2), (4, 3), (4, 4)\}.$

   Show that R is a partial ordering. Draw its Hasse diagram.

5. Draw the Hasse diagram for the divisibility relation on each of the following sets:

   (i) S = {3, 6, 12, 24, 48}

   (ii) S = {2, 3, 6, 12, 24, 36}

6. Consider the subsets {2, 3}, {4, 6} and {3, 6} is the poset {1, 2, 3, 4, 5, 6}, find for each subsets, if exist

   (a) upper bound and lower bound.

   (b) greatest lower bound and least upper bound.

7. If an element of a distributive lattice has a complement then show that the complement is unique.

8. Let L be a bounded lattice with at least two elements. Show that no element of L is its own universe.

9. Is the poset A = {2, 3, 6, 12, 24, 36, 72} under the relation of divisibility a lattice?

10. Show that a subset of a linearly ordered poset is a sublattice.

3

DETERMINANTS

INTRODUCTION

Consider the following two equations:

$$a_1x + b_1y = 0$$
$$a_2x + b_2y = 0$$

$$\therefore \qquad -\frac{a_1}{b_1} = \frac{y}{x} = -\frac{a_2}{b_2}$$

Eliminating x and y we get $-\frac{a_1}{b_1} = -\frac{a_2}{b_2}$

or $a_1b_2 - a_2b_1 = 0$

We shall experts the above eliminate in the form of

$$\begin{vmatrix} a_1 & b_1 \\ a_2 & b_2 \end{vmatrix} = 0 \qquad \text{...(1)}$$

(1) is called a determinant of second order and its value is $a_1b_2 - a_2b_1 = 0$

Let us non eliminate x, y, z from the following equations:

$$a_1x + b_1y + c_1 = 0$$
$$a_2x + b_2y + c_2 = 0$$
$$a_3x + b_3y + c_3 = 0$$

Solving eqn. (2) and (3) by cross multiplication, we have

$$\frac{x}{b_2c_3 - b_3c_2} = \frac{y}{(a_2c_3 - a_3c_2)} = \frac{y}{a_2c_3 - a_3b_2}$$

Substituting the value of x_1 and z in (1), we get

$$K[a_1 (b_2c_3 - b_3c_2) - b_1 (a_2c_3 - c_3a_2) + c_1 (a_2b_3 - a_3b_2)] = 0$$

$\therefore$ The eliminant is $a_1(b_2c_3 - b_3c_2) - (a_2c_3 - c_3a_2) + c_1(a_2b_3 - a_3b_2) = 0$

$$\begin{vmatrix} a_1 & b_1 & c_1 \\ a_2 & b_2 & c_2 \\ a_3 & b_3 & c_3 \end{vmatrix} = 0 \qquad ...(2)$$

eqn. (2) is called the determinant of order (3).

PRODUCT OF DETERMINANTS

Complementary Minor of a Determinant

Definition: *If B is r × r sub-matrix of an n × n matrix A, then the determinant E' of A formed by removing the rows and columns of A containing the elements of B is called the complementary minor of B.*

In the matrix $\begin{bmatrix} a_1 & b_1 & c_1 & d_1 \\ a_2 & b_2 & c_2 & d_2 \\ a_3 & b_3 & c_3 & d_3 \\ a_4 & b_4 & c_4 & d_4 \end{bmatrix}$

the complementary minor of the det. $\begin{vmatrix} a_1 & b_1 \\ a_2 & b_2 \end{vmatrix}$ is $\begin{vmatrix} c_3 & d_3 \\ c_4 & d_4 \end{vmatrix}$;

the complementary minor of $\begin{vmatrix} b_1 & c_1 & d_1 \\ b_2 & c_2 & d_2 \\ b_3 & c_3 & d_3 \end{vmatrix}$ is a_4

and the complementary minor of $\begin{vmatrix} b_2 & c_2 \\ b_3 & c_3 \end{vmatrix}$ is $\begin{vmatrix} a_1 & d_1 \\ a_4 & d_4 \end{vmatrix}$

LAPLACE'S EXPANSION OF A DETERMINANT BY THE MINORS OF FIRST R COLUMNS

If $| B_i |$ is r × r minor of an n × n matrix A formed by the elements of the first r columns of A and $| B' |$ is the complementary minor of $| B_i |$, then

$$| A | = \Sigma \pm | B_i |, | B'_i |,$$

where the summation is extended over all the possible r × r minor of A which can be formed from the elements of the first r columns and + or – sign is taken according as an even or odd number of interchanges of adjacent rows of A is required to bring the submatrix B_i into the first r rows of A.

Theorem:

If A is an $n \times n$ matrix and E is an elementary matrix obtained from the identity matrix I_n, then

$$|EA| = |AE| = |E| . |A| = |A| |E|.$$

Proof:

$\because I_n$ is the $n \times n$ identity matrix.

$\therefore |I_n| = 1.$ ...(i)

Let E_a, E_b, E_c be three elementary matrices.

Then $|E_a| = -|I_n|$

$\therefore$ From (i) we get $|E_a| = -1$ ...(ii)

Again $|E_b| = c|I_n|$

$\therefore$ From (i) we get $|E_b| = c$ ...(iii)

Similarly $|E_c| = |I_n|$

$\therefore$ From (i) we get $|E_c| = 1$ ...(iv)

Now, as the matrix $E_a A$ can be obtained by interchanging two rows of the matrix A, so we have from (ii)

$$|E_a| = -|A| = |E_a| . |A| \quad \text{...(v)}$$

Similarly, the product E_bA can be obtained by applying second elementary row operation on the matrix A, so we have from (iii)

$$|E_bA| = c|A| = |E_b| . |A| \quad \text{...(vi)}$$

Again the product $E_c A$ can be obtained by applying third elementary row operation on the matrix A, so we have from (iv)

$$|E_c A| = |A| = |E_c| . |A| \quad \text{...(vii)}$$

Hence from (v). (vi) and (vii) we conclude that

$$|EA| = |E| . |A|,$$

where E is any one of the elementary matrices.

In similar way we can prove that

$$|AE| = |A| . |E|,$$

CANONICAL FORM (OR NORMAL FORM OF A MATRIX)

Every non-zero $m \times n$ matrix A can be reduced by means of elementary transformation (*i.e.,* elementary row and column operations) to the form

$$\begin{bmatrix} I & 0 \\ 0 & 0 \end{bmatrix}'$$

where I_r is the r × r identity matrix and the remaining sub-matrices are zero matrices.

The above form is called the *canonical form or orthogonal form or normal form* of the matrix A.

Definition: *If A and B be two m × n matrices, then B ~ A if and only if B = SAT, where S is an m × m non-singular matrix and T is an n × n non-singular matrix.*

The following two properties of the above relation are fundamental:

1. **Symmetry:** If A ~ B, then B ~ A for if A = PBQ,

   then $\quad B \sim P^{-1} AQ^{-1}$, where P^{-1} and Q^{-1} are non-singular matrices.

2. **Reflexivity:** Every matrix A is equivalent to itself since we can write A = IAI, so that P = I = Q.

Theorem 1:

$| A_1 A_2 | = | A_1 | . | A_2 |$, where $| A_2 |$ and $| A_3 |$ are two determinants.

Proof:

Let C_1 and C_2 be the canonical form of the matrices A_1 and A_2 *i.e.,* $A_1 \sim C_1$ and $A_2 \sim C_2$.

If $\quad A_1 \sim C_1$, we have

$A_1 \sim S\, C_1\, T$, where S and T are non-singular matrices.

$\Rightarrow \quad A_1 = E_r \ldots E_2 E_1 C_1 D_1 D_2 \ldots D_s$,

where E_i and D_i are elementary matrices.

Similarly $A_2 = F_i \ldots F_2 F_1 C_2 K_1 K_2 \ldots K_s$,

were F_i and K_i are elementary matrices.

$\therefore A_1 A_2 = E_r \ldots E_2 E_1 C_1 D_1 D_2 \ldots D_s F_t \ldots F_2 F_1 C_2 K_2 \ldots K_s$

We have

$$| A_1A_2 | = | E_r \ldots E_2 E_1 | . | C_1 D_1 D_2 \ldots D_s F_t \ldots F_2 F_1 C_2 | . | K_1 \ldots K_s | \quad \ldots(i)$$

Now the following cases arise.

Case 1: Let A_1 be a singular matric.

Then C_1 has at least one row of zero.

$\therefore | C_1 D_1 D_2 \ldots D_s F_t \ldots F_2 F_1 C_2 | = 0$

since the matrix $C_1 D_1 D_2 \ldots D_s F_t \ldots F_2 F_1 C_2$ has a row of zero.

$\therefore$ From (i) above we have $| A_1 A_2 | = 0$.

Case 2: If A_2 is a singular matrix. Then C_2 has at least one column of zero, hence as in Case I above

$$| C_1 D_1 D_2 \; D_s F_t \ldots F_1 C_2 | = 0$$

$\therefore$ From (i) above we have $| A_1 A_2 | = 0$.

Case 3: If either A_1 or A_2 is singular, then

$$| A_1 | . | A_2 | = 0$$

Hence $\quad | A_1 A_2 | = 0 = | A_1 | . | A_2 |$.

Case 4: If A_1 and A_2 are non-singular matrices. Then C_1 and C_2 are identity matrices. We have

$| A_1 A_2 | = | E_r \ldots E_2 E_1 C_1 D_1 D_2 \ldots D_s | . | F_t \ldots F_2 F_1 C_2 K_1 \ldots K_s|$

$$= | A_1 | . | A_2 |$$

$\therefore$ From all the above cases it is clear that

$| A_1 A_2 | = | A_1 | . | A_2 |$.

Cor.

$| A_1 A'_2 | = | A_1 | . | A_2 |$, where A'_2 is the transpose of A_2.

Proof:

$| A'_2 | = | A_2 |$, $\qquad \because A'_2$ is the transpose of A_2.

$\therefore | A_1 A'_2 | = | A_1 | | A'_2 |$,

$= | A_1 | | A_2 |$, $\qquad \because | A'_2 | = | A_2 |$.

The corollary leads to the row by row rule of multiplication of the determinants as given in Examples below:

Theorem 2:

If C_{ij} be the cofactor of a_{ij} in the $n \times n$, matrix $A = [a_{ij}]$ then $| C_{ij} | = \{ | a_{jk} | \}^{n-1}$

Proof:

If $A = [a_{ij}]$, then

$A' = [a'_{ki}]$, where A' is the transpose of A and $a'_{ki} = a_{ik}$.

Now $A' . [C_{ij}] = [a'_{ki}] [C_{ij}] = [b_{kj}]$, say. ...(i)

where $b_{kj} = \sum_{i=1}^{n} a'_{ki} C_{ij} = \sum_{i=1}^{n} a_{ij} C_{ij}, \quad \because a'_{ki} = a_{ik}$

We know that

$$b_{kj} = \sum_{i=1}^{n} a_{ik} C_{ij} = 0, \text{ if } j \neq k$$

$$= | A |, \text{ if } j = k.$$

$\therefore$ From (i) we conclude that for the product A', $[C_{ij}]$ *i.e.,* $[b_{ki}]$ all the diagonal terms (for which j = k) are | A |, whereas the nondiagonal terms (for which j ≠ k) are zero.

i.e.,, $$A'.[C_{ij}] = \begin{bmatrix} |A| & 0 & \ldots & 0 \\ 0 & |A| & \ldots & 0 \\ \ldots & \ldots & \ldots & \ldots \\ \ldots & \ldots & \ldots & \ldots \\ 0 & 0 & \ldots & |A| \end{bmatrix}$$

Hence $$| A' . [C_{ij}] | = \begin{bmatrix} |A| & 0 & \ldots & 0 \\ 0 & |A| & \ldots & 0 \\ \ldots & \ldots & \ldots & \ldots \\ \ldots & \ldots & \ldots & \ldots \\ 0 & 0 & \ldots & |A| \end{bmatrix}$$

$\Rightarrow | A' | . | C_{ij} | = \{| A |\}^n \quad \because \quad | A_1 . A_2 | = |A_1| . | A_2 |$

$\Rightarrow | A | . | C_{ij} | = \{| A |\}^n, \quad \because \quad | A' | = | A |$

$\Rightarrow | C_{ij} | = \{| A |\}^{n-1}$

Theorem 3:

An n × n matrix A is non-singular (or invertible) if and only if the determinant | A | ≠ 0.

Proof:

If C be the canonical form of the matrix A, then

$$C \sim A$$

Therefore $C = S A T$

where S and T are non-singular.

Hence $A = S^{-1} C T^{-1}$

$\Rightarrow \qquad A = E_r \ldots E_2 E_1 C D_1 D_2 \ldots D_s,$

where E_1 and D_1 are elementary matrices.

By the successive application, we have

$$| A | = | E_r | \ldots | E_2 | | E_1 | | C | | D_1 | | D_2 \ldots | D_s |$$

$\therefore$ If $| A | = 0$, then $| C | = 0$, as $| E_i | \neq 0$ and $| D_i | \neq 0$.

If $| C | = 0$, then it has at least one row of zero.

$\therefore$ The rank of matrix A is less than n

i.e., the matrix A is singular.

If the matrix A is non-singular, then

$C = I_n$, where I_n is the $n \times n$ identity matrix.

i.e., $\quad | C | = | I_n | = 1$

$\therefore$ From (i) above, we have $| A | \neq 0$.

MINOR OF AN ELEMENT

Definition:

If M_{ij} be the $(n - 1) \times (n - 1)$ sub-matrix of the matrix $A = [a_{ij}]$ obtained by removing the ith row and jth column, then the determinant $| M_{ij} |$ is defined as the minor of the element a_{ij} in the determinant $| a_{ij} |$ of order n.

Theorem 1:

The cofactor C_{ij} of the element a_{ij} in the determinant $| a_{ij} |$ is given by $C_{ij} = (-1)^{i+j} | M_{ij} |$.

Proof:

Let us first of all prove the case $C_{11} = (-1)^2 | M_{11} |$ *i.e.*, $C_{11} = | M_{11} |$.

The terms in C_{11} are composed of elements taken from the $(n - 1) \times (n - 1)$ sub-matrix M_{11} of A.

The general term of $a_{11} C_{11} = \pm a_{11} a_{i22} a_{i33} \ldots a_{inn}$, where $i_2, i_3, \ldots i_n$ are $2, 3, \ldots, n$ is some order.

This term can also be obtained from the trace of matrix A *i.e.*, the product of the elements of the diagonal of the matrix A *i.e.*, $a_{11} a_{22} a_{33} \ldots a_{nn}$, by operating on its row subscripts by the permutation $p = \begin{pmatrix} 1 & 2 & 3 & \ldots & n \\ i_1 & i_2 & i_3 & \ldots & i_n \end{pmatrix}$, where $i_2, i_3, \ldots, i_n$ are defined as above.

Thus the permutation p may be regarded as a permutation on the symbols $2, 3, \ldots, n$. Hence all the terms of $a_{11} C_{11}$ can be obtained by running p through the $(n - 1)!$ permutations on the symbols $2, 3, \ldots, n$ keeping 1 fixed.

Thus, the terms of C_{11} can be obtained by operating on the row subscripts of the elements of the product $a_{22}\ a_{33}\ .\ .\ .\ a_{nn}$, which is the product of the elements of the diagonal of M_{11} *i.e.,* the trace of M_{11}.

Hence $C_{11} = | M_{11} |$.

Now let us prove $C_{ij} = (-1)^{i+j} | M_{11} |$.

Move the jth column of the matrix A to the first column by performing (j – 1) successive interchanges of adjacent columns and move the ith row of the matrix A to the first row by performing (i – 1) successive interchanges of adjacent rows. Then the element a_{ij} is in the first row and first column of the resulting matrix B, say. The sub-matrix of B obtained by removing the first row and first column is the sub-matrix M_{ij} of the matrix A. Hence a_{ij} $| M_{ij} |$ is the term of $| B |$ containing a_{ij}.

Also we know that, if two rows or two columns of a determinant $| A |$ are interchanged, the new determinant $= - | A |$.

$$\therefore \text{ We have } | B | = (-1)^{i-1+j-i} | A |$$

$$= (-1)^{i+j} (-1)^{-2} | A |$$

$$= (-1)^{i+j} | A |, \qquad \because \quad (-1)^{-2} = 1$$

$$\Rightarrow \qquad | A | = (-1)^{i+j} | B |$$

Equating the coefficient of a_{ij} from both sides, we have

$C_{ij} = (-1)^{i+j} | M_{ij} |$.

Theorem 2:

If C_{ij} is the cofactor of a_{ij} in the determinant $| A | = | a_{ij} |$ of order n, then (i) the sum of the products of the elements of the ith row with the cofactors of the corresponding elements of the kth row is zero provided $i \neq k$.

(ii) Also the sum of the products of the elements of the jth column with the cofactors of the corresponding elements of the kth column is zero provided $j \neq k$,

i.e., (i) $$\sum_{j=1}^{n} c_{ij}\, C_{kj} = 0, \text{ if } i \neq k$$

and (ii) $$\sum_{j=1}^{n} c_{ij}\, C_{ik} = 0, \text{ if } j \neq k$$

Proof:

(i) The given determinant $| A | = \sum_{j=1}^{n} a_{kj} C_{kj}$.

Now replace the kth row by ith row, then we have the new determinant

$$= \sum_{j=1}^{n} a_{ij} C_{kj}.$$

But the kth and ith rows of the new determinant are identical, hence its values is zero.

$$\sum_{j=1}^{n} a_{ij} C_{kj} = 0$$

(ii) The given determinant $| A | = \sum_{i=1}^{n} a_{ik} C_{ik}$

Now replace the kth column by jth column, then we have the new determinant $= \sum_{i=1}^{n} a_{ij} C_{ik}$.

But the kth and jth columns of the new determinant are identical, hence its value is zero.

$$\therefore \quad \sum_{i=1}^{n} a_{ij} C_{ik} = 0.$$

AN IMPORTANT PROPERTY OF THE DETERMINANT

(a) *If A_i, the ith row of a determinant $| A | = | a_{ij} |$ of order n be replaced by $A_j + cA_k$, where c is a scalar and A_k denotes the kth row of the determinant $| A |$, then the value of the determinant remains unaltered.*

Proof:

The determinant $| A | = \sum_{j=1}^{n} a_{ij} C_{ij}$.

Replacing A_i by $A_i + c\, A_k$ we get the new determinant $| B |$, say

$$\text{Then} \quad | B | = \sum_{j=1}^{n} (a_{ij} + c\, a_{kj})\, C_{ij}$$

$$= \sum_{j=1}^{n} a_{ij} C_{ij} + c \sum_{j=1}^{n} a_{kj} C_{ij}$$

$$= | A | + c.0$$

i.e., $\quad | B | = | A |$.

(b) *If C_i, the ith column of determinant $| A | = | a_{ij} |$ of order n be replaced by $C_i + \lambda\, C_k$, where λ is a scalar and C_k denotes the kth column of $| A |$, then the value of the determinant remains unaltered.*

Proof is similar to part (a) above.

In the following examples R_1. R_2, R_3, . . . stand for first, second third, . . . rows and C_1, C_2,C_3, . . . stand for first, second, third . . . columns.

DETERMINANT OF A SQUARE MATRIX

Let us consider a square matrix A of order n × n given by

$$A = \begin{bmatrix} a_{11} & a_{12} & a_{13} & \cdots & \cdots & a_{1n} \\ a_{21} & a_{22} & a_{23} & \cdots & \cdots & a_{2n} \\ a_{31} & a_{32} & a_{33} & \cdots & \cdots & a_{3n} \\ \cdots & \cdots & \cdots & \cdots & \cdots & \cdots \\ a_{n1} & a_{n2} & a_{n3} & \cdots & \cdots & a_{nn} \end{bmatrix}$$

The product of the elements in the principal diagonal is

$$a_{11}\, a_{22}\, a_{33} \ldots a_{nn}.$$

This is also called the *trace of the matrix.*

Now obtain n! terms of the above type by operating on the row-subscripts of the elements of the above expression by n! permutations $p = \begin{pmatrix} 1 & 2 & 3 & \ldots & n \\ i_1 & i_2 & i_3 & \ldots & i_n \end{pmatrix}$, where i_1, i_2, i_3, . . . i_n are one of the n! permutations of the integers 1, 2, 3, . . ., n.

The sum of n! signed terms thus obtained is defined as the *determinants of the matrix* A and is denoted by | A | or $| a_{ij} |$.

Therefore the determinant of the square matrix A = $[a_{ij}]$ of order n × n is given by

$$| a_{ij} | = \Sigma \pm a_{\alpha 1}\, a_{\beta 2}\, a_{\gamma 3}\, a_{\delta 4} \ldots a_{k_n}$$

where + or – sign is taken where α, β, γ, δ, . . ., k is an even or odd permutation of 1, 2, 3, . . ., n, and the summation extends over n! permutations of the row subscripts 1, 2, 3, ...

Note: The determinant of a square matrix of order n is known as a determinant of order n.

DETERMINANT OF ORDER TWO

Let us consider a square matrix A = $\begin{bmatrix} a_{11} & a_{12} \\ a_{21} & a_{22} \end{bmatrix}$ of order 2 × 2.

Then 2! permutations on two symbols 1 and 2 are

$$I = \begin{pmatrix} 1 & 2 \\ 1 & 2 \end{pmatrix},\ p = \begin{pmatrix} 1 & 2 \\ 2 & 1 \end{pmatrix}.$$

The product of the elements of the principal diagonal are $a_{11}\ a_{22}$.

Operating on the row subscripts of $a_{11}\ a_{22}$ by the permutation I we get $+ a_{11}\ a_{22}$, prefixing + sign as the permutation I is even and by the permutation p we get $-a_{21}\ a_{12}$, prefixing – sign as the permutation is odd.

Hence $\begin{vmatrix} a_{11} & a_{12} \\ a_{21} & a_{22} \end{vmatrix} = a_{11}\ a_{22} - a_{21}\ a_{12}$

DETERMINANT OF ORDER THREE

Let us consider a square matrix $A = \begin{bmatrix} a_{11} & a_{12} & a_{13} \\ a_{21} & a_{22} & a_{23} \\ a_{31} & a_{32} & a_{33} \end{bmatrix}$, of order 3 × 3.

Then 3! permutations on three symbols 1, 2, and 3 are

$$I = \begin{pmatrix} 1 & 2 & 3 \\ 1 & 2 & 3 \end{pmatrix};\ p_1 = \begin{pmatrix} 1 & 2 & 3 \\ 1 & 3 & 2 \end{pmatrix};\ p_2 = \begin{pmatrix} 1 & 2 & 3 \\ 2 & 3 & 1 \end{pmatrix};$$

$$p_3 = \begin{pmatrix} 1 & 2 & 3 \\ 2 & 1 & 3 \end{pmatrix};\ p_4 = \begin{pmatrix} 1 & 2 & 3 \\ 3 & 2 & 1 \end{pmatrix};\ \text{and}\ p_5 = \begin{pmatrix} 1 & 2 & 3 \\ 3 & 2 & 1 \end{pmatrix};$$

Operating on the row subscripts of $a_{11}\ a_{22}\ a_{33}$ by the permutation I, p_1, p_2, . . ., p_5 we have successively

(a) $+a_{11}\ a_{22}\ a_{33}$, prefixing + as permutation I is even,

(b) $-a_{11}\ a_{32}\ a_{23}$, prefixing – as permutation p_1 is odd,

(c) $+a_{21}\ a_{32}\ a_{13}$, prefixing + sign as permutation p_2 is even,

(d) $-a_{21}\ a_{12}\ a_{33}$, prefixing + sign as permutation p_3 is odd,

(e) $+a_{31}\ a_{12}\ a_{23}$, prefixing + sign as permutation p_4 is even,

(f) $-a_{31}\ a_{22}\ a_{13}$, prefixing – sign as permutation p_5 is odd.

Hence we have

$$\begin{bmatrix} a_{11} & a_{12} & a_{13} \\ a_{21} & a_{22} & a_{23} \\ a_{31} & a_{32} & a_{33} \end{bmatrix} = a_{11}\ a_{23}\ a_{33} - a_{11}\ a_{32}\ a_{23} + a_{21}\ a_{32}\ a_{13} - a_{21}\ a_{12}\ a_{33}$$

$$+ a_{31}\ a_{12}\ a_{23} - a_{31}\ a_{22}\ a_{13}$$

$$= a_{11}\ (a_{22}\ a_{33} - a_{23}\ a_{32}) - a_{13}\ (a_{21}\ a_{33} - a_{23}\ a_{21}) + a_{13}\ (a_{21}\ a_{22} - a_{22}\ a_{31})$$

$$= a_{11} \begin{vmatrix} a_{22} & a_{23} \\ a_{32} & a_{33} \end{vmatrix} - a_{12} \begin{vmatrix} a_{21} & a_{23} \\ a_{31} & a_{33} \end{vmatrix} + a_{13} \begin{vmatrix} a_{21} & a_{22} \\ a_{31} & a_{32} \end{vmatrix}.$$

COFACTOR OF AN ELEMENT

Definition: If in the expansion of a determinant $|a_{ij}|$, all the terms containing a_{ij} as a factor are collected and their sum be denoted by $a_{ij} C_{ij}$ then the factor C_{ij} is defined as the cofactor of the element a_{ij}.

From the above definition we find that if $[a_{ij}]$ be the $n \times n$ matrix whose determinant is $|a_{ij}|$ then if from $[a_{ij}]$ the element of its ith row and jth column are removed, the terms of C_{ij} are then composed of elements from the remaining $(n-1) \times (n-1)$ sub-matrix M_{ij} of $[a_{ij}]$.

Hence, the determinant $|a_{ij}|$ can be expressed as a function of the elements of ith row, by collecting all the terms containing $a_{i1}, a_{i2}, a_{i3}, \ldots, a_{in}$ and finding their sum.

i.e.,
$$|a_{ij}| = a_{i1} C_{i1} + a_{i2} C_{i2} + \ldots + a_{in} C_{in}$$
$$= \sum_{j-1}^{n} a_{ij} C_{ij},$$

which is the expansion of the determinant $|a_{ij}|$ by the elements of the ith row and their cofactors.

In a similar manner we can expand the determinant $|a_{ij}|$ by the elements of the kth column and their cofactor and write as $a_{ij}| = \sum_{j-k}^{n} a_{ik} C_{ik}$

or

The determinant that is left by cancelling the row and columns intersecting at a particular constituent, when the particular constituent is brought in the top left hand corner of a determinant is called its cofactor and is denoted by corresponding capital letter.

PROPERTIES OF DETERMINANTS

Prop. 1:

If $A = [a_{ij}]$ is an $n \times n$ matrix then $|A'| = |A|$, where A' is the transpose of the matrix A.

or

The value of a determinant is not altered by changing the rows into column and column into rows.

Proof:

If $A = [a_{ij}]$, then $A' = [a'_{ij}]$, where $a'_{ij} = a_{ij}$.

Now the product of the elements of the principal diagonal of

$$A' = a'_{11}\, a'_{22}\, a'_{33} \ldots a'_{nn}.$$

Operating on the row subscripts of the elements of this product by the permutation $p = \begin{pmatrix} 1 & 2 & 3 & n \\ i_1 & i_2 & i_3 \ldots i_n \end{pmatrix}$, where $i_1, t_2, i_3, \ldots$ are 1, 2, 3, . . . in some order, we have $\pm\, a'_{i11}\, a'_{i22}\, a'_{i33} \ldots a'_{inn}$ as a term of | A' | plus or minus sign to be taken according as p is even or odd.

Now as $a'_{ij} = a_{ji}$, so we have

$a'_{i11}\, a'_{i22}\, a_{i33} \ldots a'_{inn} = a_{1/1}\, a_{2/2}\, a_{3/3} \ldots a_{n/n}$

The term $a_{1/1}\, a_{2/2} \ldots a_{n/n}$ can be obtained from the term $a_{i1/1}\, a_{i2/2}\, a_{i3/3} \ldots a_{in/n}$ by operating on its row subscripts the permutation $p' = \begin{pmatrix} i_1 & i_2 & i_3 \ldots i_n \\ 1 & 2 & 3 & n \end{pmatrix} = p^{-1}$.

Hence p' is even or odd according as p is even or odd. Therefore the term $\pm a'_{i11}\; a'_{i22}\; a'_{i33} \ldots a'_{inn}$ of | A' | is also a term of | A |.

We can thus prove that every one of the n! terms of | A' | is a term of | A |. Hence the property.

For example, If $|A| = \begin{vmatrix} a_1 & a_2 & a_3 \\ b_1 & b_2 & b_3 \\ c_1 & c_2 & c_3 \end{vmatrix}$

$$= a_1 \begin{vmatrix} b_2 & b_3 \\ c_2 & c_3 \end{vmatrix} - a_2 \begin{vmatrix} b_1 & b_3 \\ c_1 & c_3 \end{vmatrix} + a_3 \begin{vmatrix} b_1 & b_2 \\ c_1 & c_2 \end{vmatrix}$$

$$= a_1b_2c_3 - a_1b_3c_2 - a_2b_1c_3 + a_2b_3c_1 + a_3b_1c_2 - a_3b_2c_1.$$

Then $|A'| = \begin{vmatrix} a_1 & b_1 & c_1 \\ a_2 & b_2 & c_2 \\ a_3 & b_3 & c_3 \end{vmatrix}$

$$= a_1 \begin{vmatrix} b_2 & c_2 \\ b_3 & c_3 \end{vmatrix} - b_1 \begin{vmatrix} a_2 & c_2 \\ a_3 & c_3 \end{vmatrix} + c_1 \begin{vmatrix} a_2 & b_2 \\ a_3 & b_3 \end{vmatrix}$$

$$= a_1b_2c_3 - a_1b_3c_2 - a_2b_1c_3 + a_3b_1c_2 + a_2b_3c_1 - a_3b_2c_1.$$

$\therefore$ | A' | = | A |.

Prop. 2:

If B is obtained from A by interchanging two raws (or columns) then | B | = – | A |.

or

It two adjacent rows or columns of a determinant are interchanged the sign of the determinant is changed whereas its numerical value remarks the same.

Proof:

Let us suppose that sth and rth columns of the determinant A are interchanged where $s < 1$.

Let A = $\begin{bmatrix} a_{11} & a_{12} & \cdots & a_{1s} & \cdots & a_{1t} & \cdots & a_{1n} \\ a_{21} & a_{22} & \cdots & a_{2s} & \cdots & a_{2t} & \cdots & a_{2n} \\ \cdots & \cdots & \cdots & \cdots & \cdots & \cdots & \cdots & \cdots \\ a_{i1} & a_{i2} & \cdots & a_{is} & \cdots & a_{it} & \cdots & a_{in} \\ \cdots & \cdots & \cdots & \cdots & \cdots & \cdots & \cdots & \cdots \\ a_{n1} & a_{n2} & \cdots & a_{ns} & \cdots & a_{nt} & \cdots & a_{nn} \end{bmatrix}$

Then the trace of A *i.e.,* the product of the elements of the principal diagonal of A = $a_{11}\, a_{22} \ldots a_{ss} \ldots a_{tt} \ldots a_{nn}$.

If the sth and rth columns are interchanged the product of the elements of the principal diagonal of B

$= a_{11}\, a_{22} \ldots a_{ss} \ldots a_{tt} \ldots a_{nn}$

In order to have a term of | A |, let us operate on the row subscripts of the trace of A by the permutation

$$p = \begin{pmatrix} 1 & 2 & \ldots & s & \ldots & t & \ldots & n \\ i_1 & i_2 & \ldots & i_s & \ldots & i_t & \ldots & i_n \end{pmatrix}$$

Then we have $a_{i11}\, a_{i22} \ldots a_{iss} \ldots a_{itt} \ldots a_{inn}$.

This term can also be obtained from the trace of B by operating on its row subscripts by the permutation

$$p' = \begin{pmatrix} 1 & 2 & \ldots & s & \ldots & t & \ldots & n \\ i_1 & i_2 & \ldots & i_t & \ldots & i_s & \ldots & i_n \end{pmatrix}$$

Here we observe that $p' = p\,(i_s\, i_t)$, since $(i_s\, i_t)$ is a transposition. Therefore p' is odd or even according as p is even or odd. Hence the term $\pm a_{i11}\, a_{i22} \ldots a_{iss} \ldots a_{itt} \ldots a_{inn}$ is also a term of | B | but with its sign changed. Thus every one of the n! terms of | A | is a term of | B | but with sign changed.

Here | B | = – | A |.

Example:

Let | A | = $\begin{vmatrix} a_1 & a_2 & a_3 \\ b_1 & b_2 & b_3 \\ c_1 & c_2 & c_3 \end{vmatrix}$

Let the determinant | B | be formed by interchanging second and third columns.

Then | B | = $\begin{vmatrix} a_1 & a_2 & a_3 \\ b_1 & b_2 & b_3 \\ c_1 & c_2 & c_3 \end{vmatrix}$

Expanding | B | by the elements of its first row, we get

$$|B| = a_1 \begin{vmatrix} b_3 & b_2 \\ c_3 & c_2 \end{vmatrix} - a_3 \begin{vmatrix} b_1 & b_2 \\ c_1 & c_2 \end{vmatrix} + a_2 \begin{vmatrix} b_1 & b_3 \\ c_1 & c_3 \end{vmatrix}$$

$$= a_1(b_3c_2 - b_2c_3) - a_3(b_1c_2 - b_2c_1) + a_2(b_1c_3 - b_3c_1)$$

$$= -[a_1(b_2c_3 - b_3c_2) - a_2(b_1c_3 - b_3c_1) + a_3(b_1c_2 - b_2c_1)$$

$$= -\left[a_1 \begin{vmatrix} b_2 & b_3 \\ c_2 & c_3 \end{vmatrix} - a_2 \begin{vmatrix} b_1 & b_3 \\ c_1 & c_3 \end{vmatrix} + a_3 \begin{vmatrix} b_1 & b_2 \\ c_1 & c_2 \end{vmatrix}\right]$$

$$= -\begin{vmatrix} a_1 & a_2 & a_3 \\ b_1 & b_2 & b_3 \\ c_1 & c_2 & c_3 \end{vmatrix} = -|A|.$$

Prop. 3:

If the elements of ith row (or ith column) of the determinant $|a_{ij}|$ *are multiplied by a scalar c then the resulting determinant is* $c\,|a_{ij}|$.

Proof:

We can write $|a_{ij}| = \sum_{j=1}^{n} a_{ij}\, C_{ij} \ldots$

In this case, the resulting determinant (when the elements of ith row are multiplied by c)

$$= \sum_{j=1}^{n} ca_{ij}\, C_{ij} = c \sum_{j=1}^{n} a_{ij}\, C_{ij} = c\,|a_{ij}|.$$

Similarly we can prove that statement when elements of kth column are multiplied by c.

Prop. 4:

If two rows or two columns of a determinant | *A* | *are identical then* | *A* | = *0.*

Proof:

In Prop. 2 above we have proved that if any two rows or columns of a determinant are interchanged then the value of the determinant changes in sign only.

Thus, if the two *identical* columns (or rows) of a determinant are interchanged, then the determinant does not change but its sign only changes.

Hence | A | = – | A | *i.e.*, | A | + | A | = 0 *i.e.*, | A | = 0.

Example:

Evaluate $\begin{vmatrix} a_1 & a_2 & a_1 \\ b_1 & b_2 & b_1 \\ c_1 & c_2 & c_1 \end{vmatrix}$

Expanding the determinant with respect to the first row, we have the given determinant

$$= a_1 \begin{vmatrix} b_2 & b_1 \\ c_2 & c_1 \end{vmatrix} - a_2 \begin{vmatrix} b_1 & b_1 \\ c_1 & c_1 \end{vmatrix} + a_1 \begin{vmatrix} b_1 & b_2 \\ c_1 & c_2 \end{vmatrix}$$

$$= a_1(b_2c_1 - b_1c_2) - a_2(b_1c_1 - b_1c_1) + a_1(b_1c_2 - b_2c_1)$$

$$= a_1b_2c_1 - a_1b_1c_2 - a_2(0) + a_1b_1c_2 - a_1b_2c_1 = 0.$$

SOLVED EXAMPLES

Example 1:

Prove that $\begin{vmatrix} b+c & c+a & a+b \\ q+r & r+p & p+q \\ y+z & z+x & x+y \end{vmatrix} = 2 \begin{vmatrix} a & b & c \\ p & q & r \\ x & y & z \end{vmatrix}$

Solution:

We have

$$\begin{vmatrix} b+c & c+a & a+b \\ q+r & r+p & p+q \\ y+z & z+x & x+y \end{vmatrix}$$

$$= \begin{vmatrix} b & c+a & a+b \\ q & r+p & p+q \\ y & z+x & x+y \end{vmatrix} + \begin{vmatrix} c & c+a & a+b \\ r & r+p & p+q \\ z & z+x & x+y \end{vmatrix}$$

$$= \begin{vmatrix} b & c+a & a \\ q & r+p & p \\ y & z+x & x \end{vmatrix} + \begin{vmatrix} b & c+a & b \\ q & r+p & q \\ y & z+x & y \end{vmatrix} + \begin{vmatrix} c & c & a+b \\ r & r & p+q \\ z & z & x+y \end{vmatrix}$$

$$+ \begin{vmatrix} c & a & a+b \\ r & p & p+q \\ z & x & x+y \end{vmatrix}$$

$$= \begin{vmatrix} b & c & a \\ q & r & p \\ y & z & x \end{vmatrix} + \begin{vmatrix} b & a & a \\ q & p & p \\ y & x & x \end{vmatrix} + \begin{vmatrix} c & a & a \\ r & p & p \\ z & x & x \end{vmatrix}$$

$$+\begin{vmatrix} c & a & b \\ r & p & q \\ z & x & y \end{vmatrix},$$ second and third deter minants vanish as two coumns in each are identical.

$$= \begin{vmatrix} b & c & a \\ q & r & p \\ y & z & x \end{vmatrix} + \begin{vmatrix} c & a & b \\ r & p & q \\ z & x & y \end{vmatrix},$$ second and third deter minants vanish as two columns in each are identical.

$$= -\begin{vmatrix} b & a & c \\ q & p & r \\ y & x & z \end{vmatrix} - \begin{vmatrix} a & c & b \\ p & r & q \\ x & z & y \end{vmatrix},$$ interchanging C_2 and C_3 in first determin ant and C_1 and C_2 in second.

$$= \begin{vmatrix} a & b & c \\ p & q & r \\ x & y & z \end{vmatrix} + \begin{vmatrix} a & b & c \\ p & q & r \\ x & y & z \end{vmatrix},$$ interchanging C_1 and C_2 in first and C_2 and C_3 in second determinat.

$$= 2\begin{vmatrix} a & b & c \\ p & q & r \\ x & y & z \end{vmatrix}$$

Example 2(a):

Prove that $$\begin{vmatrix} 0 & -c & b & -l \\ c & 0 & -a & -m \\ -b & a & 0 & -n \\ x & y & z & 0 \end{vmatrix} = (al + bm + cn)\,(ax + by + cz)$$

Solution:

We have

$$= \frac{1}{a}\begin{vmatrix} 0 & -ac & ab & -al \\ c & 0 & -a & -m \\ -b & a & 0 & -n \\ x & y & z & 0 \end{vmatrix},$$ taking 1/a common from R_1

$$= \frac{1}{a}\begin{vmatrix} 0 & 0 & 0 & -al - bm - cn \\ c & 0 & -a & -m \\ -b & a & 0 & -n \\ x & y & z & 0 \end{vmatrix},$$ replacing R_1 by $R_1 + b\,R_2 + c\,R_3$

$$= \frac{(al + bm + cn)}{a} \begin{vmatrix} c & 0 & -a \\ -b & a & 0 \\ x & y & z \end{vmatrix}$$, expanding with respect to R_1

$$= \frac{(al + bm + cn)}{a^2} \begin{vmatrix} ac & 0 & -a \\ -ab & a & 0 \\ ax & y & z \end{vmatrix}$$, taking 1/a common from C_1.

$$= \frac{(al + bm + cn)}{a^2} \begin{vmatrix} 0 & 0 & -a \\ 0 & a & 0 \\ ax + by + cz & y & z \end{vmatrix}$$, replacing C_1 by $C_1 + b\,C_2 + c\,C_2$

$$= \frac{(al + bm + cn)(ax + by + cz)}{a^2} \begin{vmatrix} 0 & -a \\ a & 0 \end{vmatrix},$$

$$= (al + bm + cn)\,(ax + by + cz).$$

Example 2(b):

Evaluate $\begin{vmatrix} 1 & 1 & 1 \\ a & b & c \\ a^2 & b^2 & c^2 \end{vmatrix}$

Solution:

The given determinant

$$= \begin{vmatrix} 1 & 0 & 0 \\ a & b - a & c - a \\ a^2 & b^2 - a^2 & c^2 - a^2 \end{vmatrix}$$, replacing R_2 and R_3 by $R_3 - R_1$ and $R_3 - R_1$ respectively.

$$= \begin{vmatrix} b - a & c - a \\ b^2 - a^2 & c^2 - a^2 \end{vmatrix}$$, expanding with respect to R_1

$$= (b - a)\,(c - a) \begin{vmatrix} 1 & 1 \\ b + a & c + a \end{vmatrix}$$, taking common factors out

$= (b - a)(c - a)[(c + a) - (b + a)]$

$= (a - b)(c - a)(b - c).$

Example 2(c):

In the determinant $|A| = \begin{vmatrix} a_1 & b_1 & c_1 \\ a_2 & b_2 & c_2 \\ a_3 & b_3 & c_3 \end{vmatrix}$ *prove that*

$a_1A_2 + b_1B_2 + c_1C_2 = 0,\ b_1C_1 + b_2C_2 + b_3C_3 = 0$ *and*

$c_1B_1 + c_2B_2 + c_3B_3 = 0$, *where capital letters denote the cofactors of the corresponding small letters.*

Also prove that

$a_1A_1 + b_1B_1 + c_1C_1 = |A| = a_2A_2 + b_2B_2 + c_2C_2 = a_3A_3 + b_3B_3 + c_3C_3$

Solution:

In the det $|A| = \begin{vmatrix} a_1 & b_1 & c_1 \\ a_2 & b_2 & c_2 \\ a_3 & b_3 & c_3 \end{vmatrix}$, we have

$A_1 = \begin{vmatrix} b_2 & c_2 \\ b_3 & c_3 \end{vmatrix};\ A_2 = -\begin{vmatrix} b_1 & c_1 \\ b_3 & c_3 \end{vmatrix};\ B_1 = -\begin{vmatrix} a_2 & c_2 \\ a_3 & c_3 \end{vmatrix};\ B_2 = \begin{vmatrix} a_1 & c_1 \\ a_3 & c_3 \end{vmatrix};$

$B_3 = -\begin{vmatrix} a_1 & c_1 \\ a_2 & c_2 \end{vmatrix};\ C_1 = \begin{vmatrix} a_2 & b_2 \\ a_3 & b_3 \end{vmatrix};\ C_2 = -\begin{vmatrix} a_1 & b_1 \\ a_3 & b_3 \end{vmatrix};\ A_3 = \begin{vmatrix} b_1 & c_1 \\ b_2 & c_2 \end{vmatrix};$

$C_3 = \begin{vmatrix} a_1 & b_1 \\ a_2 & b_2 \end{vmatrix}$

$\therefore\ a_1A_2 + b_1B_2 + c_1C_2$

$= -a_1\begin{vmatrix} b_1 & c_1 \\ b_3 & c_3 \end{vmatrix} + b_1\begin{vmatrix} a_1 & c_1 \\ a_3 & c_3 \end{vmatrix} - c_1\begin{vmatrix} a_1 & b_1 \\ a_3 & b_3 \end{vmatrix}$

$= -a_1(b_1c_3 - b_3c_1) + b_1(a_1c_3 - a_3c_1) - c_1(a_1b_3 - a_3b_1)$

$= 0$, on simplifying

$b_1C_1 + b_2C_2 + b_3C_3 = b_1\begin{vmatrix} a_2 & b_2 \\ a_3 & b_3 \end{vmatrix} - b_2\begin{vmatrix} a_1 & b_1 \\ a_3 & b_3 \end{vmatrix} + b_3\begin{vmatrix} a_1 & b_1 \\ a_2 & b_2 \end{vmatrix}$

$= b_1(a_2b_3 - a_3b_2) - b_2(a_1b_3 - a_3b_1) + b_3(a_1b_2 - a_2b_1)$

$= 0$, on simplifying.

In a similar way the remaining part can also be proved.

Also $|A| = \begin{vmatrix} a_1 & b_1 & c_1 \\ a_2 & b_2 & c_2 \\ a_3 & b_3 & c_3 \end{vmatrix}$

$$= a_1 \begin{vmatrix} b_2 & c_2 \\ b_3 & c_3 \end{vmatrix} - b_2 \begin{vmatrix} a_2 & c_2 \\ a_3 & c_3 \end{vmatrix} + c_1 \begin{vmatrix} a_2 & b_2 \\ a_3 & b_3 \end{vmatrix} \qquad \text{...(i)}$$

expanding with respect to first row.

Again $a_1A_1 + b_1B_1 + c_1C_1$

$$= a_1 \begin{vmatrix} b_2 & c_2 \\ b_3 & c_3 \end{vmatrix} + b_2 \left\{ - \begin{vmatrix} a_2 & c_2 \\ a_3 & c_3 \end{vmatrix} \right\} + c_1 \begin{vmatrix} a_2 & b_2 \\ a_3 & b_3 \end{vmatrix}$$

$$= a_1 \begin{vmatrix} b_2 & c_2 \\ b_3 & c_3 \end{vmatrix} - b_2 \begin{vmatrix} a_2 & c_2 \\ a_3 & c_3 \end{vmatrix} + c_1 \begin{vmatrix} a_2 & b_2 \\ a_3 & b_3 \end{vmatrix}$$

$= |A|$, from (i)

Similarly, $a_2A_2 + b_2B_2 + c_2C_2$

$$= a_2 \left\{ - \begin{vmatrix} b_1 & c_1 \\ b_3 & c_3 \end{vmatrix} \right\} + b_2 \begin{vmatrix} a_1 & c_1 \\ a_3 & c_3 \end{vmatrix} + c_2 \left\{ - \begin{vmatrix} a_1 & b_1 \\ a_3 & b_3 \end{vmatrix} \right\}$$

$$= -a_2 \begin{vmatrix} b_1 & c_1 \\ b_3 & c_3 \end{vmatrix} + b_2 \begin{vmatrix} a_1 & c_1 \\ a_3 & c_3 \end{vmatrix} - c_2 \begin{vmatrix} a_1 & b_1 \\ a_3 & b_3 \end{vmatrix} \qquad \text{...(ii)}$$

Also $|A| = \begin{vmatrix} a_1 & b_1 & c_1 \\ a_2 & b_2 & c_2 \\ a_3 & b_3 & c_3 \end{vmatrix}$

$$= -a_2 \begin{vmatrix} b_1 & c_1 \\ b_3 & c_3 \end{vmatrix} + b_2 \begin{vmatrix} a_1 & c_1 \\ a_3 & c_3 \end{vmatrix} - c_2 \begin{vmatrix} a_1 & b_1 \\ a_3 & b_3 \end{vmatrix}, \qquad \text{...(iii)}$$

expanding with respect to second row.

∴ From (ii) and (iii), we get

$$a_2A_2 + b_2B_2 + c_2C_2 = |A|.$$

In a similar way we can prove that $a_2A_2 + b_2B_2 + c_2C_2 = |A|$, by expanding $|A|$ with respect to third row.

Example 2(d):

Find the value of $\begin{vmatrix} a & h & g \\ h & b & f \\ g & f & c \end{vmatrix}$

Solution:

$$\begin{vmatrix} a & h & g \\ h & b & f \\ g & f & c \end{vmatrix} = a\,\{b'c - f'f\} - h\,\{hc - f'g\} + g\,\{hf - bg\}$$

$$= abc - af^2 - ch^2 + fgh + fgh - bg^2$$

$$= abc + 2fgh - af^2 - bg^2 - ch^2.$$

Example 3:

Evaluate $\begin{vmatrix} a & -a & -a & -a \\ b & -b & -b & -b \\ c & -c & -c & -c \\ d & -d & -d & -d \end{vmatrix}$

Solution:

Since three columns of the given determinant are identical, so the value of the determinant is zero.

Example 4:

Evaluate $\begin{vmatrix} 1 & 2 & 3 \\ 4 & 5 & 6 \\ 7 & 8 & 9 \end{vmatrix}$

Solution:

$$\begin{vmatrix} 1 & 2 & 3 \\ 4 & 5 & 6 \\ 7 & 8 & 9 \end{vmatrix} = 1\,\{5.9 - 6.8\} - 2\,\{4.9 - 6.7\} + 3\,\{4.8 - 5.7\},$$

$$= 1\,\{45 - 48\} - 2\{36 - 42\} + 3\,\{32 - 35\}$$

$$= -3 + 12 - 9 = 0.$$

Example 5(a):

Evaluate $\begin{vmatrix} 1 & 1 & 1 & 1 \\ 1 & 1+x & 1 & 1 \\ 1 & 1 & 1+y & 1 \\ 1 & 1 & 1 & 1+z \end{vmatrix}$

Solution:

We have

$$= \begin{vmatrix} 1 & 1 & 1 & 1 \\ 0 & x & 0 & 0 \\ 0 & 0 & y & 0 \\ 0 & 0 & 0 & z \end{vmatrix},$$ replacing R_2, R_3 and R_4 by $R_4 - R_1$, $R_3 - R_1$ and $R_4 - R_1$ respectively.

$$= \begin{vmatrix} x & 0 & 0 \\ 0 & y & 0 \\ 0 & 0 & z \end{vmatrix},$$ expanding with respect to C_1

$$= x \begin{vmatrix} y & 0 \\ 0 & z \end{vmatrix},$$ expanding with respect to R_1

$= xyz.$

Example 5(b):

Show that $\begin{vmatrix} (b+c)^2 & a^2 & a^2 \\ b^2 & (c+a)^2 & b^2 \\ c^2 & c^2 & (a+b)^2 \end{vmatrix} = 2abc\,(a+b+c)^2$

Solution:

We have

$$= \begin{vmatrix} (b+c)^2 - a^2 & 0 & a^2 \\ 0 & (c+a)^2 - b^2 & b^2 \\ c^2 - (a+b)^2 & c^2 - (a+b)^2 & (a+b)^2 \end{vmatrix},$$ replacing C_1 and C_2 by $C_1 - C_2$ and $C_2 - C_3$ respectively.

$$= \begin{vmatrix} (b+c+a)(b+c-a) & 0 & a^2 \\ 0 & (c+a+b)(c+a-b) & b^2 \\ (c+a+b)(c-a-b) & (c+a+b)(c-a-b) & (a+b)^2 \end{vmatrix},$$

$$= (a+b+c)^2 \begin{vmatrix} b+c-a & 0 & a^2 \\ 0 & c+a-b & b^2 \\ c-a-b & c-a-b & (a+b)^2 \end{vmatrix},$$ taking out the common factors from C_1 and C_2

$$= (a + b + c)^2 \begin{vmatrix} b + c - a & 0 & a^2 \\ 0 & c + a - b & b^2 \\ -2b & -2a & 2ab \end{vmatrix},$$ replacing R_2 by $R_3 - R_1 - R_2$

$$= (a + b + c)^2 \begin{vmatrix} b + c & a^2/b & a^2 \\ b^2/a & c + a & b^2 \\ 0 & 0 & 2ab \end{vmatrix},$$ replacing C_1 and C_2 by $C_1 + \frac{1}{a} C_2$ and $C_2 + \frac{1}{b} C_2$ respectively. **(Note)**

$$= 2ab\,(a + b + c)^2 \begin{vmatrix} b + c & a^2/b \\ b^2/a & c + a \end{vmatrix},$$ expanding with respect to R_2

$= 2ab\,(a + b + c)^2\,[(b + c)\,(c + a) - (b^2/a)\,(a^2/b)]$

$= 2ab\,(a + b + c)^2\,[bc + ba + c^2 + ca - ab]$

$= 2ab\,(a + b + c)^2\,[b + a + c] = 2abc\,(a + b + c)^2.$

Example 5(c):

Show that $$\begin{vmatrix} 1 & 1 & 1 \\ bc(b + c) & ca\,(c + a) & ab\,(a + b) \\ c^2c^2 & c^2a^2 & a^2b^2 \end{vmatrix}$$

$- abc\,(a - b)\,(b - c)\,(c - a)\,(a + b + c).$

Solution:

We have

$$= \begin{vmatrix} 1 & 1 & 1 \\ b^2c + bc^2 & c^2a + ca^2 & a^2b + ab^2 \\ b^2c^2 & c^2a^2 & a^2b^2 \end{vmatrix}$$

$$= \begin{vmatrix} 1 & 0 & 0 \\ b^2c + bc^2 & c(a - b)\,(a + b + c) & b(a - c)\,(a + b + c) \\ c^2c^2 & c^2(a - b)\,(a + b) & b^2(a - c)\,(a + c) \end{vmatrix},$$

replacing C_2, C_3 by $C_2 - C_1$, $C_3 - C_1$

$$= \begin{vmatrix} c(a-b)(a+b+c) & b(a-c)(a+b+c) \\ c^2(a-b)(a+b) & b^2(a-c)(a+c) \end{vmatrix},$$

expanding with respect to R_1

$$= c(a-b)\,b(a-c) \begin{vmatrix} a+b+c & a+b+c \\ c(a+b) & b(b+c) \end{vmatrix},$$

$$= bc(a-b)(a-c)(a+b+c) \begin{vmatrix} 1 & 1 \\ ca+cb & ba+bc \end{vmatrix},$$

taking out $a + b + c$ common from R_1

$$= bc(a-b)(a-c)(a+b+c) \begin{vmatrix} 1 & 0 \\ ca+cb & ba-bc \end{vmatrix},$$

replacing C_2 by $C_2 - C_1$

$$= bc(a-b)(a-c)(a+b+c)(ab-ca)$$

$$= -abc(a-b)(b-c)(c-a)(a+b+c)$$

Example 6:

Evaluate $\begin{vmatrix} a & b & ax+by \\ b & c & bx+cy \\ ax+by & bx+cy & 0 \end{vmatrix}$

Solution:

We have

$$= \begin{vmatrix} a & b & 0 \\ b & c & 0 \\ ax+by & bx+cy & -x(ax+by) - y(bx+cy) \end{vmatrix},$$ replacing C_2 by $C_3 - x\,C_1 - y\,C_2$

$$= -(ax^2 + 2bxy + cy^2) \begin{vmatrix} a & b \\ b & c \end{vmatrix},$$ expanding with respect to C_2

$$= -(ax^2 + 2bxy + cy^2)(ac - b^2).$$

Example 7:

Evaluate $\begin{vmatrix} a^2 & a^2-(b-c)^2 & bc \\ b^2 & b^2-(c-a)^2 & ca \\ c^2 & c^2-(a-b)^2 & ab \end{vmatrix}$

Solution:

We have

$$= \begin{vmatrix} a^2 & -(b-c)^2 & bc \\ b^2 & -(c-a)^2 & ca \\ c^2 & -(a-b)^2 & ab \end{vmatrix}, \text{ replacing } C_2 \text{ by } C_2 - C_1$$

$$= -\begin{vmatrix} a^2 & (b^2+c^2)-2bc & bc \\ b^2 & (c^2-a^2)-2ca & ca \\ c^2 & (a^2+b^2)-2ab & ab \end{vmatrix}$$

$$= -\begin{vmatrix} a^2 & b^2+c^2 & bc \\ b^2 & c^2+a^2 & ca \\ c^2 & a^2+b^2 & ab \end{vmatrix}, \text{ replacing } C_2 \text{ by } C_2 + 2C_3$$

$$= -\begin{vmatrix} a^2 & b^2+c^2+a^2 & bc \\ b^2 & c^2+a^2+b^2 & ca \\ c^2 & a^2+b^2+c^2 & ab \end{vmatrix}, \text{ replacing } C_2 \text{ by } C_2 + C_1$$

$$= -(a^2+b^2+c^2)\begin{vmatrix} a^2 & 1 & bc \\ b^2 & 1 & ca \\ c^2 & 1 & ab \end{vmatrix}, \text{ taking out the common factor from } C_2$$

$$= -(a^2+b^2+c^2)\begin{vmatrix} a^2 & 1 & bc \\ b^2-a^2 & 0 & ca-bc \\ c^2-a^2 & 0 & ab-bc \end{vmatrix}, \text{ replacing } R_2 \text{ and } R_3 \text{ by } R_2 - R_1 \text{ and } R_3 - R_1 \text{ respectively.}$$

$$= (a^2 + b^2 + c^2) \begin{vmatrix} b^2 - a^2 & c(a-b) \\ c^2 - a^2 & b(a-c) \end{vmatrix},$$ expanding with respect to C_2

$$= (a - b)(a - c)(a^2 + b^2 + c^2) \begin{vmatrix} -(b+a) & c \\ -(c+a) & b \end{vmatrix},$$

taking out the common factors

$$= (a - b)(a - c)(a^2 + b^2 + c^2)[-b(b + a) + (c + a)]$$
$$= (a - b)(a - c)(a^2 + b^2 + c^2)[-b^2 - ab + c^2 + ac]$$
$$= (a - b)(a - c)(a^2 + b^2 + c^2)[(c^2 - b^2) + a(c - b)]$$
$$= (a - b)(a - c)(a^2 + b^2 + c^2)(c - b)(a + b + c)$$
$$= (a - b)(a - c)(a + b + c)(a^2 + b^2 + c^2).$$

Example 8:

Evaluate $$\begin{vmatrix} 1^2 & 2^2 & 3^2 & 4^2 \\ 2^2 & 3^2 & 4^2 & 5^2 \\ 3^2 & 4^2 & 5^2 & 6^2 \\ 4^2 & 5^2 & 6^2 & 7^2 \end{vmatrix}$$

Solution:

We have

$$= \begin{vmatrix} 1 & 4 & 9 & 16 \\ 4 & 9 & 16 & 25 \\ 9 & 16 & 25 & 36 \\ 16 & 25 & 36 & 49 \end{vmatrix}$$

$$= \begin{vmatrix} 1 & 0 & 0 & 0 \\ 4 & -7 & -20 & -39 \\ 9 & -20 & -56 & -108 \\ 16 & -39 & -108 & -207 \end{vmatrix},$$ replacing C_2, C_3, C_4 by $C_2 - 4C_1$, $C_2 - 9C_1$ and $C_4 - 16C_1$ respctively.

$$= (-1)^2 \begin{vmatrix} 7 & 20 & 39 \\ 20 & 56 & 108 \\ 39 & 108 & 207 \end{vmatrix},$$

$$= \begin{vmatrix} 7 & -1 & -1 \\ 20 & -4 & -4 \\ 39 & -9 & -9 \end{vmatrix},$$ replacing C_2 and C_3 by $C_2 - 3C_1$ and $C_3 - 2C_3$ respctively.

= 0, as two columns are identical.

Example 9(a):

Prove that

$$\begin{vmatrix} a^2+\lambda & ab & ac & ad \\ bd & b^2+\lambda & bc & bd \\ ca & cb & c^2+\lambda & cd \\ da & db & dc & d^2+\lambda \end{vmatrix} = \lambda^2 (a^2+b^2+c^2+d^2+\lambda)$$

Solution:

We have

$$= abcd \begin{vmatrix} a+\frac{\lambda}{a} & a & a & a \\ b & b+\frac{\lambda}{b} & b & b \\ c & c & c+\frac{\lambda}{c} & c \\ d & d & d & d+\frac{\lambda}{d} \end{vmatrix},$$ taking out a, b, c, d common from C_1, C_2, C_3 and C_4 respectively.

$$= abcd \begin{vmatrix} a+\frac{\lambda}{a} & -\frac{\lambda}{a} & -\frac{\lambda}{a} & -\frac{\lambda}{a} \\ b & \frac{\lambda}{b} & 0 & 0 \\ c & 0 & \frac{\lambda}{c} & 0 \\ d & 0 & 0 & \frac{\lambda}{d} \end{vmatrix},$$ replacing C_2, C_3 and C_4 by $C_2 - C_1$, $C_3 - C_1$ and $C_4 - C_1$ respectively.

$$= abcd \begin{vmatrix} a & 0 & 0 & -\frac{\lambda}{a} \\ b & \frac{\lambda}{b} & 0 & 0 \\ c & 0 & \frac{\lambda}{c} & 0 \\ d+\frac{\lambda}{d} & -\frac{\lambda}{d} & -\frac{\lambda}{d} & \frac{\lambda}{d} \end{vmatrix},$$ replacing C_1, C_2 and C_3 by $C_1 + C_4$, $C_2 - C_4$ and $C_3 - C_4$ respectively.

$$= a^2 bcd \begin{vmatrix} \frac{\lambda}{b} & 0 & 0 \\ 0 & \frac{\lambda}{c} & 0 \\ -\frac{\lambda}{d} & -\frac{\lambda}{d} & \frac{\lambda}{d} \end{vmatrix} + \lambda\ bcd \begin{vmatrix} b & \frac{\lambda}{b} & 0 \\ 0 & 0 & \frac{\lambda}{c} \\ d+\frac{\lambda}{d} & -\frac{\lambda}{d} & -\frac{\lambda}{d} \end{vmatrix},$$

expanding with respect to R_1

$$= a^2 bcd.\frac{\lambda}{b} \begin{vmatrix} \frac{\lambda}{c} & 0 \\ -\frac{\lambda}{d} & \frac{\lambda}{d} \end{vmatrix} + \lambda\ bcd.b \begin{vmatrix} 0 & \frac{\lambda}{c} \\ -\frac{\lambda}{d} & -\frac{\lambda}{d} \end{vmatrix}$$

$$- \lambda bcd.\frac{\lambda}{b} \begin{vmatrix} c & \frac{\lambda}{c} \\ d+\frac{\lambda}{d} & -\frac{\lambda}{d} \end{vmatrix},$$

expanding each determinant with respect to R_1

$$= \lambda\ a^2\ cd \left(\frac{\lambda^2}{cd}\right) + \lambda b^2\ cd \left(\frac{\lambda^2}{cd}\right) - \lambda^2\ cd \left(-\frac{\lambda c}{d} - \frac{\lambda d}{c} - \frac{\lambda^2}{cd}\right)$$

$$= \lambda^2\ a^2 + \lambda^2\ b^2 + \lambda^2\ cd \left(\frac{\lambda c^2 + \lambda d^2 + \lambda^2}{cd}\right)$$

$$= \lambda^2\ a^2 + \lambda^2\ b^2 + \lambda^2\ (c^2 + d^2 + \lambda) = \lambda^2\ (a^2 + b^2 + c^2 + d^2 + \lambda)$$

Example 9(b):

Prove that $\begin{vmatrix} a & b & b & b \\ a & b & a & a \\ a & a & b & a \\ b & b & b & a \end{vmatrix} = -(b-a)^4$

Solution:

We have

$$= \begin{vmatrix} a & b-a & b-a & b-a \\ a & b-a & 0 & 0 \\ a & 0 & b-a & 0 \\ b & 0 & 0 & a-b \end{vmatrix},$$ replacing C_2, C_3 and C_4 by $C_2 - C_1$, $C_3 - C_1$ and $C_4 - C_1$ respectively.

$$= -(b-a)\begin{vmatrix} a & b-a & 0 \\ a & 0 & b-a \\ b & 0 & 0 \end{vmatrix} + (a-b)\begin{vmatrix} a & b-a & b-a \\ a & b-a & 0 \\ a & 0 & b-a \end{vmatrix},$$

expanding with respect to C_1

$$= -(b-a)\,b\begin{vmatrix} b-a & 0 \\ 0 & b-a \end{vmatrix} + (a-b)\begin{vmatrix} 0 & 0 & b-a \\ a & b-a & 0 \\ a & 0 & b-a \end{vmatrix},$$

expanding first determinant with respect to R_2 and in the second determinant replacing R_1 by $R_1 - R_2$.

$$= -b(b-a)^2 + (a-b)(b-a)\begin{vmatrix} a & b-a \\ a & 0 \end{vmatrix},$$

expanding the second det. with respect to R_1

$$= -b(b-a)^3 + (a-b)(b-a)[(0 - a(b-a)]$$

$$= -b(b-a)^3 + a(b-a)^2 = -(b-a)^3(b-a)$$

$$= -(b-a)^4.$$

Example 9(c):

Evaluate $\begin{vmatrix} 1 & bc+ad & b^2c^2+a^2d^2 \\ 1 & ca+bd & c^2a^2+b^2d^2 \\ 1 & ab+cd & a^2b^2+c^2d^2 \end{vmatrix}$

Solution:

We have

$$= \begin{vmatrix} 1 & bc+ad & b^2c^2+a^2d^2 \\ 0 & ca+bd-bc-ad & c^2a^2+b^2d^2-b^2c^2-a^2d^2 \\ 0 & ab+cd \quad bc \quad ad & a^2b^2+c^2d^2-b^2c^2-c^2d^2 \end{vmatrix},$$

replacing R_2 and R_3 by $R_2 - R_1$ and $R_3 - R_1$ respectively.

$$= \begin{vmatrix} ca - bc + bd - ad & c^2a^2 - b^2c^2 + b^2d^2 - a^2d^2 \\ ab - bc + cd - ad & a^2b^2 - b^2c^2 + c^2d^2 - a^2d^2 \end{vmatrix},$$

expanding with respect to C_1

$$= \begin{vmatrix} (c - d)(a - b) & (c^2 - d^2)(a^2 - b^2) \\ (b - d)(a - c) & (b^2 - d^2)(a^2 - c^2) \end{vmatrix},$$ factorising the elements

$$= (c - d)(a - b)(b - d)(a - c) \begin{vmatrix} 1 & (c + d)(a + b) \\ 1 & (b + d)(a + c) \end{vmatrix},$$

taking out the common factors

$$= (c - d)(a - b)(b - d)(a - c) \begin{vmatrix} 1 & ca + bc + da + db \\ 0 & ba + dc - ca - db \end{vmatrix},$$

replacing R_2 by $R_3 - R_1$

$$= (c - d)(a - b)(b - d)(a - c)(ba + dc - ca - db)$$
$$= (c - d)(a - b)(b - d)(a - c)(a - d)(b - c)$$

Example 10:

Prove that

$$\begin{vmatrix} 1 & 0 & x & 0 & x \\ 0 & 1 & 0 & x & 0 \\ x & 0 & x+1 & 0 & x \\ 0 & x & 0 & 1 & 0 \\ x & 0 & x & 0 & 1 \end{vmatrix} = (x - 1)^2 (x + 1)(1 + 2x - x^2)$$

Solution:

The given determinant

$$= \begin{vmatrix} 1 & 0 & 0 & 0 & 0 \\ 0 & 1 & 0 & x & 0 \\ x & 0 & x + 1 - x^2 & 0 & -1 \\ 0 & x & 0 & 1 & 0 \\ x & 0 & x - x^2 & 0 & 1 - x \end{vmatrix},$$ replacing C_3 and C_5 by $C_3 - x\,C_1$ and $C_5 - C_3$ respectively.

$$= \begin{vmatrix} 1 & 0 & x & 0 \\ 0 & x+1-x^2 & 0 & -1 \\ x & 0 & 1 & 0 \\ 0 & x-x^2 & 0 & 1-x \end{vmatrix}$$, expanding with respect to R_1

$$= \begin{vmatrix} 1 & 0 & 0 & 0 \\ b & x+1-x^2 & 0 & -1 \\ x & 0 & 1-x^2 & 0 \\ 0 & x-x^2 & 0 & 1-x \end{vmatrix}$$, replacing C_3 by $C_2 - x\,C_1$

$$= \begin{vmatrix} x+1-x^2 & 0 & -1 \\ 0 & 1-x^2 & 0 \\ x(1-x) & 0 & 1-x \end{vmatrix}$$, expanding with respect to R_1

$$= -(1-x) \begin{vmatrix} 0 & 1-x^2 & 0 \\ x+1-x^2 & 0 & -1 \\ x & 0 & 1 \end{vmatrix}$$, interchanging R_1 and R_2 and taking out $(1-x)$ common from R_2

$$= (1-x)(1-x^2) \begin{vmatrix} x+1-x^2 & -1 \\ x & 1 \end{vmatrix}$$, expanding with respect to R_1

$$= (1-x)^2 (1+x) [(x+1-x^2)\,1 - (-1).x]$$
$$= (x-1)^2 (1+x) (2x+1-x^2).$$

Example 11:

Evaluate $\begin{vmatrix} 13 & 16 & 19 \\ 14 & 17 & 20 \\ 15 & 18 & 21 \end{vmatrix}$

Solution:

We have

$$= \begin{vmatrix} 13 & 16 & 3 \\ 14 & 17 & 3 \\ 15 & 18 & 3 \end{vmatrix}$$, replacing C_2 by $C_3 - C_2$

$$= \begin{vmatrix} 13 & 3 & 3 \\ 14 & 3 & 3 \\ 15 & 3 & 3 \end{vmatrix}, \text{ replacing } C_2 \text{ by } C_3 - C_1$$

= 0, since two columns are identical.

Example 12:

Evaluate $\begin{vmatrix} 1 & 2 & 5 \\ 2 & 3 & 1 \\ -1 & 1 & 1 \end{vmatrix}$

Solution:

The given determinant

$$= \begin{vmatrix} 1 & 3 & 6 \\ 2 & 5 & 3 \\ -1 & 0 & 0 \end{vmatrix}, \text{ replacing } C_2 \text{ and } C_3 \text{ by } C_2 + C_1 \text{ and } C_3 + C_1 \text{ respctively.}$$

$$= \begin{vmatrix} 0 & 3 & 6 \\ 0 & 5 & 3 \\ -1 & 0 & 0 \end{vmatrix}, \text{ replacing } R_1 \text{ by } R_1 + R_3 \text{ and } R_2 \text{ by } R_2 + 2R_3$$

$$= -1 \times \begin{vmatrix} 3 & 6 \\ 5 & 3 \end{vmatrix}, \text{ expanding with respect to first column}$$

$$= -[3(3) - 5(6)] = 21.$$

Example 13:

Evaluate $\begin{vmatrix} 1 & a & b+c \\ 1 & b & c+a \\ 1 & c & a+b \end{vmatrix}$

Solution:

We have

$$= \begin{vmatrix} 1 & a & a+b+c \\ 1 & b & b+c+a \\ 1 & c & c+a+b \end{vmatrix}, \text{ replacing } C_3 \text{ by } C_3 + C_2$$

$$= (a+b+c) \begin{vmatrix} 1 & a & 1 \\ 1 & b & 1 \\ 1 & c & 1 \end{vmatrix}, \text{ taking out } (a+b+c) \text{ common from } C_3$$

= 0, since two columns are identical.

Example 14(a):

Evaluate $\begin{vmatrix} 1 & 2 & 3 & 4 \\ 2 & 3 & 4 & 1 \\ 3 & 4 & 1 & 2 \\ 4 & 1 & 2 & 3 \end{vmatrix}$

Solution:

The given determinant

$= \begin{vmatrix} 10 & 2 & 3 & 4 \\ 10 & 3 & 4 & 1 \\ 10 & 4 & 1 & 2 \\ 10 & 1 & 2 & 3 \end{vmatrix}$, replacing C_1 by $C_1 + C_2 + C_3 + C_4$

$= 10 \begin{vmatrix} 1 & 2 & 3 & 4 \\ 1 & 3 & 4 & 1 \\ 1 & 4 & 1 & 2 \\ 1 & 1 & 2 & 3 \end{vmatrix}$, taking out 10 common from C_1

$= 10 \begin{vmatrix} 1 & 2 & 3 & 4 \\ 0 & 1 & 1 & -3 \\ 0 & 2 & -2 & -2 \\ 0 & -1 & -1 & -1 \end{vmatrix}$, replacing R_2, R_3, R_4 by $R_2 - R_1$, $R_3 - R_1$ and $R_4 - R_1$ respectively

$= 10 \begin{vmatrix} 1 & 1 & -3 \\ 2 & -2 & -2 \\ -1 & -1 & -1 \end{vmatrix}$, expanding with respect to C_1

$= -20 \begin{vmatrix} 1 & 1 & -3 \\ 1 & -1 & -1 \\ 1 & 1 & 1 \end{vmatrix}$, taking out 2 common from R_2 and -1 common from R_3

$= -20 \begin{vmatrix} 1 & 1 & -3 \\ 0 & -2 & 2 \\ 0 & 0 & 4 \end{vmatrix}$, replacing R_2 and R_3 by $R_2 - R_1$ and $R_3 - R_1$ respectively

$= -20 \begin{vmatrix} -2 & 2 \\ 0 & 4 \end{vmatrix}$, expanding with respect to C_1

$= -20\ [(-2).4 - 0.2] = -20\ [-8] = 160.$

Example 14(b):

Evaluate $\begin{vmatrix} 1 & 1 & 1 \\ 1 & 1+x & 1 \\ 1 & 1 & 1+y \end{vmatrix}$

Solution:

We have

$$= \begin{vmatrix} 1 & 1 & 1 \\ 0 & x & 0 \\ 0 & 0 & y \end{vmatrix}, \text{ replacing } R_3 \text{ by } R_2 - R_1 \text{ and } R_3 \text{ by } R_3 - R_1$$

$$= 1 \times \begin{vmatrix} x & 0 \\ 0 & y \end{vmatrix}, \text{ expanding with respect to the first column.}$$

$= xy.$

Example 14(c):

Evaluate $\begin{vmatrix} -4 & 1 & 1 & 1 & 1 \\ 1 & -4 & 1 & 1 & 1 \\ 1 & 1 & -4 & 1 & 1 \\ 1 & 1 & 1 & -4 & 1 \\ 1 & 1 & 1 & 1 & -4 \end{vmatrix}$

Solution:

We have

$$= \begin{vmatrix} 0 & 1 & 1 & 1 & 1 \\ 0 & -4 & 1 & 1 & 1 \\ 0 & 1 & -4 & 1 & 1 \\ 0 & 1 & 1 & -4 & 1 \\ 0 & 1 & 1 & 1 & -4 \end{vmatrix}, \text{ replacing } C_1 \text{ by } C_1 + C_2 + C_3 + C_4 + C_5$$

= 0, expanding with respect to elements of first column.

Example 15:

Evaluate $\begin{vmatrix} 3 & 2 & 1 & 4 \\ 15 & 29 & 2 & 14 \\ 16 & 19 & 3 & 17 \\ 33 & 39 & 8 & 38 \end{vmatrix}$

Solution:

The given determinant

$$= \begin{vmatrix} 0 & 0 & 1 & 0 \\ 9 & 25 & 2 & 6 \\ 7 & 13 & 3 & 5 \\ 9 & 23 & 8 & 6 \end{vmatrix},$$ replacing C_1, C_2 and C_4 by $C_1 - 3C_3$, $C_2 - 2C_3$, $C_4 - 4C_3$ respectively.

$$= \begin{vmatrix} 9 & 25 & 6 \\ 7 & 13 & 5 \\ 9 & 23 & 6 \end{vmatrix}, \text{ expanding with respect to } R_1$$

$$= \begin{vmatrix} 0 & 2 & 0 \\ 7 & 13 & 5 \\ 9 & 23 & 6 \end{vmatrix}, \text{ replacing } R_1 \text{ by } R_1 - R_3$$

$$= -2 \begin{vmatrix} 7 & 5 \\ 9 & 6 \end{vmatrix}, \text{ expanding with respect to } R_1$$

$$= -2\,[42 - 45] = (-2) \times (-3) = 6.$$

Example 16:

Show that

$$\begin{vmatrix} a + b + 2c & a & b \\ c & b + c + 2a & b \\ c & a & c + a + 2b \end{vmatrix} = 2\,(a + b + c)^2$$

Solution:

The given determinant

$$= \begin{vmatrix} 2a + 2b + 2c & a & b \\ 2a + 2b + 2c & b + c + 2a & b \\ 2a + 2b + 2c & a & c + a + 2b \end{vmatrix}, \text{ replacing } C_1 \text{ by } C_1 + C_2 + C_3$$

$$= (2a + 2b + 2c) \begin{vmatrix} 1 & a & b \\ 1 & b + c + 2a & b \\ 1 & a & c + a + 2b \end{vmatrix}, \text{ taking out } 2a + 2b + 2c \text{ common}$$

$$= 2(a + b + c) \begin{vmatrix} 1 & a & b \\ 0 & b + c + a & b \\ 0 & a & c + a + b \end{vmatrix}, \text{ replacing } R_2 \text{ and } R_3 \text{ by } R_2 - R_1 \text{ and } R_3 - R_1 \text{ respectively}$$

$$= 2(a + b + c) \begin{vmatrix} b + c + a & 0 \\ 0 & c + a + b \end{vmatrix}, \text{ expanding with respect to 1st column.}$$

$= 2(a + b + c)\ [(b + c + a)\ (c + a + b)]$

$= 2\ (a + b + c)^2.$

Example 17:

Evaluate $\begin{vmatrix} 1 & 1 & 1 & 1 & 1 \\ 1 & 2 & 3 & 4 & 5 \\ 1 & 3 & 6 & 10 & 15 \\ 1 & 4 & 10 & 20 & 35 \\ 1 & 5 & 15 & 35 & 69 \end{vmatrix}$

Solution:

We have

$= \begin{vmatrix} 1 & 0 & 0 & 0 & 0 \\ 1 & 1 & 2 & 3 & 4 \\ 1 & 2 & 5 & 9 & 14 \\ 1 & 3 & 9 & 19 & 34 \\ 1 & 4 & 14 & 34 & 68 \end{vmatrix}$, replacing C_2, C_3, C_4 and C_5 by $C_2 - C_1, C_3 - C_1, C_4 - C_1$ and $C_5 - C_1$ respectively

$= \begin{vmatrix} 1 & 2 & 3 & 4 \\ 2 & 5 & 9 & 14 \\ 3 & 9 & 19 & 34 \\ 4 & 14 & 34 & 68 \end{vmatrix}$, expanding with respect to first row.

$= \begin{vmatrix} 1 & 0 & 0 & 0 \\ 2 & 1 & 3 & 6 \\ 3 & 3 & 10 & 22 \\ 4 & 6 & 22 & 52 \end{vmatrix}$, replacing C_2, C_3 and C_4 by $C_2 - 2C_1, C_3 - 3C_1$ and $C_4 - 4C_1$ respectively.

$= \begin{vmatrix} 1 & 3 & 6 \\ 3 & 10 & 22 \\ 6 & 22 & 52 \end{vmatrix}$, expanding with respect to first row

$= \begin{vmatrix} 1 & 0 & 0 \\ 3 & 1 & 4 \\ 6 & 4 & 16 \end{vmatrix}$, replacing C_2 and C_3 by $C_3 - 3C_1$ and $C_3 - 6C_1$ respectively.

$= \begin{vmatrix} 1 & 4 \\ 4 & 16 \end{vmatrix}$, expanding with respect to first row.

$= 1\ (16) - 4\ 4 = 0.$

Example 18:

Show that $= \begin{vmatrix} 1 & 1 & 1 & 1 \\ \alpha & \beta & \gamma & \delta \\ \beta+\gamma & \gamma+\delta & \delta+\alpha & \alpha+\beta \\ \delta & \alpha & \beta & \gamma \end{vmatrix} = 0$

Solution:

We have

$$= \begin{vmatrix} 1 & 1 & 1 & 1 \\ \alpha & \beta & \gamma & \delta \\ \beta+\gamma & \gamma+\delta & \delta+\alpha & \alpha+\beta \\ \alpha+\beta+\gamma+\delta & \alpha+\beta+\gamma+\delta & \alpha+\beta+\gamma+\delta & \alpha+\beta+\gamma+\delta \end{vmatrix},$$

replacing R_4 by $R_2 + R_3 + R_4$

$$= (\alpha+\beta+\gamma+\delta) \begin{vmatrix} 1 & 1 & 1 & 1 \\ \alpha & \beta & \gamma & \delta \\ \beta+\gamma & \gamma+\delta & \delta+\alpha & \alpha+\beta \\ 1 & 1 & 1 & 1 \end{vmatrix},$$

taking out $(\alpha + \beta + \gamma + \delta)$ common from R_4

= 0, since two row are identical.

Example 19(a):

Evaluate $\begin{vmatrix} a-b-c & 2a & 2a \\ 2b & b-c-a & 2b \\ 2c & 2c & c-a-b \end{vmatrix}$

Solution:

The given determinant

$$= \begin{vmatrix} a+b+c & b+c+a & a+b+c \\ 2b & b-c-a & 2b \\ 2c & 2c & c-a-b \end{vmatrix},$$ replacing R_1 by $R_1 + R_2 + R_3$

$$= (a + b + c)\begin{vmatrix} 1 & 1 & 1 \\ 2b & b - c - a & 2b \\ 2c & 2c & c - a - b \end{vmatrix},$$ taking out (a + b + c) common.

$$= (a + b + c)\begin{vmatrix} 1 & 0 & 0 \\ 2b & - a - b - c & 0 \\ 2c & 0 & - a - b - c \end{vmatrix},$$ replacing C_2 and C_3 by $C_2 = C_1$ and $C_3 - C_1$ respectively.

$$= (a + b + c)\begin{vmatrix} - a - b - c & 0 \\ 0 & - a - b - c \end{vmatrix},$$ expanding with respect to first row.

$$= (a + b + c)\,(- a - b - c)\,(- a - b - c) = (a + b + c)^2.$$

Example 19(b):

Prove that $\begin{vmatrix} a^3 & 3a^2 & 3a & 1 \\ a^2 & a^2 + 2a & 2a + 1 & 1 \\ a & 2a + 1 & a + 2 & 1 \\ 1 & 3 & 3 & 1 \end{vmatrix} = (a - 1)^6$

Solution:

The given determinant

$$= \begin{vmatrix} a^3 - 3a^2 + 3a - 1 & 3a^2 & 3a & 1 \\ 0 & a^2 + 2a & 2a + 1 & 1 \\ 0 & 2a + 1 & a + 2 & 1 \\ 0 & 3 & 3 & 1 \end{vmatrix},$$

replacing C_1 by $C_1 - C_2 + C_3 - C_4$

$$= (a^3 - 3a^2 + 3a - 1)\begin{vmatrix} a^2 + 2a & 2a + 1 & 1 \\ 2a + 1 & a + 2 & 1 \\ 3 & 3 & 1 \end{vmatrix},$$ expanding with respect to C_1

$$= (a-1)^3 \begin{vmatrix} (a^2+2a)-(2a+1)+1 & 2a+1 & 1 \\ (2a+1)-(a+2)+1 & a+2 & 1 \\ 3-3+1 & 3 & 1 \end{vmatrix},$$

replacing C_1 by $C_1 - C_2 + C_3$

$$= (a-1)^3 \begin{vmatrix} a^2 & 2a+1 & 1 \\ a & a+2 & 1 \\ 1 & 3 & 1 \end{vmatrix}, \text{ on simplifying}$$

$$= (a-1)^3 \begin{vmatrix} a^2-1 & 2a-2 & 0 \\ a-1 & a-1 & 0 \\ 1 & 3 & 1 \end{vmatrix},$$ replacing R_1 and R_2 by $R_1 - R_2$ and $R_2 - R_3$ respectively.

$$= (a-1)^3 \begin{vmatrix} a^2-1 & 2(a-1) \\ a-1 & a-1 \end{vmatrix}, \text{ expanding with respect to } C_3$$

$$= (a-1)^3(a-1)(a-1) \begin{vmatrix} a+1 & 2 \\ 1 & 1 \end{vmatrix}, \text{ taking out common factors.}$$

$$= (a-1)^3\,[(a+1)-2] = (a-1)^6.$$

Example 19(c):

Evaluate $\begin{vmatrix} y+z & x & y \\ z+x & z & x \\ x+y & y & z \end{vmatrix}$

Solution:

We have

$$= \begin{vmatrix} 2x+2y+2z & x+y+z & x+y+z \\ z+x & z & x \\ x+y & y & z \end{vmatrix},$$ replacing R_1 by $R_1 + R_2 + R_3$

$$= (x + y + z)\begin{vmatrix} 2 & 1 & 1 \\ z + x & z & x \\ x + y & y & z \end{vmatrix},$$

$$= (x + y + z)\begin{vmatrix} 0 & 1 & 1 \\ 0 & z & x \\ x - y & y & z \end{vmatrix}, \text{ replacing } C_1 \text{ by } C_1 - C_2 - C_3$$

$$= (x + y + z)(x - z)\begin{vmatrix} 1 & 1 \\ z & x \end{vmatrix}, \text{ expanding with respect to first column.}$$

$$= (x + y + z)(x - z) = (x + y + z)(x - z)^2.$$

Example 20:

Evaluate $\begin{vmatrix} b + c & a + b & a \\ c + a & b + c & b \\ a + b & c + a & c \end{vmatrix}$

Solution:

We have

$$= \begin{vmatrix} 2a + 2b + 2c & 2a + 2b + 2c & a + b + c \\ c + a & b + c & b \\ a + b & c + a & c \end{vmatrix}, \text{ replacing } R_1 \text{ by } R_1 + R_2 + R_3$$

$$= (a + b + c)\begin{vmatrix} 2 & 2 & 1 \\ c + a & b + c & b \\ a + b & c + a & c \end{vmatrix}, \text{ taking out } (a + b + c) \text{ common.}$$

$$= (a + b + c)\begin{vmatrix} 0 & 2 & 1 \\ a - b & b + c & b \\ b - c & c + a & c \end{vmatrix}, \text{ replacing } C_1 \text{ by } C_1 - C_2$$

$$= (a + b + c)\begin{vmatrix} 0 & 0 & 1 \\ a - b & c - b & b \\ b - c & a - c & c \end{vmatrix}, \text{ replacing } C_2 \text{ by } C_2 - 2C_2$$

$= (a + b + c)\,[(a - b)(a - c) - (b - c)(c - b)]$

$= (a + b + c)\,[a^2 - ac - ba + bc + b^2 + c^2 - 2bc]$

$= (a + b + c)(a^2 + b^2 + c^2 - ab - bc - ca)$

$= a^2 + b^2 + c^2 - 3abc.$

Example 21:

Evaluate $\begin{vmatrix} a & 1 & 1 & 1 \\ 1 & a & 1 & 1 \\ 1 & 1 & a & 1 \\ 1 & 1 & 1 & a \end{vmatrix}$

Solution:

We have

$= \begin{vmatrix} a+3 & 1 & 1 & 1 \\ a+3 & a & 1 & 1 \\ a+3 & 1 & a & 1 \\ a+3 & 1 & 1 & a \end{vmatrix}$, replacing C_1 by $C_1 + C_2 + C_3 + C_4$

$= (a + 3)\begin{vmatrix} 1 & 1 & 1 & 1 \\ 1 & a & 1 & 1 \\ 1 & 1 & a & 1 \\ 1 & 1 & 1 & a \end{vmatrix}$, taking out $(a + 3)$ common from C_1

$= (a + 3)\begin{vmatrix} 1 & 1 & 1 & 1 \\ 0 & a-1 & 0 & 0 \\ 0 & 0 & a-1 & 0 \\ 0 & 0 & 0 & a-1 \end{vmatrix}$, replacing R_2, R_3, R_4 by $R_2 - R_1, R_3 - R_1$ and $R_4 - R_1$ respectively.

$= (a + 3)\begin{vmatrix} a-1 & 0 & 0 \\ 0 & a-1 & 0 \\ 0 & 0 & a-1 \end{vmatrix}$, expanding with respect to C_1

$= (a + 3)(a - 1)\begin{vmatrix} a-1 & 0 \\ 0 & a-1 \end{vmatrix}$, expanding with respect to R_1

$= (a + 3)(a - 1)\,[(a - 1)(a - 1) = 0.0]$

$= (a + 3)(a - 1)^3.$

Example 22:

Prove that

$$\begin{vmatrix} 1+a_1 & a_2 & a_3 & a_4 \\ a_1 & 1+a_2 & a_3 & a_4 \\ a_1 & a_2 & 1+a_3 & a_4 \\ a_1 & a_2 & a_3 & 1+a_4 \end{vmatrix} = 1 + a_1 + a_2 + a_3 + a_4$$

Solution:

We have

$$= \begin{vmatrix} 1+a_1+a_2+a_3+a_4 & a_2 & a_3 & a_4 \\ 1+a_1+a_2+a_3+a_4 & 1+a_2 & a_3 & a_4 \\ 1+a_1+a_2+a_3+a_4 & a_2 & 1+a_3 & a_4 \\ 1+a_1+a_2+a_3+a_2 & a_2 & a_3 & 1+a_4 \end{vmatrix}$$, replacing C_1 by $C_1 + C_2 + C_3 + C_4$

$$= (1 + a_1 + a_2 + a_3 + a_4) \begin{vmatrix} 1 & a_2 & a_3 & a_4 \\ 1 & 1+a_2 & a_3 & a_4 \\ 1 & a_2 & 1+a_3 & a_4 \\ 1 & a_2 & a_3 & 1+a_4 \end{vmatrix}$$, taking out $(1 + a_1 + a_2 + a_3 + a_4)$ common from C_1

$$= (1 + a_1 + a_2 + a_3 + a_4) \begin{vmatrix} 1 & a_2 & a_3 & a_4 \\ 0 & 1 & 0 & 0 \\ 0 & 0 & 1 & 0 \\ 0 & 0 & 0 & 1 \end{vmatrix}$$, replacing R_2, R_3 and R_4 by $R_2 - R_1$, $R_3 - R_1$ and $R_4 - R_1$ respectively.

$$= (1 + a_1 + a_2 + a_3 + a_4) \begin{vmatrix} 1 & 0 & 0 \\ 0 & 1 & 0 \\ 0 & 0 & 1 \end{vmatrix}$$, expanding with respect to C_1

$$= (1 + a_1 + a_2 + a_3 + a_4) \begin{vmatrix} 1 & 0 \\ 0 & 1 \end{vmatrix}$$, expanding with respect to R_1

$= (1 + a_1 + a_2 + a_3 + a_4)$.

Example 23:

Evaluate $\begin{vmatrix} 1+a & 1 & 1 & 1 \\ 1 & 1+a & 1 & 1 \\ 1 & 1 & 1+a & 1 \\ 1 & 1 & 1 & 1+a \end{vmatrix}$

Solution:

We have

$$= \begin{vmatrix} 4+a & 1 & 1 & 1 \\ 4+a & 1+a & 1 & 1 \\ 4+a & 1 & 1+a & 1 \\ 4+a & 1 & 1 & 1+a \end{vmatrix}, \text{ replacing } C_1 \text{ by } C_1 + C_2 + C_3 + C_4$$

$$= (4+a) \begin{vmatrix} 1 & 1 & 1 & 1 \\ 1 & 1+a & 1 & 1 \\ 1 & 1 & 1+a & 1 \\ 1 & 1 & 1 & 1+a \end{vmatrix}, \text{ taking out } (4+a) \text{ common from first column}$$

$$= (4+a) \begin{vmatrix} 1 & 0 & 0 & 0 \\ 1 & a & 0 & 0 \\ 1 & 0 & a & 0 \\ 1 & 0 & 0 & a \end{vmatrix}, \text{ replacing } C_1, C_2 \text{ and } C_4 \text{ by } C_2 - C_1, C_3 - C_1 \text{ and } C_4 \quad C_1 \text{ respectively.}$$

$$= (4+a) \begin{vmatrix} a & 0 & 0 \\ 0 & a & 0 \\ 0 & 0 & a \end{vmatrix}, \text{ expanding with respect to } R_1.$$

$$= (4+a) \begin{vmatrix} a & 0 \\ 0 & a \end{vmatrix}, \text{ expanding with respect to } R_1.$$

$= (4 + a)\ a\ (a^2) = (4 + a)\ a^2.$

Example 24:

Evaluate $\begin{vmatrix} 1 & x & 1 & y \\ x & 1 & y & 1 \\ 1 & y & 1 & x \\ y & 1 & x & 1 \end{vmatrix}$

Solution:

We have

$$= \begin{vmatrix} x+y+2 & x+y+2 & x+y+2 & x+y+2 \\ x & 1 & y & 1 \\ 1 & y & 1 & x \\ y & 1 & x & 1 \end{vmatrix}, \text{ replacing } R_1 \text{ by } R_1+R_2+R_3+R_4$$

$$= (x+y+2) \begin{vmatrix} 1 & 1 & 1 & 1 \\ x & 1 & y & 1 \\ y & 1 & x & 1 \\ y & 1 & x & 1 \end{vmatrix}, \text{ taking out } (x + y + 2) \text{ common from } R_1$$

$$= (x+y+2) \begin{vmatrix} 1 & 0 & 0 & 0 \\ x & 1-x & y-x & 1-x \\ 1 & y-1 & 0 & x-1 \\ y & 1-y & x-y & 1-y \end{vmatrix}, \text{ replacing } C_2, C_3 \text{ and } C_4 \text{ by } C_2-C_1, C_3-C_1 \text{ and } C_4-C_1 \text{ respectively.}$$

$$= (x+y+2) \begin{vmatrix} 1-x & y-x & 1-x \\ y-1 & 0 & x-1 \\ 1-y & x-y & 1-y \end{vmatrix}, \text{ expanding with respect to } R_1$$

$$= (x+y+2) \begin{vmatrix} 0 & y-x & 1-x \\ y-x & 0 & x-1 \\ 0 & x-y & 1-y \end{vmatrix}, \text{ replacing } C_1 \text{ by } C_1 - C_2$$

$$= -(x+y+2)(y-x) \begin{vmatrix} y-x & 1-x \\ x-y & 1-y \end{vmatrix}, \text{ expanding with respect to } C_1$$

$$= -(x+y+2)(y-x)(y-x) \begin{vmatrix} -1 & 1-x \\ -1 & 1-y \end{vmatrix}, \text{ taking out } (y-x) \text{ common}$$

$= -(x + y + 2)(y - x)^2 [(1 - y) - (-1)(1 - x)]$
$= -(x + y + 2)(x - y)^2 (2 - x - y)$
$= (x + y + 2)(x - y)^2 (x + y - 2).$

Example 25(a):

Evaluate $\begin{vmatrix} 5 & 7 & 10 & 14 \\ 2 & 3 & 7 & 6 \\ 3 & 3 & 6 & 9 \\ 5 & 6 & 11 & 20 \end{vmatrix}$

Solution:

The given determinant

$$= \begin{vmatrix} 0 & 1 & -1 & -6 \\ 2 & 3 & 7 & 6 \\ 1 & 0 & -1 & 3 \\ 5 & 6 & 11 & 20 \end{vmatrix}, \text{ replacing } R_1 \text{ and } R_3 \text{ by } R_1 - R_4 \text{ and } R_3 - R_2 \text{ respectively.}$$

$$= \begin{vmatrix} 0 & 0 & -1 & 0 \\ 2 & 10 & 7 & 24 \\ 1 & -1 & -1 & 5 \\ 5 & 17 & 11 & 56 \end{vmatrix}, \text{ replacing } C_2 \text{ and } C_4 \text{ by } C_2 + C_3 \text{ and } C_4 + 6C_2 \text{ respectively.}$$

Now expand with respect to 1st row and proceed.

Example 25(b):

Evaluate $\begin{vmatrix} x & a & a & a \\ a & x & a & a \\ a & a & x & a \\ a & a & a & x \end{vmatrix}$

Solution:

We have

$$= \begin{vmatrix} x + 3a & a & a & a \\ x + 3a & x & a & a \\ x + 3a & a & x & a \\ x + 3a & a & a & x \end{vmatrix}, \text{ replacing } C_1 \text{ by } C_1 + C_2 + C_3 + C_4$$

$$= (x + 3a)\begin{vmatrix} 1 & a & a & a \\ 1 & x & a & a \\ 1 & a & x & a \\ 1 & a & a & x \end{vmatrix}, \text{ taking out } (x + 3a) \text{ common}$$

$$= (x + 3a)\begin{vmatrix} 1 & a & a & a \\ 0 & x-a & 0 & 0 \\ 0 & 0 & x-a & 0 \\ 0 & 0 & 0 & x-a \end{vmatrix},$$ replacing R_2, R_3 and R_4 by $R_2 - R_1$, $R_3 - R_1$ and $R_4 - R_1$ respectively.

$$= (x + 3a)\begin{vmatrix} x-a & 0 & 0 \\ 0 & x-a & 0 \\ 0 & 0 & x-a \end{vmatrix}, \text{ expanding with respect to } C_1$$

$$= (x + 3a)(x - a)\begin{vmatrix} x-a & 0 \\ 0 & x-a \end{vmatrix}, \text{ expanding with respect to } C_1$$

$$= (x + 3a)\ (x - a)\ (x - a)\ (x - a) = (x + 3a)\ (x - a)^2.$$

Example 25(c):

Evaluate $\begin{vmatrix} x & a & b & c & 1 \\ d & x & f & h & 1 \\ d & e & x & k & 1 \\ d & e & g & x & 1 \\ d & e & g & m & 1 \end{vmatrix}$

Solution:

The given determinant

$$= \begin{vmatrix} x-d & a-x & b-f & c-h & 0 \\ 0 & x-e & f-x & h-k & 0 \\ 0 & 0 & x-g & k-x & 0 \\ 0 & 0 & 0 & x-m & 0 \\ d & e & g & m & 1 \end{vmatrix},$$

replacing R_1, R_2, R_3 and R_4 by $R_1 - R_2$, $R_2 - R_3$, $R_3 - R_4$ and $R_{4 - R}5$ respectively.

$$= (x - m) \begin{vmatrix} x - d & a - x & b - f & 0 \\ 0 & x - e & f - x & 0 \\ 0 & 0 & x - g & 0 \\ d & e & g & 1 \end{vmatrix},$$

expanding with respect to R_4

$$= (x - m) \begin{vmatrix} x - d & a - x & b - f \\ 0 & x - e & f - x \\ 0 & 0 & x - g \end{vmatrix},$$ expanding with respect to C_4

$$= (x - m)(x - d) \begin{vmatrix} x - c & f - x \\ 0 & x - g \end{vmatrix},$$ expanding with respect to C_1

$$= (x - m)(x - d)(x - c)(x - g).$$

Example 26:

Evaluate $\begin{vmatrix} b + c & a - c & a - b \\ b - c & c + a & b - a \\ c - b & c - a & a + b \end{vmatrix}$

Solution:

$$= \begin{vmatrix} 2b & 2a & 0 \\ b - c & c + a & b - a \\ 0 & 2c & 2b \end{vmatrix},$$ replacing R_1 and R_2 by $R_1 + R_2$ and $R_3 + R_2$ respectively.

$$= 4 \begin{vmatrix} b & a & 0 \\ b - c & c + a & b - a \\ 0 & c & b \end{vmatrix},$$ taking out 2 common from R_1 and R_2

$$= 4 \begin{vmatrix} b & a & 0 \\ -c & c & b - a \\ 0 & c & b \end{vmatrix},$$ replacing R_2 by $R_2 - R_1$

$$= 4\left[b\begin{vmatrix} c & b-a \\ c & b \end{vmatrix} + c\begin{vmatrix} a & 0 \\ c & b \end{vmatrix}\right],$$ expanding with respect to C_1

$$= 4\left[b\begin{vmatrix} c & b-a \\ 0 & a \end{vmatrix} + c\begin{vmatrix} a & 0 \\ c & b \end{vmatrix}\right],$$ replacing R_2 by $R_2 - R_1$ in first det er min ant.

$$= 4\,[b\,(ca) + c\,(ab)] = 4\,[2abc] = 8abc.$$

Example 27(a):

Evaluate $$\begin{vmatrix} b+c & a & a \\ b & c+a & b \\ c & c & a+b \end{vmatrix}$$

Solution:

We have

$$= \begin{vmatrix} b+c & 0 & 0 \\ b & c+a-b & b \\ c & c-a-b & a+b \end{vmatrix},$$ replacing C_2 by $C_2 - C_3$

$$= (b+c)\begin{vmatrix} c+a-b & b \\ c-a-b & a+b \end{vmatrix} + a\begin{vmatrix} b & c+a-b \\ c & c-a-b \end{vmatrix},$$

expanding with respect to R_1

$$= (b+c)\begin{vmatrix} c+a-b & b \\ -2a & a \end{vmatrix} + a\begin{vmatrix} b & c+a-b \\ c-b & -2a \end{vmatrix},$$

Replacing R_2 by $R_2 - R_1$ in each determinant.

$$= (b+c)\,[a\,(c+a-b) - b\,(-2a)] + a\,[b - (-2a) - (c-b)\,(c+a-b)]$$

$$= (b+c)\,(ac + a^2 + ab) + a\,(-2ab - c^2 - ca + bc + bc + ab - b^2)$$

$$= abc + a^2b + ab^2 + ac^2 + a^2c + abc - 2a^2b - ac^2 - ca^2 + 2abc + a^2b - ab^2$$

$$= 4\ 4abc.$$

Example 27(b):

Show that $$\begin{vmatrix} 1 & a & bc \\ 1 & b & ca \\ 1 & c & ab \end{vmatrix} = (b-c)\,(c-a)\,(a-b)$$

Solution:

The given determinant

$$= \begin{vmatrix} 1 & a & bc \\ 0 & b-a & ca-bc \\ 0 & c-a & ab-bc \end{vmatrix}, \text{ replacing } R_2 \text{ and } R_3 \text{ by } R_2 - R_1 \text{ and } R_3 - R_1 \text{ respective.}$$

$$= \begin{vmatrix} b-a & -c(b-a) \\ c-a & -b(c-a) \end{vmatrix}, \text{ expanding with respect to } C_1$$

$$= (b-a)(c-a) \begin{vmatrix} 1 & -c \\ 1 & -c \end{vmatrix}, \text{ taking common factors out}$$

$$= (b-a)(c-a)(-b+c)$$

$$= (a-b)(b-c)(c-a).$$

Example 27(c):

Evaluate $\begin{vmatrix} 1 & bc & a(b+c) \\ 1 & ca & b(c+a) \\ 1 & ab & c(a+b) \end{vmatrix}$

Solution:

The given determinant

$$= \begin{vmatrix} 1 & bc & a(b+c) \\ 0 & c(a-b) & c(b-a) \\ 0 & b(a-c) & b(c-a) \end{vmatrix}, \text{ replacing } R_2 \text{ and } R_3 \text{ by } R_2 - R_1 \text{ and } R_3 - R_1 \text{ respectively.}$$

$$= \begin{vmatrix} c(a-b) & c(b-a) \\ b(a-c) & b(c-a) \end{vmatrix}, \text{ expanding with respect to } C_1$$

$$- (a-b)(a-c) \begin{vmatrix} c & -c \\ b & -b \end{vmatrix}, \text{ taking out } (a-b) \text{ and } (a-c) \text{ common from } R_1 \text{ and } R_2 \text{ respectively}$$

$$= -(a-b)(a-c) \begin{vmatrix} c & c \\ b & b \end{vmatrix} = 0, \text{ two columns belong identical.}$$ **Ans.**

Example 28:

Prove that

$$\begin{vmatrix} 1 & a^2 + bc & a^3 \\ 1 & b^2 + ca & b^3 \\ 1 & c^2 + ab & c^3 \end{vmatrix} = -(b-c)(c-a)(a-b)(a^2+b^2+c^2)$$

Solution:

The given determinant

$$= \begin{vmatrix} 1 & a^2 & a^3 \\ 1 & b^2 & b^3 \\ 1 & c^2 & c^3 \end{vmatrix} + \begin{vmatrix} 1 & bc & a^3 \\ 1 & ca & b^3 \\ 1 & ab & c^3 \end{vmatrix},$$ breaking into two determinante

$$= \begin{vmatrix} 1 & 1 & 1 \\ a^2 & b^2 & c^2 \\ a^3 & b^3 & c^3 \end{vmatrix} + \begin{vmatrix} 1 & 1 & 1 \\ bc & ca & ab \\ a^3 & b^3 & c^3 \end{vmatrix},$$ interchanging rows and columns ...(i)

The first determinant

$= (a - b)(b - c)(c - a)(ab + bc + ca)$ ...(ii)

The second determinant

$$= \begin{vmatrix} 1 & 0 & 0 \\ ba & ca - bc & ab - bc \\ a^2 & b^3 - a^3 & c^3 - a^3 \end{vmatrix},$$ replacing C_2 and C_3 by $C_2 - C_1$ and $C_3 - C_1$

$$= \begin{vmatrix} c(a-b) & b(a-c) \\ (b-a)(b^2+ab+a^2) & (c-a)(c^2+ac+a^2) \end{vmatrix},$$

expanding with respect to R_1

$$= (a-b)(a-c)\begin{vmatrix} c & b \\ -(b^2+ab+a^2) & -(c^2+ac+a^2) \end{vmatrix},$$

taking out common factors from C_1 and C_2

$$= (a-b)(a-c)\begin{vmatrix} c-b & b \\ -(b^2-c^2+ab-ac) & -(c^2+ac+a^2) \end{vmatrix},$$

replacing C_1 by $C_1 - C_2$

$$= (a-b)(a-c)\begin{vmatrix} (c-b) & b \\ (c-b)(c+b+a) & -(c^2+ac+a^2) \end{vmatrix}$$

$$= (a-b)(a-c)(c-b)\begin{vmatrix} 1 & b \\ a+b+c & -(c^2+ca+a^2) \end{vmatrix},$$

taking $(c-b)$ common from C_1.

$$= (a-b)(b-c)(c-a)[-(c^2+ca+a^2) - b(a+b+c)]$$

$$= -(a-b)(b-c)(c-a)(a^2+b^2+c^2+ab+bc+ca) \qquad \ldots(iii)$$

Substituting values from (ii) and (iii) in (i), we find the given determinant

$$= (a-b)(b-c)(c-a)(ab+bc+ca)$$
$$- (a-b)(b-c)(c-a)(a^2+b^2+c^2+ab+bc+ca)]$$
$$= (a-b)(b-c)(c-a)(a^2+b^2+c^2).$$

Example 29:

Evaluate $\begin{vmatrix} a & x & y & a \\ x & 0 & 0 & y \\ y & 0 & 0 & x \\ a & y & z & a \end{vmatrix}$

Solution:

The given determinant

$$= \begin{vmatrix} 0 & x-y & y & a \\ x-y & 0 & 0 & y \\ y-x & 0 & 0 & x \\ a & y+x & x & a \end{vmatrix},$$ replacing C_1 and C_2 by $C_1 - C_4$ and $C_2 + C_3$ respectively.

$$= (x-y)(x+y)\begin{vmatrix} 0 & 1 & y & a \\ 1 & 0 & 0 & y \\ -1 & 0 & 0 & x \\ 0 & 1 & x & a \end{vmatrix},$$ taking out $x-y$ and $x+y$ common from C_1 and C_2 respectively.

$$= (x^2-y^2)\begin{vmatrix} 0 & 1 & y & a \\ 1 & 0 & 0 & y \\ 0 & 0 & 0 & x+y \\ 0 & 0 & x-y & a \end{vmatrix},$$ replacing R_3 and R_4 by $R_2 + R_3$ and $R_4 - R_1$ respectively.

$$= -(x^2 - y^2)\begin{vmatrix} 1 & y & a \\ 0 & 0 & x+y \\ 0 & x-y & 0 \end{vmatrix}, \text{ expanding with respect to } C_1$$

$$= -(x^2 - y^2)\begin{vmatrix} 0 & x+y \\ x-y & 0 \end{vmatrix}$$

$$= -(x^2 - y^2)\,[-(x + y)(x - y)] = (x^2 - y^2)^2.$$

Example 30:

Evaluate $\begin{vmatrix} 1 & 1 & 1 \\ a^2 & b^2 & c^2 \\ a^3 & b^3 & c^3 \end{vmatrix}$

Solution:

We have

$$= \begin{vmatrix} 1 & 0 & 0 \\ a^2 & b^2 - a^2 & c^2 - a^2 \\ a^3 & b^3 - a^3 & c^3 - a^3 \end{vmatrix}, \text{ replacing } C_2 \text{ and } C_3 \text{ by } C_2 - C_1 \text{ and } C_3 - C_1 \text{ respectively.}$$

$$= \begin{vmatrix} b^2 - a^2 & c^2 - a^2 \\ b^3 - a^3 & c^3 - a^3 \end{vmatrix}, \text{ expanding with respect to } R_1$$

$$= (b - a)(c - a)\begin{vmatrix} b + a & c + a \\ b^2 + ab + a^2 & c^2 + ac + a^2 \end{vmatrix},$$

taking out common factors of C_1, C_2

$$= (b - a)(c - a)\begin{vmatrix} b - c & c + a \\ b^2 + ab - c^2 - ca & c^2 + ca + a^2 \end{vmatrix},$$

replacing C_1 by $C_1 - C_2$

$$= (b - a)(c - a)\begin{vmatrix} b-c & c + a \\ (b^2 - c^2) + a(b-c) & c^2 + ca + a^2 \end{vmatrix}$$

$$= (b - a)(c - a)\begin{vmatrix} 1 & c + a \\ b + c + a & c^2 + ca + a^2 \end{vmatrix}$$

$$= -(a - b)(b - c)(c - a)\,[1\,(c^2 + ca + a^2) - (c + a)(b + c + a)$$

$= -(a-b)(b-c)(c-a)$

$[c^2 + ca + a^2 - cb - c^2 - ac - ab - ac - a^2]$

$= -(a-b)(b-c)(c-a)[-ab - bc - ca]$

$= (a-b)(b-c)(c-a)(ab + bc + ca).$

Example 31:

Show that $\begin{vmatrix} a & b & c \\ a^2 & b^2 & c^2 \\ a^3 & b^3 & c^3 \end{vmatrix} = abc\,(a-b)(b-c)(c-a)$

Solution:

The given determinant

$= abc \begin{vmatrix} 1 & 1 & 1 \\ a & b & c \\ a^2 & b^2 & c^2 \end{vmatrix}$, taking out a, b and c common from C_1, C_2 and C_3 respectively.

Example 32(a):

Evaluate $\begin{vmatrix} 1 & 1 & 1 \\ a & b & c \\ a^3 & b^3 & c^3 \end{vmatrix}$

Solution:

The given determinant

$= \begin{vmatrix} 1 & 0 & 0 \\ a & b-a & c-a \\ a^3 & b^3-a^3 & c^3-a^3 \end{vmatrix}$, replacing C_2, C_3 by $C_3 - C_1$ and $C_3 - C_1$ respectively.

$= \begin{vmatrix} b-a & c-a \\ b^3-a^3 & c^3-a^3 \end{vmatrix}$, expanding with respect to R_1

$= (b-a)(c-a) \begin{vmatrix} 1 & 1 \\ b^2+ab+a^2 & c^2+ac+a^2 \end{vmatrix}$,

taking common factors out from C_1 and C_2

$= (b - a)(c - a)[(c^2 + ac + a^2) - (b^2 + ab + a^2)]$

$= (b - a)(c - a)[(c^2 - b^2) + a(c - b)]$

$= (b - a)(c - a)(c - b)[(c + b) + a]$

$= (a - b)(b - c)(c - a)(a + b + c).$

Example 32(b):

If a, b, c are all different and

$\begin{vmatrix} a & a^2 & 1+a^3 \\ b & b^2 & 1+b^3 \\ c & c^2 & 1+c^3 \end{vmatrix} = 0$, *prove that* $1 + abc = 0$.

Solution:

The given determinant

$= \begin{vmatrix} a & a^2 & 1 \\ b & b^2 & 1 \\ c & c^2 & 1 \end{vmatrix} + \begin{vmatrix} a & a^2 & a^3 \\ b & b^2 & b^3 \\ c & c^2 & c^3 \end{vmatrix}$, breaking into two determinants.

$= \begin{vmatrix} a & b & c \\ a^2 & b^2 & c^2 \\ 1 & 1 & 1 \end{vmatrix} + \begin{vmatrix} a & b & c \\ a^2 & b^2 & c^2 \\ a^3 & b^3 & c^3 \end{vmatrix}$, interchanging rows and columns in each determinant.

The value of given determinant $= (1 + abc)[(a - b)(b - c)(c - a)]$

Example 32(c):

Show that

$\begin{vmatrix} x & y & z \\ x^2 & y^2 & z^2 \\ yz & zx & xy \end{vmatrix} = \begin{vmatrix} 1 & 1 & 1 \\ x^2 & y^2 & z^2 \\ x^3 & y^3 & z^3 \end{vmatrix}$

$= (y - x)(z - x)(x - y)(yz + zx + xy).$

Solution:

$\begin{vmatrix} x & y & z \\ x^2 & y^2 & z^2 \\ yz & zx & xy \end{vmatrix} = \frac{1}{xyz}\begin{vmatrix} x^2 & y^2 & z^2 \\ x^3 & y^3 & z^3 \\ xyz & xyz & xyz \end{vmatrix}$, multiplying C_1, C_2 C_3 by x, y, z respectively

$$= \frac{1}{xyz}.xyz\begin{vmatrix} x^2 & y^2 & z^2 \\ x^3 & y^3 & z^3 \\ 1 & 1 & 1 \end{vmatrix}$$, taking out xyz common from R_2

$$= \begin{vmatrix} 1 & 1 & 1 \\ x^2 & y^2 & z^2 \\ x^3 & y^3 & z^3 \end{vmatrix}$$, interchanging R_2 and R_3 and then R_1 and R_2

For the 2nd part do your self.

Example 32(d):

Prove that $$\begin{vmatrix} x+a & b & c & d \\ a & x+b & c & d \\ a & b & x+c & d \\ a & b & c & x+d \end{vmatrix}$$

$= x^2 (x + a + b + c + d)$.

Solution:

We have

$$= \begin{vmatrix} x+a+b+c+d & b & c & d \\ x+a+b+c+d & x+b & c & d \\ x+a+b+c+d & b & x+c & d \\ x+a+b+c+d & b & c & x+d \end{vmatrix}$$, replacing C_1 by $C_1 + C_2 + C_3 + C_4$

$$= (x+a+b+c+d)\begin{vmatrix} 1 & b & c & d \\ 1 & x+b & c & d \\ 1 & b & x+c & d \\ 1 & b & c & x+d \end{vmatrix}$$, taking out $(x+a+b+c+d)$ common from C_1

$$= (x+a+b+c+d)\begin{vmatrix} 1 & b & c & d \\ 0 & x & 0 & 0 \\ 0 & 0 & x & 0 \\ 0 & 0 & 0 & x \end{vmatrix}$$, replacing R_2, R_3, R_4 by $R_2 - R_1$, $R_3 - R_1$ and $R_4 - R_1$ respectively.

$$= (x + a + b + c + d)\begin{vmatrix} x & 0 & 0 \\ 0 & x & 0 \\ 0 & 0 & x \end{vmatrix}$$, expanding with respect to C_1

$$= (x + a + b + c + d) \begin{vmatrix} x & 0 \\ 0 & x \end{vmatrix},$$

expanding with respect to C_1

$$= x\,(x + a + b + c + d)\, x^2$$
$$= x^3\,(x + a + b + c + d).$$

Example 32(e):

Expand the determinant $\begin{vmatrix} a & h & g \\ h & b & f \\ g & f & c \end{vmatrix}$ *by the elements of 1st row.*

Solution:

Elements of firs row are a, h, g.

Let A, H, and G denote the cofactors of a, h, g.

Then $A = \begin{vmatrix} b & f \\ f & c \end{vmatrix}$; $H = -\begin{vmatrix} h & f \\ g & c \end{vmatrix}$ and $G = \begin{vmatrix} h & b \\ g & f \end{vmatrix}$

Hence $\begin{vmatrix} a & h & g \\ h & b & f \\ g & f & c \end{vmatrix} = aA + hH + gG$

$$= a\begin{vmatrix} b & f \\ f & c \end{vmatrix} - h\begin{vmatrix} h & f \\ g & c \end{vmatrix} + g\begin{vmatrix} h & b \\ g & f \end{vmatrix}$$
$$= a\,(bc - f^2) - h\,(ch - fg) + g\,(hf - bg)$$
$$= abc + 2fgh - af^2 - bg^2 - ch^2.$$

Example 33:

Evaluate $\begin{vmatrix} 1+a & 1 & 1 & 1 \\ 1 & 1+b & 1 & 1 \\ 1 & 1 & 1+c & 1 \\ 1 & 1 & 1 & 1+d \end{vmatrix}$

Solution:

We have

$$= abcd \begin{vmatrix} \frac{1}{a}+1 & \frac{1}{a} & \frac{1}{a} & \frac{1}{a} \\ \frac{1}{b} & \frac{1}{b}+1 & \frac{1}{b} & \frac{1}{b} \\ \frac{1}{c} & \frac{1}{c} & \frac{1}{c}+1 & \frac{1}{c} \\ \frac{1}{d} & \frac{1}{d} & \frac{1}{d} & \frac{1}{d}+1 \end{vmatrix},$$ taking a, b, c, d common from R_1, R_2, R_3 and R_4 respectively.

$$= abcd\left(\frac{1}{a}+\frac{1}{b}+\frac{1}{c}+\frac{1}{d}+1\right) \begin{vmatrix} 1 & 1 & 1 & 1 \\ \frac{1}{b} & \frac{1}{b}+1 & \frac{1}{b} & \frac{1}{b} \\ \frac{1}{c} & \frac{1}{c} & \frac{1}{c}+1 & \frac{1}{c} \\ \frac{1}{d} & \frac{1}{d} & \frac{1}{d} & \frac{1}{d}+1 \end{vmatrix},$$

replacing R_1 by $R_1 + R_2 + R_3 + R_4$ and taking $\left(\frac{1}{a}+\frac{1}{b}+\frac{1}{c}+\frac{1}{d}+1\right)$ common from R_1

$$= abcd\left(\frac{1}{a}+\frac{1}{b}+\frac{1}{c}+\frac{1}{d}+1\right) \begin{vmatrix} 1 & 0 & 0 & 0 \\ \frac{1}{b} & 1 & 0 & 0 \\ \frac{1}{c} & 0 & 1 & 0 \\ \frac{1}{d} & 0 & 0 & 1 \end{vmatrix},$$

replacing C_2, C_3 and C_4 by $C_2 - C_1$, $C_3 - C_1$ and $C_4 - C_1$ respectively.

$$= abcd\left(\frac{1}{a}+\frac{1}{b}+\frac{1}{c}+\frac{1}{d}+1\right) \begin{vmatrix} 1 & 0 & 0 \\ 0 & 1 & 0 \\ 0 & 0 & 0 \end{vmatrix},$$ expanding with respect to R_1

$$= abcd\left(\frac{1}{a}+\frac{1}{b}+\frac{1}{c}+\frac{1}{d}+1\right) \begin{vmatrix} 1 & 0 \\ 0 & 1 \end{vmatrix},$$ expanding with respect to R_1

$$= abcd\left(\frac{1}{a}+\frac{1}{b}+\frac{1}{c}+\frac{1}{d}+1\right)$$

Example 34:

For a fixed positive integer n, if

$$D = \begin{vmatrix} n! & (n+1)! & (n+2)! \\ (n+1)! & (n+2)! & (n+3)! \\ (n+2)! & (n+3)! & (n+4)! \end{vmatrix}$$

then show that $\left(\dfrac{D}{(n!)^3} - 3\right)$ *is divisible by n.*

Solution:

Taking n! ((n + 1) !) {(n + 2) !} common from C_1 (C_2) (C_3) we get

$$D = n!\,(n+1)!\,(n+2)! \begin{vmatrix} 1 & 1 & 1 \\ n+1 & n+2 & n+3 \\ (n+1)(n+2) & (n+2)(n+3) & (n+3)(n+4) \end{vmatrix}$$

Applying $C_3 \to C_3 - C_2$ and $C_2 \to C_2 - C_1$ we get

$$D = n!\,(n+1)!\,(n+2)! \begin{vmatrix} 1 & 0 & 0 \\ n+1 & 1 & 1 \\ (n+1)(n+2) & 2n+4 & 2n+6 \end{vmatrix}$$

$= n!\,(n + 1)!\,(n + 2)!\,(2n + 6 - 2n - 4)$

$D = 2(n!)\,(n + 1)!\,(n + 2)!$

$$\Rightarrow \quad \frac{D}{(n!)^3} - 4 = 2(n + 1)\,(n + 1)\,(n + 2) - 4$$

$$= 2n(n^2 + 4n + 5)$$

$$\Rightarrow \quad \left(\frac{D}{(n!)^3} - 3\right) \text{ is divisible by n.}$$

Example 35:

Express $\begin{vmatrix} b^2 + c^2 & ab & ca \\ ab & c^2 + a^2 & bc \\ ca & bc & a^2 + b^2 \end{vmatrix}$

as the square of a determinant.

Hence evaluate.

Solution:

The element in first row and first column is $b^2 + c^2$ which can be written as 0.0 + c.c + b.b.

So by trial and error method, we get the given determinant

$$= \begin{vmatrix} 0 & c & b \\ c & 0 & a \\ b & a & 0 \end{vmatrix} \times \begin{vmatrix} o & c & b \\ c & 0 & a \\ b & a & 0 \end{vmatrix} = \begin{vmatrix} 0 & c & b \\ c & 0 & a \\ b & a & 0 \end{vmatrix}^2$$

Now $\begin{vmatrix} o & c & b \\ c & 0 & a \\ b & a & 0 \end{vmatrix}$

$$= -c \begin{vmatrix} c & a \\ b & 0 \end{vmatrix} + b \begin{vmatrix} c & 0 \\ b & a \end{vmatrix}$$, expanding with respect to R_1.

$= -$ c[c.0 $-$ a.b] + b [c.a $-$ b.0] = 2abc.

$\therefore$ The given determinant

$$= \begin{vmatrix} 0 & c & b \\ c & 0 & a \\ b & a & 0 \end{vmatrix}^2 = (2abc)^2 = 4a^2b^2c^2.$$

Example 36(a):

Show that

$$\Delta = \begin{vmatrix} \sin 3\alpha & \sin 2\alpha & \sin \alpha \\ \sin 3\beta & \sin 2\beta & \sin \beta \\ \sin 3\gamma & \sin 2\gamma & \sin \gamma \end{vmatrix}$$

$$= 2^6 \sin\alpha \sin\beta \sin\gamma \sin\frac{\alpha+\beta}{2} \sin\frac{\beta+\gamma}{2}$$

Solution:

We know that $\sin 3\theta = 3 \sin \theta - 4 \sin^3 \theta$ and $\sin 2\theta = 2 \sin \theta \cos \theta$. Applying $C_1 \to C_1 - 3C_3$ we get

$$\Delta = \begin{vmatrix} -4\sin^3\alpha & 2\sin\alpha\cos\alpha & \sin\alpha \\ -4\sin^3\beta & 2\sin\beta\cos\beta & \sin\beta \\ -4\sin^3\gamma & 2\sin\gamma\cos\gamma & \sin\gamma \end{vmatrix}$$

$$= 8\sin\alpha\,\sin\beta\sin\gamma \begin{vmatrix} -\sin^2\alpha & \cos\alpha & 1 \\ -\sin^2\beta & \cos\beta & 1 \\ -\sin^2\gamma & \cos\gamma & 1 \end{vmatrix}$$

Applying $R_2 \to R_2 - R_1$ and $R_3 \to R_3 - R_1$ we get

$$\Delta = 8\sin\alpha\,\sin\beta\sin\gamma \begin{vmatrix} -\sin^2\alpha & \cos\alpha & 1 \\ \sin^2\alpha - \sin^2\beta & \cos\beta - \cos\alpha & 0 \\ \sin^2\alpha - \sin^2\gamma & \cos\gamma - \cos\alpha & 1 \end{vmatrix}$$

$$= 8 \sin\alpha \sin\beta \sin\gamma \begin{vmatrix} \cos^2\beta - \cos^2\alpha & \cos\beta - \cos\alpha \\ \cos^2\gamma - \cos^2\alpha & \cos\gamma - \cos\alpha \end{vmatrix}$$

$$= 8 \sin\alpha \sin\beta \sin\gamma(\cos\beta - \cos\alpha)(\cos\gamma - \cos\alpha) \begin{vmatrix} \cos\beta + \cos\alpha & 1 \\ \cos\gamma + \cos\alpha & 1 \end{vmatrix}$$

$$= 8 \sin\alpha \sin\beta \sin\gamma (\cos\beta - \cos\alpha)$$
$$(\cos\gamma - \cos\alpha)\ (\cos\beta - \cos\gamma)$$

$$= 2^6 \sin\alpha \sin\beta \sin\gamma \sin\frac{\alpha-\beta}{2} \sin\frac{\alpha+\beta}{2} \sin$$
$$\frac{\alpha-\gamma}{2} \times \sin\frac{\alpha+\gamma}{2} \sin\frac{\gamma-\beta}{2} \sin\frac{\beta+\gamma}{2}$$

$$= 2^6 \sin\alpha \sin\beta \sin\gamma \sin\frac{\beta+\gamma}{2} \sin\frac{\gamma+\alpha}{2} \sin$$
$$\frac{\gamma+\beta}{2} \times \sin\frac{\beta-\alpha}{2} \sin\frac{\gamma-\alpha}{2} \sin\frac{\gamma-\beta}{2}$$

Example 36(b):

Express $\begin{vmatrix} (a-x)^2 & (b-x)^2 & (c-x)^2 \\ (a-y)^2 & (b-y)^2 & (c-y)^2 \\ (a-z)^2 & (b-z)^2 & (c-z)^2 \end{vmatrix}$

as the product of two determinants.

Solution:

We have

$$= \begin{vmatrix} a^2-2ax+x^2 & b^2-2bx+x^2 & c^2-2cx+x^2 \\ a^2-2ay+y^2 & b^2-2by+y^2 & c^2-2cy+y^2 \\ a^2-2az+z^2 & b^2-2bz+z^2 & c^2-2cz+z^2 \end{vmatrix}$$

The element in the first row and first column is $a^2 - 2ax + x^2$, which can be written as $1 (a^2) + (-2x) (a) + x^2 (1)$.

This suggests that the first row of the required determinants are $1, -2x, x^2$ and $a^2, a, 1$.

Example 37:

Expand $\begin{vmatrix} 3 & 2 & 1 & 4 \\ 5 & 29 & 2 & 14 \\ 16 & 19 & 3 & 17 \\ 33 & 39 & 8 & 38 \end{vmatrix}$

by Laplace's expansion by the minors of the first two columns.

Solution:

All the possible minors of the first two columns and their complementary minors and given by :

$$|B_1| = \begin{vmatrix} 3 & 2 \\ 15 & 29 \end{vmatrix}, \quad |B'_1| = \begin{vmatrix} 3 & 17 \\ 8 & 38 \end{vmatrix}$$

$$|B_2| = \begin{vmatrix} 3 & 2 \\ 16 & 19 \end{vmatrix}, \quad |B'_2| = \begin{vmatrix} 2 & 14 \\ 8 & 38 \end{vmatrix}$$

$$|B_3| = \begin{vmatrix} 3 & 2 \\ 33 & 39 \end{vmatrix}, \ |B'_3| = \begin{vmatrix} 2 & 14 \\ 3 & 17 \end{vmatrix}$$

$$|B_4| = \begin{vmatrix} 15 & 29 \\ 16 & 19 \end{vmatrix}, \ |B'_4| = \begin{vmatrix} 1 & 4 \\ 8 & 38 \end{vmatrix}$$

$$|B_5| = \begin{vmatrix} 15 & 29 \\ 33 & 39 \end{vmatrix}, \ |B'_5| = \begin{vmatrix} 1 & 4 \\ 3 & 17 \end{vmatrix}$$

$$|B_6| = \begin{vmatrix} 16 & 19 \\ 33 & 39 \end{vmatrix}, \ |B'_6| = \begin{vmatrix} 1 & 4 \\ 2 & 14 \end{vmatrix}$$

$\therefore$ The given determinant

$$= \begin{vmatrix} 13 & 2 \\ 15 & 29 \end{vmatrix} . \begin{vmatrix} 3 & 17 \\ 8 & 38 \end{vmatrix} - \begin{vmatrix} 3 & 2 \\ 16 & 19 \end{vmatrix} . \begin{vmatrix} 2 & 14 \\ 8 & 38 \end{vmatrix}$$

$$+ \begin{vmatrix} 3 & 2 \\ 33 & 39 \end{vmatrix} . \begin{vmatrix} 2 & 14 \\ 2 & 17 \end{vmatrix} + \begin{vmatrix} 15 & 29 \\ 16 & 19 \end{vmatrix} . \begin{vmatrix} 1 & 4 \\ 8 & 38 \end{vmatrix}$$

$$- \begin{vmatrix} 15 & 29 \\ 33 & 39 \end{vmatrix} . \begin{vmatrix} 1 & 14 \\ 3 & 17 \end{vmatrix} + \begin{vmatrix} 16 & 19 \\ 33 & 39 \end{vmatrix} . \begin{vmatrix} 1 & 4 \\ 2 & 14 \end{vmatrix}$$

Example 38:

Expand $\begin{vmatrix} a & x & y & a \\ x & 0 & 0 & y \\ y & 0 & 0 & x \\ a & y & x & a \end{vmatrix}$

by Laplace's expansion by the minors of the first two columns. Hence evaluate it.

Solution:

All the possible minors of the first two columns and their complementary minors are given by

$$|B_1| = \begin{vmatrix} a & x \\ x & 0 \end{vmatrix}; \; |B'_1| = \begin{vmatrix} 0 & x \\ x & a \end{vmatrix};$$

$$|B_2| = \begin{vmatrix} a & x \\ y & 0 \end{vmatrix}; \; |B'_2| = \begin{vmatrix} 0 & y \\ x & a \end{vmatrix};$$

$$|B_3| = \begin{vmatrix} a & x \\ y & a \end{vmatrix}; \; |B'_2| = \begin{vmatrix} 0 & y \\ 0 & x \end{vmatrix};$$

$$|B_4| = \begin{vmatrix} x & 0 \\ y & 0 \end{vmatrix}; \; |B'_4| = \begin{vmatrix} y & a \\ x & a \end{vmatrix};$$

$$|B_5| = \begin{vmatrix} x & 0 \\ a & y \end{vmatrix}; \; |B'_5| = \begin{vmatrix} y & 0 \\ 0 & x \end{vmatrix};$$

and $$|B_6| = \begin{vmatrix} y & 0 \\ a & y \end{vmatrix}; \; |B'_6| = \begin{vmatrix} y & a \\ 0 & y \end{vmatrix};$$

Therefore the given determinant

$$= \begin{vmatrix} a & x \\ x & 0 \end{vmatrix} . \begin{vmatrix} 0 & x \\ x & a \end{vmatrix} - \begin{vmatrix} a & x \\ y & 0 \end{vmatrix} . \begin{vmatrix} 0 & y \\ x & a \end{vmatrix} + \begin{vmatrix} a & x \\ a & y \end{vmatrix} . \begin{vmatrix} 0 & y \\ 0 & x \end{vmatrix}$$

$$+ \begin{vmatrix} x & 0 \\ y & 0 \end{vmatrix} . \begin{vmatrix} y & a \\ x & a \end{vmatrix} - \begin{vmatrix} x & 0 \\ a & y \end{vmatrix} . \begin{vmatrix} y & a \\ 0 & x \end{vmatrix} + \begin{vmatrix} y & 0 \\ a & y \end{vmatrix} . \begin{vmatrix} y & a \\ 0 & y \end{vmatrix} \quad ...(i)$$

The submatrix B_2 requires one interchange of rows viz. of second and third rows to bring it into the first two rows therefore –sign is put before the product $|B_2| \, |B'_2|$. Again the submatrix B_2 requires two interchanges of rows to bring fourth row to the position of second row *i.e.*, to bring B_2 into first two rows, therefore +sign is put before the product $|B_3| \, |B'_3|$.

Similarly, B_4 requires two interchanges, B_5 requires three inter changes and B_3 requires four interchanges, hence +, – and + signs are put before $|B_4| \, |B'_4|$, $|B'_5| \, |B'_5|$ and $|B_6| \, |B'_6|$ respectively.

Hence from (i) we have (expanding the determinants) the given determinant

$= (-x^2)(-x^2) - (-xy) + (ay - ax)(0) + 0; (ay - ax) - (xy),(xy) + (y^2)(y^2)$
$= x^4 - 2x^2y^2 + y^4 = (x^2 - y^2)^2.$

Example 39:

Expand $\begin{vmatrix} a & b & c & d \\ e & f & g & h \\ 0 & 0 & i & k \\ 0 & 0 & l & m \end{vmatrix}$ *by Laplace's expansion by the minors of the first two columns.*

Solution:

All the possible minors of the first two columns and their complementary minors are given by:

$$| B_1 | = \begin{vmatrix} a & b \\ e & f \end{vmatrix}; | B'_1 | = \begin{vmatrix} j & k \\ l & m \end{vmatrix};$$

$| B_2 | = \begin{vmatrix} a & b \\ 0 & 0 \end{vmatrix} = 0$, hence $| B'_2 |$ need not be calculated

Similarly $| B_3 | = \begin{vmatrix} a & b \\ 0 & 0 \end{vmatrix} = 0; | B_4 | = \begin{vmatrix} e & f \\ 0 & 0 \end{vmatrix} = 0;$

$| B_5 | = \begin{vmatrix} e & f \\ 0 & 0 \end{vmatrix} = 0$ and $| B_6 | = \begin{vmatrix} 0 & 0 \\ 0 & 0 \end{vmatrix} = 0$ and therefore their complementary minors need not be calculated.

Then the given determinant by Laplace's Expansion method

$$= \begin{vmatrix} a & b \\ e & f \end{vmatrix} . \begin{vmatrix} j & k \\ l & m \end{vmatrix}$$

Example 40:

If $u = ax + by + cz$, $v = ay + bz + cx$, $w = az + bx + cy$, *prove that*

$$\begin{vmatrix} a & b & c \\ b & c & a \\ c & a & b \end{vmatrix} \times \begin{vmatrix} x & y & z \\ y & z & x \\ z & x & y \end{vmatrix} = u^2 + v^2 + w^2 - 3uvw.$$

Solution:

By row-by-row multiplication, the product of the given determinants

$$= \begin{vmatrix} ax + by + cz & ay + bz + cx & az + bx + cy \\ bx + cy + az & by + cz + ax & bz + cx + ay \\ cx + ay + bz & cy + az + bx & cz + ax + by \end{vmatrix}$$

$$= \begin{vmatrix} u & v & w \\ w & u & v \\ v & w & u \end{vmatrix}, \text{ since } ax + by + cz = u \text{ etc. (given)}$$

$$= u\begin{vmatrix} u & v \\ w & u \end{vmatrix} - v\begin{vmatrix} w & v \\ v & u \end{vmatrix} + w\begin{vmatrix} w & u \\ v & w \end{vmatrix}$$

$$= u(u^2 - vw) - v(uw - v^2) + w(w^2 - uv)$$
$$= u^2 + v^2 + w^2 - 3uvw.$$

Example 41:

Expand $\begin{vmatrix} a & 1 & 0 & 0 & 0 \\ b & a & 1 & 0 & 0 \\ 0 & b & a & 1 & 0 \\ 0 & 0 & b & a & 1 \\ 0 & 0 & 0 & b & a \end{vmatrix}$ *by Laplace's expansion by the minors of the first two columns. Hence evaluate it.*

Solution:

All the possible minors of the first two columns and their complementary minors are given by:

$$|B_1| = \begin{vmatrix} a & 1 \\ b & a \end{vmatrix}, \ |B'_1| = \begin{vmatrix} a & 1 & 0 \\ b & a & 1 \\ 0 & b & a \end{vmatrix}$$

$$|B_2| = \begin{vmatrix} a & 1 \\ 0 & b \end{vmatrix}, \ |B'_2| = \begin{vmatrix} 1 & 0 & 0 \\ b & a & 1 \\ 0 & b & a \end{vmatrix}$$

$$|B_3| = \begin{vmatrix} b & a \\ 0 & b \end{vmatrix}, \; |B'_3| = \begin{vmatrix} 0 & 0 & 0 \\ b & a & 1 \\ 0 & b & a \end{vmatrix} = 0$$

All other minors of the first two columns are equal to zero as they have at least one row of zero.

Hence the given determinant

$$= \begin{vmatrix} a & 1 \\ 0 & b \end{vmatrix} \cdot \begin{vmatrix} a & 1 & 0 \\ b & a & 1 \\ 0 & b & a \end{vmatrix} - \begin{vmatrix} a & 1 \\ 0 & b \end{vmatrix} \cdot \begin{vmatrix} 1 & 0 & 0 \\ b & a & 1 \\ 0 & b & a \end{vmatrix} \qquad \ldots(i)$$

Now $\begin{vmatrix} a & 1 & 0 \\ b & a & 1 \\ 0 & b & a \end{vmatrix} = \begin{vmatrix} a & 1 \\ b & a \end{vmatrix} .a - \begin{vmatrix} b & 1 \\ 0 & a \end{vmatrix} .1 + \begin{vmatrix} b & a \\ 0 & b \end{vmatrix} .0,$

expanding by the minors of first two columns.

$= (a^2 - b)\, a - (ab) = a^3 - 2ab$

$\therefore$ From (i), the given determinant

$$= \begin{vmatrix} a & 1 \\ b & a \end{vmatrix} .(a^3 - 2ab) - \begin{vmatrix} a & 1 \\ 0 & b \end{vmatrix} . \begin{vmatrix} a & 1 \\ b & a \end{vmatrix},$$

expanding the last determinant with respect to R_1.

$= (a^2 - b)\,(a^2 - 2ab) - (ab)\,(a^2 - b) = (a^2 - b)\,[a^2 + 3ab]$

$= a\,(a^2 - b)\,(a^2 - 3b)$

Example 42:

Show that $\Delta = \begin{vmatrix} (b+c)^2 & a^2 & bc \\ (c+a)^2 & b^2 & ca \\ (a+b)^2 & c^2 & ab \end{vmatrix}$

$= (a^2 + b^2 + c^2)\,(a + b + c)\,(a - b)\,(b - c)(c - a)$

Solution:

Applying $C_1 \rightarrow C_1 + C_2 - 2C_3$, we get

$$\Delta = \begin{vmatrix} a^2 + b^2 + c^2 & a^2 & bc \\ a^2 + b^2 + c^2 & b^2 & ca \\ a^2 + b^2 + c^2 & c^2 & ab \end{vmatrix}$$

Taking $(a^2 + b^2 + c^2)$ common from C_1 and applying $R_2 \to R_2 - R_1$ and $R_3 \to R_3 - R_1$, we get

$$\Delta = (a^2 + b^2 + c^2) \begin{vmatrix} 1 & a^2 & bc \\ 0 & b^2 - a^2 & ca - bc \\ 0 & c^2 - a^2 & ab - bc \end{vmatrix}$$

$$= (a^2 + b^2 + c^2)(b - a)(c - a) \begin{vmatrix} 1 & a^2 & bc \\ 0 & b + a & -c \\ 0 & c + a & -b \end{vmatrix}$$

Expanding along C_1, we get

$$\Delta = (a^2 + b^2 + c^2)(b - a)(c - a)[c(c + a) - b(b + a)]$$

$$= (a^2 + b^2 + c^2)(a + b + c)(a - b)(b - c)(c - a)$$

Example 43:

$$\textit{If } \Delta(x) = \begin{vmatrix} a_1 + x & b_1 + x & c_1 + x \\ a_2 + x & b_2 + x & c_2 + x \\ a_3 + x & b_3 + x & c_3 + x \end{vmatrix}$$

show that $\Delta''(x) = 0$ and that $\Delta(x) = \Delta(0) + Sx$, where S denotes the sum of all the cofactors of the elements in $\Delta(0)$.

Solution:

We have

$$\Delta'(x) = \begin{vmatrix} 1 & b_1 + x & c_1 + x \\ 1 & b_2 + x & c_2 + x \\ 1 & b_3 + x & c_3 + x \end{vmatrix} + \begin{vmatrix} a_1 + x & 1 & c_1 + x \\ a_2 + x & 1 & c_2 + x \\ a_3 + x & 1 & c_3 + x \end{vmatrix} + \begin{vmatrix} a_1 + x & b_1 + x & 1 \\ a_2 + x & b_2 + x & 1 \\ a_3 + x & b_3 + x & 1 \end{vmatrix}$$

Applying $C_2 \to C_{2-x} C_1$ and $C_3 \to C_{3-x} C_1$ in the first; $C_1 \to C_{1-x} C_2$ and $C_3 - C_{3-x} C_2$ in the second; and $C_1 \to C_{1-x} C_3$ and $C_2 \to C_{2-x} C_3$ in the third determinant of $\Delta'(x)$, we get

$$\Delta'(x) = \begin{vmatrix} 1 & b_1 & c_1 \\ 1 & b_2 & c_2 \\ 1 & b_3 & c_3 \end{vmatrix} + \begin{vmatrix} a_1 & 1 & c_1 \\ a_2 & 1 & c_2 \\ a_3 & 1 & c_3 \end{vmatrix} + \begin{vmatrix} a_1 & b_1 & 1 \\ a_2 & b_2 & 1 \\ a_3 & b_3 & 1 \end{vmatrix} = 5$$

where S denotes the sum of all the cofactors of all the elements in $\Delta(0)$, so that $\Delta''(x) = 0$. Since $\Delta'(x) = S$, we get $\Delta(x) = xS + k$. Also, $k = \Delta(0)$. Therefore, $\Delta(x) = \Delta(0) + Sx$.

Example 44:

Prove that

$$\Delta = \begin{vmatrix} a^2 & (s-a)^2 & (s-a)^2 \\ (s-b)^2 & b^2 & (s-b)^2 \\ (s-c)^2 & (s-c)^2 & c^2 \end{vmatrix} = 2s^3 (s-a)(s-b)(s-c)$$

where $2s = a + b + c$.

Solution:

Let $s - a = \alpha$, $s - b = \beta$ and $s - c = \gamma$, so that $\alpha + \beta + \gamma = 3s - (a + b + c) = s$. Also, $\beta + \gamma = 2s - (b + c) = a$, $\gamma + \alpha = b$ and $\alpha + \beta = c$. Now Δ can be written is

$$\Delta = \begin{vmatrix} (\beta+\gamma)^2 & \alpha^2 & \alpha^2 \\ \beta^2 & (\gamma+\alpha)^2 & \beta^2 \\ \gamma^2 & \gamma^2 & (\alpha+\beta)^2 \end{vmatrix}$$

$= 2(\alpha + \beta + \gamma)^2\, \alpha\beta\gamma = 2s^3 (s - a)(s - b)(s - c)$.

Example 45:

Show that

$$\Delta = \begin{vmatrix} (b+c)^2 & a^2 & a^2 \\ b^2 & (c+a)^2 & b^2 \\ c^2 & c^2 & (a+b)^2 \end{vmatrix} = 2abc(a+b+c)^3$$

Solution:

Applying $C_2 \to C_2 - C_1$ and $C_3 \to C_3 - C_1$, we get

$$\Delta = \begin{vmatrix} (b+c)^2 & a^2-(b+c)^2 & a^2-(b+c)^2 \\ b^2 & (c+a)^2-b^2 & 0 \\ c^2 & 0 & (a+b)^2-c^2 \end{vmatrix}$$

Taking (a + b + c) common from C_2 and C_3, we obtain

$$\Delta = (a+b+c)^2 \begin{vmatrix} (b+c)^2 & a-b-c & a^2-(b+c)^2 \\ b^2 & c+a-b & 0 \\ c^2 & 0 & a+b-c \end{vmatrix}$$

Applying $R_1 \to R_1 - R_2 - R_3$, we get

$$\Delta = (a+b+c)^2 \begin{vmatrix} 2bc & -2c & -2b \\ b^2 & c+a-b & 0 \\ c^2 & 0 & a+b-c \end{vmatrix}$$

Applying $C_2 \to C_2 + (1/b)C_1$ and $C_3 \to C_3 + (1/c)C_1$, we obtain

$$\Delta = (a+b+c)^2 \begin{vmatrix} 2bc & 0 & 0 \\ b^2 & c+a & b^2/c \\ c^2 & c^2/b & a+b \end{vmatrix}$$

Expanding along R_1, we get

$\Delta = (a + b + c)^2\, 2bc\, [(c + a)(a + b) - bc]$

$\quad = (a + b + c)^2\, 2bc\, (ab + ac + bc + a^2 - bc)$

$\quad = 2bc\,(a + b + c)^3$

Example 46:

Show that $\Delta = \begin{vmatrix} (a+b)^2 & ca & cb \\ ca & (b+c)^2 & ab \\ bc & ab & (c+a)^2 \end{vmatrix} = 2abc(a+b+c)^3$

Solution:

Multiplying R_1 by c, R_2 by a and R_3 by b, we get

$$\Delta = \frac{1}{abc}\begin{vmatrix} (a+b)^2 c & c^2 a & c^2 b \\ ca^2 & (b+c)^2 a & a^2 b \\ b^2 c & ab^2 & (c+a)^2 b \end{vmatrix}$$

Taking c common from C_1, a from C_2 and b from C_3, we get

$$\Delta = \begin{vmatrix} (a+b)^2 c & c^2 & c^2 \\ a^2 & (b+c)^2 & a^2 \\ b^2 & b^2 & (c+a)^2 \end{vmatrix}$$

$= 2abc\ (a + b + c)^3$

Example 47:

Find the value of $f(\pi/6)$, where

$$f(\theta) = \begin{vmatrix} \cos^2\theta & \cos\theta\sin\theta & -\sin\theta \\ \cos\theta\sin\theta & \sin^2\theta & \cos\theta \\ \sin\theta & -\cos\theta & 0 \end{vmatrix}$$

Solution:

Applying $C_1 \to C_1 - \sin\theta\ C_3$ and $C_2 \to C_2 + \cos\theta\ C_3$, we get

$$f(\theta) = \begin{vmatrix} 1 & 0 & -\sin\theta \\ 0 & 1 & \cos\theta \\ \sin\theta & -\cos\theta & 0 \end{vmatrix}$$

Applying $R_3 \to R_3 - \sin\theta\ R_1 + \cos\theta\ R_2$, we get

$$f(\theta) = \begin{vmatrix} 1 & 0 & -\sin\theta \\ 0 & 1 & \cos\theta \\ 0 & 0 & \sin^2\theta + \cos^2\theta \end{vmatrix} = 1$$

Thus, $f(\pi/6) = 1$.

Example 48:

Show that

$$\Delta = \begin{vmatrix} a^2 & bc & ac+c^2 \\ a^2+bc & b^2 & ac \\ ab & b^2+bc & c^2 \end{vmatrix} = 4a^2b^2c^2$$

Solution:

Taking a common from C_1, b from C_2 and c from C_3, we get

$$\Delta = abc \begin{vmatrix} a & c & a+c \\ a+b & b & a \\ b & b+c & c \end{vmatrix}$$

Applying $C_1 \rightarrow C_1 + C_2 - C_3$, we get

$$\Delta = abc \begin{vmatrix} 0 & c & a+c \\ 2b & b & a \\ 2b & b+c & c \end{vmatrix}$$

Taking 2b common from C_1 and applying $R_3 \rightarrow R_3 - R_2$, we get

$$\Delta = 2ab^2c \begin{vmatrix} 0 & c & a+c \\ 1 & b & a \\ 0 & c & c-a \end{vmatrix}$$

Expanding along C_1, we get

$$\Delta = 2ab^2c(-1) \begin{vmatrix} c & a+c \\ c & c-a \end{vmatrix}$$

$$= 2\ ab^2c\ (-1)\ [c(c-a) - c(a+c)]$$

$$= 4a^2b^2c^2.$$

Example 49:

Show that

$$\Delta = \begin{vmatrix} a & b-c & c+b \\ a+c & b & c-a \\ a-b & b+a & c \end{vmatrix} = (a+b+c)(a^2+b^2+c^2)$$

Solution:

Multiplying C_1 and dividing Δ by a, we get

$$\Delta = \frac{1}{a}\begin{vmatrix} a & b-c & c+b \\ a+c & b & c-a \\ a-b & b+a & c \end{vmatrix}$$

Applying $C_1 \rightarrow C_1 + bC_2 + cC_3$, we get

$$\Delta = \frac{1}{a}\begin{vmatrix} a^2+b^2+c^2 & b-c & c+b \\ a^2+b^2+c^2 & b & c-a \\ a^2+b^2+c^2 & b+a & c \end{vmatrix}$$

Taking $a^2 + b^2 + c^2$ common from C_1 and applying $R_2 \rightarrow R_2 - R_1$ and $R_3 \rightarrow R_3 - R_1$, we get

$$\Delta = \frac{a^2+b^2+c^2}{a}\begin{vmatrix} 1 & b-c & c+b \\ 0 & c & -(a+b) \\ 0 & a+c & -b \end{vmatrix}$$

Expanding along C_1, we get

$$\Delta = \frac{a^2+b^2+c^2}{a} \; [-bc + (a + b)(a + c)]$$

$$= \frac{a^2+b^2+c^2}{a} [a^2 + ab + ac]$$

$$= (a + b + c)(a^2 + b^2 + c^2),$$

Example 50:

Show that

$$\Delta = \begin{vmatrix} b^2+c^2 & ab & ac \\ ab & c^2+a^2 & bc \\ ca & cb & a^2+b^2 \end{vmatrix} = 4a^2b^2c^2$$

Solution:

Multiplying C_1 by a, C_2 by b and C_3 by c, we get

$$\Delta = \frac{1}{abc}\begin{vmatrix} a(b^2+c^2) & ab^2 & ac^2 \\ a^2b & b(c^2+a^2) & bc^2 \\ a^2c & cb^2 & (b^2+a^2)c \end{vmatrix}$$

Taking a, b and c common from R_1, R_2 and R_3 respectively, we get

$$\Delta = \frac{abc}{abc}\begin{vmatrix} b^2+c^2 & b^2 & c^2 \\ a^2 & c^2+a^2 & c^2 \\ a^2 & b^2 & b^2+a^2 \end{vmatrix}$$

Applying $C_1 \to C_1 - C_2 - C_3$, we get

$$\Delta = \begin{vmatrix} 0 & b^2 & c^2 \\ -2c^2 & c^2+a^2 & c^2 \\ -2b^2 & b^2 & b^2+a^2 \end{vmatrix}$$

Taking 2 common from C_1, and applying $C_2 \to C_2 + C_1$ and $C_3 \to C_3 + C_1$, we get

$$\Delta = 2\begin{vmatrix} 0 & b^2 & c^2 \\ -c^2 & a^2 & 0 \\ -b^2 & 0 & a^2 \end{vmatrix}$$

Evaluating along C_1, we get

$\Delta = 2[a^2 (b^2 a^2 - 0) - b^2 (0 - a^2c^2)] = 4a^2b^2c^2$.

Example 51:

If x, y and z are unequal and

$$\begin{vmatrix} x^3 & (x+a)^3 & (x-a)^3 \\ y^3 & (y+a)^3 & (y-a)^3 \\ z^3 & (z+a)^3 & (z-a)^3 \end{vmatrix} = 0$$

prove that $a^2 (x + y + z) = 3xyz$.

Solution:

Applying $C_2 \to C_2 - C_3$, we get

$$\Delta = \begin{vmatrix} x^3 & 6x^2 a+2a^3 & (x-a)^3 \\ y^3 & 6y^2 a+2a^3 & (y-a)^3 \\ z^3 & 6z^2 a+2a^3 & (z-a)^3 \end{vmatrix}$$

Taking 2 common from C_2 and applying $C_3 \to C_3 - C_1 + C_2$, we get

$$\Delta = -2\begin{vmatrix} x^3 & 3x^2 a+a^3 & 3xa^2 \\ y^3 & 3y^2 a+a^3 & 3ya^2 \\ z^3 & 3z^2 a+a^3 & 3za^2 \end{vmatrix} = -18a^2\Delta_1 - 6a^5\Delta_2$$

where $\Delta_1 = \begin{vmatrix} x^3 & x^2 & x \\ y^3 & y^2 & y \\ z^3 & z^2 & z \end{vmatrix} = xyz\begin{vmatrix} x^2 & x & 1 \\ y^2 & y & 1 \\ z^2 & z & 1 \end{vmatrix}$

and $\Delta_2 = \begin{vmatrix} x^3 & 1 & x \\ y^3 & 1 & y \\ z^3 & 1 & z \end{vmatrix}$

Applying $R_2 \to R_2 - R_1$ and $R_3 \to R_3 - R_1$ in Δ_1 and Δ_2 we obtain

$$\Delta_1 = xyz\begin{vmatrix} x^2 & x & 1 \\ y^2 - x^2 & y-x & 0 \\ z^2 - x^2 & z-x & 0 \end{vmatrix}$$

$$= xyz(y-x)(z-x)\begin{vmatrix} y+x & 1 \\ z+x & 1 \end{vmatrix}$$

$$= xyz\,(y - x)\,(z - x)\,(y - z)$$

and $\Delta_2 = \begin{vmatrix} x^3 & 1 & x \\ y^3 - x^3 & 0 & y-x \\ z^3 - x^3 & 0 & z-x \end{vmatrix}$

$$= -(y-x)(z-x)\begin{vmatrix} y^2 + x^2 + yx & 1 \\ z^2 + x^2 + zx & 1 \end{vmatrix}$$

$= (y - x)(z - x)(z - y)(x + y + z)$

Thus $\Delta = 6a^3 (y - x)(z - x)(y - z) \quad [a^2 (x + y + z) - 3xyz]$

As $\Delta = 0$ and x, y and z are unequal we get

$$a^2 (x + y + z) - 3xyz = 0$$

$$\Rightarrow \quad a^2 (x + y + z) = 3yz$$

Example 52(a):

If a, b and c are distinct, solve the equation

$$\Delta = \begin{vmatrix} x^2 - a^2 & x^2 - b^2 & x^2 - c^2 \\ (x-a)^3 & (x-b)^3 & (x-c)^3 \\ (x+a)^3 & (x+b)^3 & (x+c)^3 \end{vmatrix} = 0$$

Solution:

Applying $R_3 \to R_3 - R_2$, we obtain

$$\Delta = \begin{vmatrix} x^2 - a^2 & x^2 - b^2 & x^2 - c^2 \\ (x-a)^3 & (x-b)^3 & (x-c)^3 \\ 6x^2a + 2a^3 & 6x^2b + 2b^3 & 6x^2c + 2c^3 \end{vmatrix}$$

Taking 2 common from R_3 and applying $R_2 \to R_2 + R_3$, we get

$$\Delta = 2\begin{vmatrix} x^2 - a^2 & x^2 - b^2 & x^2 - c^2 \\ x^3 + 3xa^2 & x^3 + 3xb^2 & x^3 + 3xc^2 \\ 3x^2a + a^3 & 3x^2b + b^3 & 3x^2c + c^3 \end{vmatrix}$$

Taking x common from R_2 and applying $C_2 \to C_2 - C_1$ and $C_3 \to C_3 - C_1$, we get

$$\Delta = 2x\begin{vmatrix} x^2 - a^2 & x^2 - b^2 & x^2 - c^2 \\ x^2 + 3a^2 & 3(b^2 - a^2) & 3(c^2 - a^2) \\ 3x^2a + a^3 & 3x^2(b-a) + b^3 - a^3 & 3x^2(c-a) + c^3 - a^3 \end{vmatrix}$$

$= 2x (b - a)(c - a) \Delta_1$

$$\text{where } \Delta_1 = \begin{vmatrix} x^2 - a^2 & -(a+b) & -(a+c) \\ x^2 + 3a^2 & 3(b+a) & 3(a+c) \\ 3x^2a + a^3 & 3x^2 + b^2 + a^2 + ba & 3x^2 + c^2 + a^2 + ac \end{vmatrix}$$

Applying $R_2 \to R_2 + 3R_1$, we get Δ_1 =

$$\begin{vmatrix} x^2 - a^2 & -(a+b) & -(a+c) \\ 4x^2 & 0 & 0 \\ 3x^2a + a^3 & 3x^2 + b^2 + a^2 + ba & 3x^2 + c^2 + a^2 + ac \end{vmatrix}$$

Expanding along R_2, we get

$$\Delta = 8x^3 (b - a) (c - a) \begin{vmatrix} a+b & a+c \\ 3x^2 + b^2 + a^2 + ba & 3x^2 + c^2 + a^2 + ac \end{vmatrix}$$

Applying $C_2 \to C_2 - C_1$, we get

$$\Delta = 8x^3 (b - a) (c - a) \begin{vmatrix} a+b & c-b \\ 3x^2 + b^2 + a^2 + ba & c^2 - b^2 + ac - ab \end{vmatrix}$$

$= 8x^3 (b - a) (c - a) (c - b) [(a + b) (a + b + c) - 3x^2 - b^2 - a^2 - ba]$

$= 8x^2 (b - a) (c - a) (c - b) [(ab + bc + ca) - 3x^2]$

$\therefore\ \Delta = 0 \Rightarrow 8(b - a) (c - a) (c - b)x^3 [(ab + bc + ca) - 3x^2] = 0$

As a, b and c are distinct, we get

$$x = 0, 0, 0, \pm \sqrt{\frac{1}{3}(bc + ca + ab)}$$

Example 52(b):

Without expanding at any stage, show that

$$\Delta = \begin{vmatrix} x^2 + x & x+1 & x-2 \\ 2x^3 + 3x - 1 & 3x & 3x-3 \\ x^2 + 2x + 3 & 2x-1 & 2x-1 \end{vmatrix} = xA + B$$

where A and B are determinants of order 3 not involving x.

Solution:

Applying $R_2 \to R_2 - R_1 - R_3$, we obtain

$$\Delta = \begin{vmatrix} x^2 + x & x+1 & x-2 \\ -4 & 0 & 0 \\ x^2 + 2x + 3 & 2x-1 & 2x-1 \end{vmatrix}$$

Applying $R_1 \to R_1 + 1/4x^3 R_2$ and $R_3 \to R_3 + 1/4x^3 R_2$, we get

$$\Delta = \begin{vmatrix} x & x+1 & x-2 \\ -4 & 0 & 0 \\ 2x+3 & 2x-1 & 2x-1 \end{vmatrix}$$

Applying $R_3 \to R_3 - 2R_1$ we get

$$\Delta = \begin{vmatrix} x & x+1 & x-2 \\ -4 & 0 & 0 \\ 3 & -3 & 3 \end{vmatrix} = xA + B$$

$$\text{where} \quad A = \begin{vmatrix} 1 & 1 & 1 \\ -4 & 0 & 0 \\ 3 & -3 & 3 \end{vmatrix} \text{ and } B \begin{vmatrix} 0 & 1 & -2 \\ -4 & 0 & 0 \\ 3 & -3 & 3 \end{vmatrix}$$

Example 52(c):

Evaluate $\begin{vmatrix} 0 & \cos x & -\sin x \\ \sin x & 0 & \cos x \\ \cos x & \sin x & 0 \end{vmatrix}^2$

Solution:

We have

$$= \begin{vmatrix} 0 & \cos x & -\sin x \\ \sin x & 0 & \cos x \\ \cos x & \sin x & 0 \end{vmatrix} \times \begin{vmatrix} 0 & \cos x & -\sin x \\ \sin x & 0 & \cos x \\ \cos x & \sin x & 0 \end{vmatrix}$$

$$= \begin{vmatrix} 0.0 + \cos x.\cos x + \sin x \sin x & 0.\sin x + \cos.0 - \sin x \cos x & 0.\cos x + \cos x \sin x - \sin x.0 \\ \sin x.0 + 0.\cos x - \cos x \sin x & \sin x \sin x + 0.0 + \cos x \cos x & \sin x.\cos x + 0 \sin x + \cos x.0 \\ \cos x.0 + \sin x \cos x + 0(-\sin x) & \cos x \sin x + \sin x.0 + 0.\cos x & \cos x.\cos x + \sin x.\sin x + 0.0 \end{vmatrix}$$

$$= \begin{vmatrix} \cos^2 x + \sin^2 x & -\sin x \cos x & \cos x \sin x \\ -\cos x \sin x & \sin^2 + \cos^2 x & \sin x \cos x \\ \sin x \cos x & \cos x \sin x & \cos^2 x + \sin^2 x \end{vmatrix}$$

$$= \begin{vmatrix} 1 & -\lambda & \lambda \\ -\lambda & 1 & \lambda \\ \lambda & \lambda & 1 \end{vmatrix}, \text{ where } \lambda \sin x \cos x$$

Example 53:

Let α be a repeated root of the quadratic equation $f(x) = 0$ and $A(x)$, $B(x)$ and $C(x)$ polynomials of degrees 3, 4 and 5 respectively. Show that

$$\begin{vmatrix} A(x) & B(x) & C(x) \\ A(\alpha) & B(\alpha) & C(\alpha) \\ A'(\alpha) & B'(\alpha) & C'(\alpha) \end{vmatrix}$$

is divisible by $f(x)$, where prime denotes the derivative.

Solution:

Since α is a repeated root of the quadratic equation $f(x) = 0$, we have $f(x) = a(x - \alpha)^2$, where α is some constant, and $a \neq 0$.

Let $$\phi(x) = \begin{vmatrix} A(x) & B(x) & C(x) \\ A(\alpha) & B(\alpha) & C(\alpha) \\ A'(\alpha) & B'(\alpha) & C'(\alpha) \end{vmatrix}$$

Note that $\phi(x)$ is a polynomial in x of degree at most 5. We have

$$\phi(\alpha) = \begin{vmatrix} A(\alpha) & B(\alpha) & C(\alpha) \\ A(\alpha) & B(\alpha) & C(\alpha) \\ A'(\alpha) & B'(\alpha) & C'(\alpha) \end{vmatrix}$$

$= 0$ [$\because$ R_1 and R_2 are identical]

Also $$\phi'(x) = \begin{vmatrix} A'(x) & B'(x) & C'(x) \\ A(\alpha) & B(\alpha) & C(\alpha) \\ A'(\alpha) & B'(\alpha) & C'(\alpha) \end{vmatrix} \Rightarrow \phi'(\alpha) = 0$$

Thus, α is a repeated root of $\phi(x) = 0$. Therefore, $\phi(x) = (x - \alpha)^2 \psi(x)$, where $\psi(x)$ is a polynomial in x of degree at most 3. This shows that $f(x) \mid \phi(x)$.

Example 54:

Show that

$$\Delta = \begin{vmatrix} a_1 & x & x \\ x & a_2 & x \\ x & x & a_3 \end{vmatrix} = xf'(x) - f(x)$$

where $f(x) = (x - a_1)(x - a_2)(x - a_3)$

Solution:

Applying $C_3 \to C_3 - C_2$ and $C_2 \to C_2 - C_1$ we get

$$\Delta = \begin{vmatrix} a_1 & x-a_1 & 0 \\ x & a_2-x & x-a_2 \\ x & x & a_3-x \end{vmatrix}$$

Expanding along R_1 we get

$$\Delta = \begin{vmatrix} a_2-x & x-a_2 \\ 0 & a_3-x \end{vmatrix} - (x-a_1)\begin{vmatrix} x & x-a_2 \\ x & a_3-x \end{vmatrix}$$

$= a_1[(a_2 - x)(a_3 - x)] - (x - a_1)\, x\, \{(a_3 - x) - (x - a_2)\}$

$= x[(x - a_1)(x - a_2) + (x - a_1)(x - a_2)] + a_1 (x - a_2)(x - a_3)$

$= x[x - a_2)(x - a_3) + (x - a_3)(x - a_1) + (x - a_1)(x - a_2)] + a_1$

$(x - a_2)(x - a_2) - x(x - a_2)(x - a_3)$

$= xf'(x) - (x - a_1)(x - a_2)(x - a_3) = xf'(x) - f(x).$

Example 55:

Show that $\begin{vmatrix} bc-a^2 & ca-b^2 & ab-c^2 \\ ca-b^2 & ab-c^2 & bc-a^2 \\ ab-c^2 & bc-a^2 & ca-b^2 \end{vmatrix}$

$$= \begin{vmatrix} a^2 & c^2 & 2ac-b^2 \\ 2ab-c^2 & b^2 & a^2 \\ b^2 & 2bc-a^2 & c^2 \end{vmatrix}$$

Solution:

Let $\Delta = \begin{vmatrix} a & b & c \\ b & c & a \\ c & a & b \end{vmatrix}$

Replacing each element of Δ by its cofactor we obtain

$$\Delta_1 = \begin{vmatrix} bc-a^2 & ca-b^2 & ab-c^2 \\ ca-b^2 & ab-c^2 & bc-a^2 \\ ab-c^2 & bc-a^2 & ca-b^2 \end{vmatrix} = \Delta^2 \qquad (1)$$

Also $\Delta^2 = \begin{vmatrix} a & b & c \\ b & c & a \\ c & a & b \end{vmatrix}^2 = \begin{vmatrix} a & b & c \\ b & c & a \\ c & a & b \end{vmatrix}\begin{vmatrix} a & -c & b \\ b & -a & c \\ c & -b & a \end{vmatrix}$

$$= \begin{vmatrix} a^2 & c^2 & 2ac-b^2 \\ 2ab-c^2 & b^2 & a^2 \\ b^2 & 2bc-a^2 & c^2 \end{vmatrix} \qquad (2)$$

From (1) and (2) we get

$$\begin{vmatrix} bc-a^2 & ca-b^2 & ab-c^2 \\ ca-b^2 & ab-c^2 & bc-a^2 \\ ab-c^2 & bc-a^2 & ca-b^2 \end{vmatrix} = \begin{vmatrix} a^2 & c^2 & 2ac-b^2 \\ 2ab-c^2 & b^2 & a^2 \\ b^2 & 2bc-a^2 & c^2 \end{vmatrix}$$

Example 56(a):

Evaluate the determinant

$$\Delta = \begin{vmatrix} \sqrt{p}+\sqrt{q} & 2\sqrt{r} & \sqrt{r} \\ \sqrt{qr}+\sqrt{2p} & r & \sqrt{2r} \\ q+\sqrt{pr} & \sqrt{qr} & r \end{vmatrix}$$

where p, q and r are positive real numbers.

Solution:

Taking $\sqrt{r}$ common from C_2 and C_3 we get

$$\Delta = r\begin{vmatrix} \sqrt{p}+\sqrt{q} & 2 & 1 \\ \sqrt{qr}+\sqrt{2p} & \sqrt{r} & \sqrt{2} \\ q+\sqrt{pr} & \sqrt{q} & \sqrt{r} \end{vmatrix}$$

Applying $C_1 \to C_1 - \sqrt{q}\,C_2 - \sqrt{p}\,C_3$ we get

$$\Delta = r\begin{vmatrix} -\sqrt{p} & 2 & 1 \\ 0 & \sqrt{r} & \sqrt{2} \\ 0 & \sqrt{q} & \sqrt{r} \end{vmatrix} = -\,r\sqrt{q}\,(r-\sqrt{2q})$$

$$= r(2\sqrt{2q} - r\sqrt{q})$$

Example 56(b):

If $a \neq p,\ b \neq q,\ c \neq r$ and

$$\Delta = \begin{vmatrix} p & b & c \\ a & q & c \\ a & b & r \end{vmatrix} = 0, \text{ then find the value of}$$

$$\frac{p}{p-a} + \frac{q}{q-b} + \frac{r}{r-c}$$

Solution:

Applying $R_2 \to R_2 - R_1$, $R_3 \to R_3 - R_1$, we get

$$\Delta = \begin{vmatrix} p & b & c \\ a-p & q-b & 0 \\ a-p & 0 & r-c \end{vmatrix}$$

$$= c\begin{vmatrix} a-p & q-b \\ a-p & 0 \end{vmatrix} + (r-c)\begin{vmatrix} p & b \\ a-p & q-b \end{vmatrix}$$

$= -c(a - p)\ (q - b) + (r - c)\ (q - b)\ p - (r - c)\ (a - p) = 0$

As $\Delta = 0$, we get

$-c(a - p)\ (q - b) + (r - c)\ (q - b)\ p - (r - c)\ (a - p)\ b = 0$

Dividing by $(p - a)\ ((q - b)\ (r - c)$ we get

$$\frac{c}{r-c}+\frac{p}{p-a}+\frac{b}{q-b}=0 \quad ...(1)$$

Let $$k=\frac{p}{p-a}+\frac{q}{q-b}+\frac{r}{r-c} \quad ...(2)$$

Subtracting (1) from (2) we get

$$k-0=\frac{q-b}{q-b}+\frac{r-c}{r-c}=1+1=2 \Rightarrow k=2.$$

Example 56(c):

Prove that $\begin{vmatrix} a & b & c \\ b & c & a \\ c & a & b \end{vmatrix}^2 = \begin{vmatrix} 2bc-a^2 & c^2 & b^2 \\ c^2 & 2ac-b^2 & a^2 \\ b^2 & a^2 & 2ab-c^2 \end{vmatrix}$

$= (a^2 + b^2 + c^2 - 3abc)^2.$

Solution:

$$\begin{vmatrix} a & b & c \\ b & c & a \\ c & a & b \end{vmatrix}^2 = \begin{vmatrix} a & b & c \\ b & c & a \\ c & a & b \end{vmatrix} \times \begin{vmatrix} a & b & c \\ b & c & a \\ c & a & b \end{vmatrix}$$

$= -\begin{vmatrix} a & b & c \\ b & c & a \\ c & a & b \end{vmatrix} \times \begin{vmatrix} a & c & b \\ b & a & c \\ c & b & a \end{vmatrix}$, interchanging C_2, C_3 of the second determin ant.

$= \begin{vmatrix} -a & b & c \\ -b & c & a \\ -c & a & b \end{vmatrix} \times \begin{vmatrix} a & c & b \\ b & a & c \\ c & b & a \end{vmatrix}$, multiplying C_1 of first determinant by -1.

$$= \begin{vmatrix} -aa+bc+cb & -ab+ba+cc & -ac+bb+ca \\ -ba+cc+ab & -bb+ca+ac & -bc+cb+aa \\ -ca+ac+bb & -cb+aa+bc & -cc+ab+a \end{vmatrix}$$

$$= \begin{vmatrix} 2bc - a^2 & c^2 & b^2 \\ c^2 & 2ac - b^2 & a^2 \\ b^2 & a^2 & 2ab - c^2 \end{vmatrix}$$

Also $\begin{vmatrix} a & b & c \\ b & c & a \\ c & a & b \end{vmatrix} = a \begin{vmatrix} c & a \\ a & b \end{vmatrix} - b \begin{vmatrix} b & a \\ c & b \end{vmatrix} + c \begin{vmatrix} b & c \\ c & a \end{vmatrix}$

$= a(cb - a^2) - b\,(b^2 - ac) + c\,(ab - c^2)$

$= abc - a^3 - b^3 + abc - c^3$

$= -\,(a^3 + b^3 + c^3 - 3abc)$

$\therefore \begin{vmatrix} a & b & c \\ b & c & a \\ c & a & b \end{vmatrix}^2 = (a^3 + b^3 + c^3 - 3abc)^2$

Example 57:

Evaluate $\begin{vmatrix} a_1 & b_1 & c_1 \\ a_2 & b_2 & c_2 \\ a_3 & b_3 & c_3 \end{vmatrix} \times \begin{vmatrix} x_1 & y_1 & z_1 \\ x_2 & y_2 & z_2 \\ x_3 & y_3 & z_3 \end{vmatrix}$

Solution:

The required product

$$= \begin{vmatrix} a_1x_1 + b_1y_1 + c_1z_1 & a_1x_2 + b_1y_2 + c_1z_2 & a_1x_3 + b_1y_3 + c_1z_3 \\ a_2x_1 + b_2y_1 + c_2z_1 & a_2x_2 + b_2y_2 + c_2z_2 & a_2x_3 + b_2y_3 + c_2z_3 \\ a_3x_1 + b_3y_1 + c_3z_1 & a_3x_2 + b_3y_2 + c_3z_2 & a_3x_3 + b_3y_3 + c_3z_3 \end{vmatrix}$$

Example 58:

Evaluate $\begin{vmatrix} 0 & c & b \\ c & 0 & a \\ b & a & 0 \end{vmatrix}^2$

Solution:

$$\begin{vmatrix} 0 & c & b \\ c & 0 & a \\ b & a & 0 \end{vmatrix}^2 = \begin{vmatrix} 0 & c & b \\ c & 0 & a \\ b & a & 0 \end{vmatrix} \times \begin{vmatrix} 0 & c & b \\ c & 0 & a \\ b & a & 0 \end{vmatrix}$$

$$= \begin{vmatrix} 0.0 + cc + b.b & 0.c + c.0 + b.a & 0.b + c.a + b.0 \\ c.0 + 0c + a.b & c.c + 00 + a.a & c.b + 0.a + a0 \\ b.0 + a.c + 0.b & b.c + a.0 + 0.a & b.b + a.a + 0.0 \end{vmatrix}$$

$$= \begin{vmatrix} c^2 + b^2 & ba & ca \\ ab & c^2 + a^2 & bc \\ ac & bc & b^2 + a^2 \end{vmatrix}$$

Example 59:

Express $\begin{vmatrix} 2bc - a^2 & c^2 & b^2 \\ c^2 & 2ca - b^2 & a^2 \\ b^2 & a^2 & 2ab - c^2 \end{vmatrix}$

as the product of two determinants.

Solution:

The element in the first row and first column is $2bc - a^2$ which can be written as

$$a\,(-a) + b\,(c) + c\,(b) \qquad \ldots\text{(i)}$$

The element in the first row and second column ins c^2 which can be written as

$$a\,(-b) + b\,(a) + c(c) \qquad \ldots\text{(ii)}$$

The element in the first row and third column is b^2 which can be written as

$$a\,(-c) + b\,(b) + c\,(a) \qquad \ldots\text{(iii)}$$

(i), (ii) and (iii) suggest that the given determinant can be written tentatively as

$$\begin{vmatrix} a & b & c \\ b & c & a \\ c & a & b \end{vmatrix} \times \begin{vmatrix} -a & c & b \\ -b & a & c \\ -c & b & a \end{vmatrix}$$

But actually multiplying these two determinants we get the given determinant. Hence these determinants are the required ones.

Example 60:

Find the product of two determinants of different orders or evaluate

$$\begin{vmatrix} \alpha_1 & \beta_1 & \gamma_1 \\ \alpha_2 & \beta_2 & \gamma_2 \\ \alpha_3 & \beta_3 & \gamma_3 \end{vmatrix} \times \begin{vmatrix} a_1 & b_1 \\ a_2 & b_2 \end{vmatrix}$$

Solution:

Here the two given determinants are of different orders, so we adopt the following method :

$$\begin{vmatrix} \alpha_1 & \beta_1 & \gamma_1 \\ \alpha_2 & \beta_2 & \gamma_2 \\ \alpha_3 & \beta_3 & \gamma_3 \end{vmatrix} \times \begin{vmatrix} a_1 & b_1 \\ a_2 & b_2 \end{vmatrix}$$

$$= \begin{vmatrix} \alpha_1 & \beta_1 & \gamma_1 \\ \alpha_2 & \beta_2 & \gamma_2 \\ \alpha_3 & \beta_3 & \gamma_3 \end{vmatrix} \times \begin{vmatrix} a_1 & b_1 & 0 \\ a_2 & b_2 & 0 \\ 0 & 0 & 1 \end{vmatrix},$$

making the two determinants of the same order

$$= \begin{vmatrix} \alpha_1 a_1 + \beta_1 b_1 + \gamma_1.0 & \alpha_1 a_2 + \beta_1 b_2 + \gamma_1.0 & \alpha_1.0 + \beta_1.0 + \gamma_1.1 \\ \alpha_2 a_1 + \beta_2 b_1 + \gamma_2.0 & \alpha_2 a_2 + \beta_2 b_2 + \gamma_2.0 & \alpha_2.0 + \beta_2.0 + \gamma_2.1 \\ \alpha_3 a_1 + \beta_3 b_1 + \gamma_3.0 & \alpha_3 a_2 + \beta_3 b_2 + \gamma_3.0 & \alpha_3.0 + \beta_3.0 + \gamma_3.1 \end{vmatrix}$$

$$= \begin{vmatrix} a_1\alpha_1 + b_1\beta_1 & a_2\alpha_1 + b_2\beta_1 & \gamma_1 \\ a_1\alpha_2 + b_1\beta_2 & a_2\alpha_2 + b_2\beta_2 & \gamma_2 \\ a_1\alpha_3 + b_1\beta_3 & a_2\alpha_3 + b_2\beta_3 & \gamma_3 \end{vmatrix}$$

Example 61:

Express $\begin{vmatrix} (1+ax)^2 & (1+ay)^2 & (1+az)^2 \\ (1+bx)^2 & (1+by)^2 & (1+bz)^2 \\ (1+cx)^2 & (1+cy)^2 & (1+cz)^2 \end{vmatrix}$

as the product of two determinants.

Solution:

The given determinant

$$= \begin{vmatrix} 1+2ax+a^2x^2 & 1+2ay+a^2y^2 & 1+2az+a^2z^2 \\ 1+2bx+b^2x^2 & 1+2by+b^2y^2 & 1+2bz+b^2z^2 \\ 1+2cx+c^2x^2 & 1+2cy+c^2y^2 & 1+2cz+c^2z^2 \end{vmatrix}$$

The element in the first row and first column is $1 + 2ax + a^2x^2$ which can be written as

$$(1)\,(1) + (2a)\,.\,(x) + (a^2)\,.\,(x^2)$$

This suggests that the first rows of the two required determinants are $1, 2a, a^2$ and $1, x, x^2$.

Hence the given determinant may be written as

$$\begin{vmatrix} 1 & 2a & a^2 \\ 1 & 2b & b^2 \\ 1 & 2c & c^2 \end{vmatrix} \times \begin{vmatrix} 1 & x & x^2 \\ 1 & y & y^2 \\ 1 & z & z^2 \end{vmatrix}$$

Example 62:

Reduce $A = \begin{bmatrix} 13 & 16 & 19 \\ 14 & 17 & 20 \\ 15 & 18 & 21 \end{bmatrix}$ *to the canonical form.*

Solution:

We have $\begin{bmatrix} 13 & 16 & 19 \\ 14 & 17 & 20 \\ 15 & 18 & 21 \end{bmatrix}$

$$A \sim \begin{bmatrix} 13 & 16 & 19 \\ 1 & 1 & 1 \\ 1 & 1 & 1 \end{bmatrix},$$ replacing R_2 and R_3 by $R_2 - R_1$ and $R_3 - R_2$ respectively.

$$\sim \begin{bmatrix} 13 & 3 & 3 \\ 1 & 0 & 0 \\ 1 & 0 & 0 \end{bmatrix},$$ replacing C_2 and C_3 by $C_2 - C_1$, $C_3 - C_2$ respectively.

$$\sim \begin{bmatrix} 13 & 3 & 0 \\ 1 & 0 & 0 \\ 1 & 0 & 0 \end{bmatrix},$$ replacing C_3 by $C_3 - C_2$.

$$\sim \begin{bmatrix} 13 & 3 & 0 \\ 1 & 0 & 0 \\ 1 & 0 & 0 \end{bmatrix},$$ replacing R_3 by $R_3 - R_2$.

$$\sim \begin{bmatrix} 3 & 13 & 0 \\ 0 & 1 & 0 \\ 0 & 0 & 0 \end{bmatrix},$$ int erchanging C_1 and C_2.

$$\sim \begin{bmatrix} 3 & 0 & 0 \\ 0 & 1 & 0 \\ 0 & 0 & 0 \end{bmatrix},$$ replacing C_2 by $C_3 - (13/3)C_1$.

$$\sim \begin{bmatrix} 1 & 0 & 0 \\ 0 & 1 & 0 \\ 0 & 0 & 0 \end{bmatrix},$$ replacing R_1 by $\frac{1}{3} R_1$.

$$\sim \begin{bmatrix} I_2 & 0 \\ 0 & 0 \end{bmatrix},$$ which is the required canonical form

Example 63:

Reduce $A = \begin{bmatrix} 1 & 2 & -1 & 4 \\ 2 & 4 & 3 & 5 \\ 1 & 2 & 3 & 5 \\ -1 & -2 & 6 & -7 \end{bmatrix}$ *to the canonical form.*

Solution:

$$A \sim \begin{bmatrix} 1 & 2 & -1 & 4 \\ 0 & 0 & 5 & -3 \\ 0 & 0 & 4 & 0 \\ 0 & 0 & 5 & -3 \end{bmatrix},$$ replacing R_2, R_3 and R_4 by $R_2 - 2R_1$, $R_3 - R_1$ and $R_4 + R_1$ respectively.

$$\sim \begin{bmatrix} 1 & 0 & 0 & 0 \\ 0 & 0 & 5 & -3 \\ 0 & 0 & 4 & 0 \\ 0 & 0 & 5 & -3 \end{bmatrix},$$ replacing C_2, C_3 and C_4 by $C_2 - 2C_1$, $C_3 + C_1$ and $C_4 - 4C_1$ respectively.

$$\sim \begin{bmatrix} 1 & 0 & 0 & 0 \\ 0 & 0 & 5 & -3 \\ 0 & 0 & 1 & 0 \\ 0 & 0 & 0 & 0 \end{bmatrix},$$ replacing R_3 and R_4 by $\frac{1}{4}R_3$ and $R_4 - R_3$ respectively.

$$\sim \begin{bmatrix} 1 & 0 & 0 & 0 \\ 0 & -3 & 5 & 0 \\ 0 & 0 & 1 & 0 \\ 0 & 0 & 0 & 0 \end{bmatrix},$$ interchanging C_2 and C_4

$$\sim \begin{bmatrix} 1 & 0 & 0 & 0 \\ 0 & -3 & 0 & 0 \\ 0 & 0 & 1 & 0 \\ 0 & 0 & 0 & 4 \end{bmatrix},$$ replacing R_2 by $R_2 - 5R_3$.

or $\sim\begin{bmatrix} 1 & 0 & 0 & 0 \\ 0 & 1 & 0 & 0 \\ 0 & 0 & 1 & 0 \\ 0 & 0 & 0 & 0 \end{bmatrix}$, replacing R_2 by $-\frac{1}{3}R_2$

$\sim\begin{bmatrix} I_3 & 0 \\ 0 & 0 \end{bmatrix}$, which is the required canonical form

Example 64:

Let $a > 0$, $d > 0$. Find the value of the determinant

$$\begin{vmatrix} \frac{1}{a} & \frac{1}{a(a+d)} & \frac{1}{(a+d)(a+2d)} \\ \frac{1}{(a+d)} & \frac{1}{(a+d)(a+2d)} & \frac{1}{(a+2d)(a+3d)} \\ \frac{1}{(a+2d)} & \frac{1}{(a+2d)(a+3d)} & \frac{1}{(a+3d)(a+4d)} \end{vmatrix}$$

Solution:

Let us denote the given determinant by Δ. Taking

$\frac{1}{a(a+d)(a+2d)}$ common from R_1,

$\frac{1}{(a+d)(a+2d)(a+3d)}$ from R_2 and

$\frac{1}{(a+2d)(a+3d)(a+4d)}$ from R_3, we get

$$\Delta = \frac{1}{a(a+d)^2(a+2d)^3(a+3d)^2(a+4d)}\Delta_1$$

where $\Delta_1 = \begin{vmatrix} (a+d)(a+2d) & a+2d & a \\ (a+2d)(a+3d) & a+3d & a+d \\ (a+3d)(a+4d) & a+4d & a+2d \end{vmatrix}$

Applying $R_3 \to R_3 - R_2$ and $R_2 \to R_2 - R_1$, we get

$$\Delta_1 = \begin{vmatrix} (a+d)(a+2d) & d & a \\ (a+2d)(2d) & d & d \\ (a+3d)(2d) & d & d \end{vmatrix}$$ Applying $R_3 \to R_3 - R_2$, we get

$$\Delta_1 = \begin{vmatrix} (a+d)(a+2d) & a+2d & a \\ (a+2d)(2d) & d & d \\ 2d^2 & 0 & 0 \end{vmatrix}$$ Expanding along R_3, we get

$$\Delta_1 = (2d)^2 \begin{vmatrix} a+2d & a \\ d & d \end{vmatrix} = (2d^2)(d)(a + 2d - a) = 4d^2$$

Thus $\Delta = \dfrac{4d^4}{a(a+d)^2(a+2d)^3(a+3d)^2(a+4d)}$

Example 65:

Show that

$$\Delta = \begin{vmatrix} \cos(\alpha-\beta) & \cos(\beta-\gamma) & \cos(\gamma-\alpha) \\ \cos(\alpha+\beta) & \cos(\beta+\gamma) & \cos(\gamma+\alpha) \\ \sin(\alpha+\beta) & \sin(\beta+\gamma) & \sin(\gamma+\alpha) \end{vmatrix}$$

$$= -2 \sin(\beta - \gamma) \sin(\gamma - \alpha) \sin(\alpha - \beta)$$

Solution:

Applying $R_1 \to R_1 - R_2$, we get

$$\Delta = \begin{vmatrix} 2\sin\alpha\sin\beta & 2\sin\beta\sin\gamma & 2\sin\gamma\sin\alpha \\ \cos(\alpha+\beta) & \cos(\beta+\gamma) & \cos(\gamma+\alpha) \\ \sin(\alpha+\beta) & \sin(\beta+\gamma) & \sin(\gamma+\alpha) \end{vmatrix}$$

Taking 2 common from R_1 and applying $R_2 \to R_2 + R_1$, we get

$$\Delta = 2\begin{vmatrix} \sin\alpha\sin\beta & \sin\beta\sin\gamma & \sin\gamma\sin\alpha \\ \cos\alpha\cos\beta & \cos\beta\cos\gamma & \cos\gamma\cos\alpha \\ \sin(\alpha+\beta) & \sin(\beta+\gamma) & \sin(\gamma+\alpha) \end{vmatrix}$$

Taking $\sin\alpha \sin\beta$ common from C_1, $\sin\beta \sin\gamma$ from C_2 and $\sin\gamma \sin\alpha$ from C_3 and using the fact that $\sin(x + y)/\sin x \sin y = \cot x + \cot y$, we get

$$\Delta = 2\sin^2\alpha\sin^2\beta\sin^2\gamma\begin{vmatrix} 1 & 1 & 1 \\ \cot\alpha\cot\beta & \cot\beta\cot\gamma & \cot\gamma\cot\alpha \\ \cot\alpha+\cot\beta & \cot\beta+\cot\gamma & \cot\gamma+\cot\alpha \end{vmatrix}$$

Applying $C_2 \to C_2 - C_1$ and $C_3 \to C_3 - C_1$, we get

$\Delta = 2\sin^2\alpha\sin^2\beta\sin^2\gamma\Delta_1$

$$\text{where } \Delta_1 = \begin{vmatrix} 1 & 0 & 1 \\ \cot\alpha\cot\beta & \cot\beta(\cot\gamma-\cot\alpha) & \cot\alpha(\cot\gamma-\cot\beta) \\ \cot\alpha+\cot\beta & \cot\gamma-\cot\alpha & \cot\gamma-\cot\beta \end{vmatrix}$$

Expanding along R_1, we get

$$\Delta = 2\sin^2\alpha\sin^2\beta\sin^2\gamma(\cot\gamma-\cot\alpha)\ (\cot\gamma-\cot\beta)\begin{vmatrix} \cot\beta & \cot\alpha \\ 1 & 1 \end{vmatrix}$$

$$= 2\sin^2\alpha\sin^2\beta\sin^2\gamma(\cot\gamma-\cot\alpha)(\cot\gamma-\cot\beta)(\cot\beta-\cot\alpha)$$

$$= 2\sin^2\alpha\sin^2\beta\sin^2\gamma\ \frac{\sin(\alpha-\gamma)\sin(\beta-\gamma)\sin(\alpha-\beta)}{\sin^2\alpha\sin^2\beta\sin^2\gamma}$$

$$= -2\sin(\beta-\gamma)\sin(\gamma-\alpha)\sin(\alpha-\beta).$$

Example 66:

By differentiation or otherwise, show that the determinant

$$\begin{vmatrix} \sin(x+\alpha) & \cos(x+\alpha) & a+x\sin\alpha \\ \sin(x+\beta) & \cos(x+\beta) & b+x\sin\beta \\ \cos(x+\gamma) & \cos(x+\beta) & c+x\sin\gamma \end{vmatrix}$$ *is independent of x.*

Solution:

Let us denote the given determinant as $\Delta(x)$. Then

$$\Delta'(x) = \begin{vmatrix} \cos(x+\alpha) & \cos(x+\alpha) & a+x\sin\alpha \\ \cos(x+\beta) & \cos(x+\beta) & b+x\sin\beta \\ \cos(x+\gamma) & \cos(x+\beta) & c+x\sin\gamma \end{vmatrix}$$

$$+\begin{vmatrix} \sin(x+\alpha) & -\sin(x+\alpha) & a+x\sin\alpha \\ \sin(x+\beta) & -\sin(x+\beta) & b+x\sin\beta \\ \sin(x+\gamma) & -\sin(x+\beta) & c+x\sin\gamma \end{vmatrix} + \begin{vmatrix} \sin(x+\alpha) & \cos(x+\alpha) & \sin\alpha \\ \sin(x+\beta) & \cos(x+\beta) & \sin\beta \\ \sin(x+\gamma) & \cos(x+\beta) & \sin\gamma \end{vmatrix}$$

The first two determinants are zero. In the third, apply $C_1 \to C_1 - (\cos x)\, C_2$ and $C_2 \to C_2 + (\sin x)\, C_3$.

$$\Delta'(x) = \begin{vmatrix} \sin x\cos\alpha & \cos x\cos\alpha & \sin\alpha \\ \sin x\cos\beta & \cos x\cos\beta & \sin\beta \\ \sin x\cos\gamma & \cos x\cos\beta & \sin\gamma \end{vmatrix}$$

$= 0 \quad [\because C_1 \text{ and } C_2 \text{ are proportional}]$

This show that D(x) is a constant and hence independent of x.

Example 67:

Show that

$$\Delta = \begin{vmatrix} -bc & b^2+bc & c^2+bc \\ a^2+ac & -ac & c^2+ac \\ a^2+ab & b^2+ab & -ab \end{vmatrix} = (bc+ca+ab)^2$$

Solution:

If any of a, b, c, is zero, the result is trivial.

Applying $R_1 \to aR_1$, $R_2 \to bR_2$ and $R_3 \to cR_3$, we get

$$\Delta = \frac{1}{abc}\begin{vmatrix} -abc & ab^2+abc & ac^2+abc \\ a^2b+abc & -abc & bc^2+abc \\ a^2c+abc & b^2c+abc & -abc \end{vmatrix}$$

Applying $C_1 \to (1/a)C_1$, $C_2 \to (1/b)C_2$ and $C_3 \to (1/c)C_3$, we get

$$\Delta = \frac{abc}{abc}\begin{vmatrix} -bc & ab+ac & ac+ab \\ ab+bc & -ac & bc+ab \\ ac+bc & bc+ac & -ab \end{vmatrix}$$

Applying $R_1 \to R_1 + R_2 + R_3$ and taking ab + bc + ca common from R_1, we get

$$\Delta = (bc+ca+ab)\begin{vmatrix} 1 & 1 & 1 \\ ab+bc & -ac & bc+ab \\ ac+bc & bc+ac & -ab \end{vmatrix}$$

Applying $C_2 \to C_2 - C_1$ and $C_3 \to C_3 - C_1$, we get $\Delta = (ab + bc + ca)$

$$\begin{vmatrix} 1 & 0 & 0 \\ ab+bc & -(ab+bc+ac) & 0 \\ ac+bc & 0 & -(ab+bc+ca) \end{vmatrix}$$

Expanding along R_1, we get $\Delta = (bc + ca + ab)^3$.

Example 68:

Expand $\begin{vmatrix} a & b & c & d \\ e & f & g & h \\ 0 & 0 & i & k \\ 0 & 0 & l & m \end{vmatrix}$ *by Laplace's expansion by the minors of the first two columns.*

Solution:

All the possible minors of the first two columns and their complementary minors are given by:

$$|B_1| = \begin{vmatrix} a & b \\ e & f \end{vmatrix}; \; |B'_1| = \begin{vmatrix} j & k \\ l & m \end{vmatrix};$$

$$|B_2| = \begin{vmatrix} a & b \\ 0 & 0 \end{vmatrix} = 0, \text{ hence } |B'_2| \text{ need not be calculated}$$

Similarly $|B_3| = \begin{vmatrix} a & b \\ 0 & 0 \end{vmatrix} = 0; \; |B_4| = \begin{vmatrix} e & f \\ 0 & 0 \end{vmatrix} = 0;$

$|B_5| = \begin{vmatrix} e & f \\ 0 & 0 \end{vmatrix} = 0$ and $|B_6| = \begin{vmatrix} 0 & 0 \\ 0 & 0 \end{vmatrix} = 0$ and therefore their complementary minors need not be calculated.

Then the given determinant by Laplace's Expansion method

$$= \begin{vmatrix} a & b \\ e & f \end{vmatrix} . \begin{vmatrix} j & k \\ l & m \end{vmatrix}$$

EXERCISES

1. *Evalute* $\begin{vmatrix} x + a & x + 2a & x + 3a \\ x + 2a & x + 3a & x + 4a \\ x + 4a & x + 5a & x + 6a \end{vmatrix}$

2. *Calculate the value of the determinant* $\begin{vmatrix} 7 & 13 & 10 & 6 \\ 5 & 9 & 7 & 4 \\ 8 & 12 & 11 & 7 \\ 4 & 10 & 6 & 3 \end{vmatrix}$

3. *Show that* $\begin{vmatrix} a-b & b-c & c-a \\ b-c & c-a & a-b \\ c-a & a-b & b-c \end{vmatrix} = 0$

4. *Show that* $\begin{vmatrix} 21 & 17 & 7 & 10 \\ 24 & 22 & 6 & 10 \\ 6 & 8 & 2 & 3 \\ 5 & 7 & 1 & 2 \end{vmatrix} = 0$

5. *Evaluate* $\begin{vmatrix} 1 & a & b+c \\ 1 & b & c+a \\ 1 & c & a+c \end{vmatrix}$

6. *Show that* $\begin{vmatrix} 29 & 26 & 22 \\ 25 & 31 & 27 \\ 63 & 54 & 46 \end{vmatrix} = 132$.

7. *Prove that* $\begin{vmatrix} -2a & a+b & c+a \\ b+a & -2b & c+b \\ c+a & c+b & -2c \end{vmatrix} = 4(a+b)(b+c)(c+a)$

4

AN INTRODUCTION TO RINGS AND FIELDS

RINGS

Definitions and Concepts : As mentioned in our goal, we would like to investigate algebraic systems whose structure imitates that of the integers.

Example 1:

*[**Z**; + , .] is a ring, where + and stand for regular addition and multiplication on **Z**. We already know that [**Z**; +] is an abelian group, so we need only check parts 2 and 3 of the definition of a ring. From elementary algebra, we know that the associative law under multiplication and the distributive laws are true for **Z**. This is our main example of an infinite ring.*

Example 2:

[$\mathbf{Z}_n$; $+_n$, $\times_n$] is a ring. The properties of modular arithmetic on $\mathbf{Z}_n$, and they give us the information we need to convince ourselves that [$\mathbf{Z}_n$; $+_n$, $\times_n$] is a ring. This example is our main example of finite rings of different orders.

POLYNOMIAL RINGS

In the previous sections we mentioned the solution of equations over different rings and field. To solve the equation $x^2 - 2 = 0$ over the field of the real numbers means to find all solutions of this equation which are in this particular field **R**. This statement can be replaced as follows: Determine all $x \in \mathbf{R}$ such that the polynomial $f(x) = x^2 - 2$ is equal to zero, or, equivalently, find all zeros of the polynomial $f(x) = x^2 - 2$ is equal to zero, or, equivalently, find all zeros of the polynomial $f(x) = x^2 - 2$ in **R**. The process of solving equations is frequently expressed via polynomials. For this reason, this section will concentrate on polynomials. We will develop concepts using the general setting of polynomials over rings since results proven over rings are true for fields (and integral domains). The reader should keep in mind that in most cases we are just formalizing concepts which he or she learned in high-school algebra over the field of reals.

Definition: P*olynomial over R. Let [R; + , ·] be a ring. A polynomial, f (x) over R is an expression of the form.*

$$f(x) = \sum_{i=0}^{n} a_i x^i = a_0 + a_1 x + a_2 x^2 + \dots + a_n x^n, n \geq 0$$

where a_0, a_1, a_2,..., $a_n \in R$ and $a_m = 0$ when $m > n$. If $a_n \neq 0$, then the degree of f (x) is n. If f (x) = 0, then the degree of f(x) is $-\infty$. If the degree of f (x) is n, we write deg (f(x)) =n.

Comment:

(1) The above definition is that of a polynomial in the single variable, or "indeterminate" x. One can also define polynomials is more than one indeterminate.

(2) The set of all polynomials in the indeterminate x with coefficient in R is denoted by R [x].

(3) Note that $R \subseteq R[x]$. The element of R are called constant polynomials.

(4) R is called the *ground ring* for R[x].

In other words, we considered sets which were closed w.r.t. a single operation. Here we shall study an important type of algebraic structure with two binary operation known as a Rings. A Rings has its origin in the set of integrals which is closed w.r.t. both the operation of ordinary addition and ordinary multiplication.

Definition: *Let R is a non-empty set equipped with two binary operations called addition and denoted by '+' and '.' respectively i.e., for all a, b, ∈ R we have a + b ∈ R and a.b ∈ R. Then this algebraic structure (R, +, .) is called a right if the following postulates are satisfied:*

1. *Addition of associative, i.e.,*

   $(a + b) + c = a + (b + c) \ \forall a, b, c \in R.$

2. *Addition is commutative, i.e.,* $a + b = b + a \ \forall a, b, \in R.$

3. *There exists an element denoted by 0 in R such that*

   $0 + a = a \ \forall a \in R.$

4. *To each element a in R there exists an element –a in R such that*

   $(-a) + a = 0.$

5. *Multiplication is associative, i.e.,*

   $a.\ (b,\ c) = (a.b)\ .\ c\ \forall\ a,\ b,\ c \in R.$

6. Multiplication is distributive with respect to addition, *i.e.*, for all a, b, c in R,

$$\left.\begin{array}{l} a.\left(b + c\right) = a.b + a.c. \\ \left(b + c\right)a = b.a + c.\ a \end{array}\right\}$$

and

Sine addition is commutative in R, therefore we shall have $0 \in R$ such that $0 + a = a + 0\ \forall\ a \in R$.

Also if $a \in R$. Then we shall have $(-a) + a = 0 = a + (-a)$. Thus, R will be an abelian group w.r.t. addition. The element $0 \in R$ will be the additive identity it's called the zero element of the *Rings*. Since in a group the identity element is unique. Therefore every ring will passes unique zero element and it will be the additive composition. We shall always denote this element by the symbol 0.

Notes:

1. When there is no confusion about the operations on R we simply say that R is a ring
2. In a ring R, the additive identity is called the zero of the ring and the additive inverse of $a \in R$ is called the negative of a to be denoted by $-a$.

Also we define $a - b = a + (-b)\ \forall\ a, b, \in R$

The equation $a + x = b$ will have a unique solution in R and it will be $x = b - a$. Obviously $a + (b - a) = a + [b + (-a)]$

$= a + [(-a) + b] = [a + (-a)] + b = [(-a) + a] + b = 0 + b = b.$

Similarly, the equation $y + a = b$ will have a unique solution in R and it will be $y = b - a$.

Both the cancellation laws will hold good for addition in R *i.e.*, for all a, b, c in R

$a + b = a + c \Rightarrow b = c$ and $b + a = c + a \Rightarrow b = c.$

If in a ring we have $a + b = 0$, then $a = -b = i$, then $a = -b$ and $b = a$.

Example 1:

*$f(x) = 3$, $g(x) = 2 - 4x + 7x^2$, and $h(x) = 2 + x^4$ are all polynomials in **Z**[x]. Their degrees are 0, 2 and 4, respectively.*

Addition and multiplication of polynomials are perform as in high school algebra. However, we must do our computations in the ring over which we are considering the polynomials.

Example 2:

In $Z_2[x]$, if $f(x) = x + 1$ and $g(x) = x + 1$, then $f(x) + g(x) = (x + 1) + (x + 1) = (1 +_2 1)x + (1 +_2 1) = 0x + 0 = 0$ and $f(x)g(x) = (x + 1) \times (x + 1) = x^2 + (1 +_2 1)x + 1 = x^2 + 1$.

However, for the same polynomials as above, f(x) and g(x) in **Z**(x), we have $f(x) + g(x) = (x + 1) + (x + 1) = (1 + 1)x + (1 + 1) = 2x + 2$ and $f(x)\,g(x) = x^2 + (1 + 1)x + 1 = x^2 + 2x + 1$.

The important fact to keep in mind is that addition and multiplication in R[x] depend solely on addition and multiplication in R. The x's merely serve the purpose of "place holders." All computations are done over the given rings.

DEGREE OF A POLYNOMIAL

Let $f(x) = a_0x^0 + a_1x + a_2x^2 + a_3x^3 + \ldots + a_nx^n + \ldots$

be a polynomial over an arbitrary ring R. We say that n is the degree of the polynomial f (x) if and only if $a_n \neq 0$ and $a_m = 0$ for all $m > n$. We shall write deg f (x) is the largest non-negative integer i for which the ith coefficient of f (x) is not 0. If in the polynomial f (x), a_0 (*i.e.*, the coefficient of x^0) is not 0 and all the other coefficients are 0, then according to our definition, the degree of f (x) will be zero. Also according to our definition, if there is no non-zero coefficient in f (x), then its degree will remain undefined. Then we do not define the degree of the zero polynomial. Also it is obvious that very non-zero polynomial will possess a unique degree.

Note: If $f(x) = a_0x^0 + a_1x + a_2x^2 + \ldots + a_n x^n + \ldots$ is a polynomial of degree n *i.e.*, if $a_n \neq 0$ and $a_m = 0$ for all $m > n$, then it is convenient to write $f(x) = \sum_{i-0}^{n} a_ix^i = a_0x^0 + a_1x + a_2x^2 + \ldots + a_nx^n$. It will remain understood that all the terms in f (x) which follow the term a_nx^n, have zero coefficients. Also we shall call a_nx^n as the *leading term* and a_nx^0 is called the *constant term* and a_0 is called the $zero^{th}$ coefficient of f (x). For example $f(x) = 2x^0 + 3x - 4x^2 + 4x^3 - 8x^4$ is a polynomial of degree 4 over the ring of integers. Here –8 is the leading coefficient and 2 is the $zero^{th}$ coefficient.

The coefficients of all terms which contain powers of x greater than 4 will be regarded as zero. Similarly $g(x) = 3\,x^0$ is a polynomial of degree zero over the ring of integers. In this polynomial the coefficients of x, x^2,

x^3,... are all equal to zero. The zero. The zero polynomial over an arbitrary ring R will be represented by $0x^0$.

Set of constant polynomials over a ring. Let R be an arbitrary ring and R [x] the set of all polynomials over r. Let R' denote the set of all polynomials over R whose coefficients are all zero except for the constant term, which may be either zero or non-zero. That is,

$$R' = \{ax^0 : a \in R\}.$$

Then R' will be called as the set of constant polynomials in R [x].

Thus all the polynomials of degree 0 as well as the zero polynomial will be called as constant polynomials.

DEGREE OF THE SUM AND THE PRODUCT OF TWO POLYNOMIALS

Theorem:

Let f (x) and g (x) be two non-zero polynomials over an arbitrary ring R. Then

1. deg [f (x) + g (x)] ≤ Max [deg f (x), deg g (x)], if f (x) g (x) ≠ 0.
2. deg [f (x) + g (x)] ≤ deg f (x), deg g (x) if f (x) g (x) ≠ 0.

Proof:

Let $f(x) = a_0x^0 + a_1x + a_2x^2 + ... + a_nx^n,\ a_n \neq 0$

and $g(x) = b_0x^0 + b_1x + b_2x^2 + ... + b_mx^m,\ a_m \neq 0$

be two element of R [x].

Here deg f (x) = n and deg g (x) = m.

From our definition of the sum of two polynomials, it is obvious that if f (x) + g (x) ≠ 0, then

$$\deg[f(x) + g(x)] = \begin{cases} \max(n,m) \text{ if } \neq m \\ n \text{ if } = m \text{ and } a_n + b_m \neq 0 \\ < n \text{ if } n = m \text{ and } a_n + b_m = 0. \end{cases}$$

Again $f(x)\,g(x) = (a_0b_0)\,x^0 + (a_0b_0 + a_1b_0)\,x + ... + a_nb_m\,x^{n+m}$.

Suppose f (x) g (x) ≠ 0. Then f (x) g (x) has a unique degree.

If $a_nb_m \neq 0$, then deg [f (x) g (x)] = n + m deg f (x) + deg g (x).

Also if $a_nb_m = 0$, then deg [f (x) g (x)] < n + m.

Cor. 1: *Suppose D is an integral domain and f (x) g (x) are two non-zero elements of D [x]. Then* *deg [f (x) g (x)] = deg f (x) + deg g (x).*

Proof:

Since $a_n \neq 0$, $b_m \neq 0$, therefore $a_n b_m \neq 0$ because in an integral domain the product of two non-zero elements cannot be zero. Hence deg [f (x) g (x)] = m + n.

Cor 2: *If F is afield and f (x), g (x) are two non-zero elements of F [x], then deg [f (x) g (x)] = deg f (x) = deg g (x).*

Proof:

Since a field is also free from zero divisors, therefore $a_n b_m \neq 0$ when $a_n \neq 0$ and $b_m \neq 0$. Hence the result.

VARIOUS TYPE OF RING

1. **Commutative Ring:** *If in a ring R, the multiplication composition is also commutative i.e., if we have a. b = b. a $\forall$ a, b $\in$ R, then R is called a commutative ring.*
2. **A Ring without Zero Divider:** *If it is not possible to find two non-zero element of R, whose product is zero i.e., R is without zero dividers when {ab = 0 $\Rightarrow$ a = 0 b = 0}*
3. **A Ring with Zero Divider:** *If it is possible to find at least two element a and b of R such that a $\neq$ 0, b $\neq$ 0 and ab = 0*
4. **Ring with Unity:** *If in a ring R there exists an element denoted by 1 such that 1. a = a = a . 1 $\forall$ a $\in$ R, then R is called a ring with unit element. The element 1 $\in$ R is called the unit element of the ring. Obviously 1 is the multiplicative identity of R. Thus if a ring possesses multiplicative identity, then it is a ring with unity.*

Note: In future we shall denote the multiplication composition in a ring R not by the symbol '.' but by multiplicative notation. Thus we shall write ab in place of a. b.

ELEMENTARY PROPERTIES OF A RING

Theorem:

If R is a ring, then for all a, b, c $\in$ R

1. a0 = 0a = 0.
2. a(–b) = – (ab) = (–a) b.
3. (–a) (–b) = ab.
4. a (b – c) = ab – ac

5. $(b - c)\ a = ba - ca$.

Proof:

1. We know that

$a0 = a(0 + 0)$ $[\because\ 0 + 0 = 0]$

$= a0 + a0.$ [by left distributive law]

$\therefore\ 0 + a0 = a0 + a0.$ $[\because\ a0 \in R$ and $0 + a0 = a0]$

We have R is a group with respect to addition, therefore applying right cancellation law for addition in R we get $0 = a0$.

Similarly we have $0a = (0 + 0)\ a$

$= 0a + 0a.$ [by right distributive law]

$\therefore\ 0 + 0a = 0a + 0a$ $[\because\ 0 + 0a = 0a]$

Applying right cancellation law for addition in R, we get $0 = 0a$.

2. We know that $a[(-b) + b + b] = a0$ $[\because\ -b + b = 0]$

$\Rightarrow a\ (-b) + ab = 0$ [by using left distributive law and the result (1)]

$\Rightarrow a\ (-b) = -\ (ab)$, since in a ring $a + b\ 0 \Rightarrow a = -\ b$.

Similarly we know that $(-a + a)\ b = 0b$

$\Rightarrow (-a)\ b + ab = 0$

$\Rightarrow (-a)\ b = -\ (ab)$,

since in a ring $a + b = 0 \Rightarrow a = -\ b$.

3. We know that $(-\ a)\ (-\ b) = -\ [(-\ a)\ b]$, since $a\ (-\ b) = -\ (ab) =$
$-\ [-(ab)]$, since $(-\ a)\ b = -\ (ab)$

$= ab$, since R is a group with respect to addition and in a group we have

$-(-\ a) = a$.

4. We have $a\ (b - a) = a\ [b + (-c)]$

$= ab + a\ (-\ c)$ [by left distributive law]

$= ab + [-(ac)]$ $[\because\ a\ (-c) - -\ (ac)]$

$= ab - ac$.

5. We know that

$(b - c)\ a = [b + (-c)]\ a$

$= ba + (-c)\ a$ [by right distributive law]

$= ba + [-(ca)] = ba - ca$.

INTEGRAL MULTIPLES OF THE ELEMENTS OF A RING

R is a ring and $a \in R$. If m is a positive integer, then we define ma = a + a + a + ... upto m terms.

Also we define 0a = 0. Here 0 on the left hand side is the integer zero and 0 on the right hand side is the zero element (additive identity) of the ring.

If m is a positive integer, then –m is a negative integer. We define (–m) a = – (ma) = – (a + a + a + ... upto m terms)

= (–a) + (–a) + (–a) + ... upto m terms = m (–a).

If m and n are any integers, we can prove that

ma + na = (m + n) a, m (na)a, and (ma) (na) = (mn) a^2.

Also we can prove that for any integer m, we have

$m (a + b) = ma + mb \ \forall\ a, b \in R.$

The students should not confuse that this is the distributive law we assumed in the postulates for a ring. It is not so. Here m is an integer and a, b are elements or R.

SOME SPECIAL TYPES OF RINGS, RINGS WITH OR WITHOUT ZERO DIVISORS

Definition: *A non-zero element a of ring R is called a zero divisor or a divisor of zero if there exists an element $b \neq 0 \in R$ such that either ab = 0.*

Ring Without Zero Divisors

A ring R is without zero divisors if the product of no two non-zero element of R is zero i.e., if

$$ab = 0 \Rightarrow a = 0 \text{ or } b = 0.$$

On the other hand we can say that, if in a ring R there exist non-zero elements a and b such that ab = 0, then R is said to be a ring with zero divisors.

Commutative Ring

A ring R is said to be commutative if the multiplication composition in R is commutative *i.e.,* of $a.b = b.a \ \forall\ a, b \in R$

Ring without Element

A ring R is said to be a ring with unity element if R has a multiplication identity, that is if a there exist an element in R denoted by 1 such that $1.a = a.1 = a \ \forall\ a \in R$

CANCELLATION LAWS IN A RING

Let R is a ring then R is an abelian group with respect to addition. For addition composition the cancellation laws hold in all rings. Therefore the question of cancellation laws holding in a ring arises only for the multiplication composition.

We say that cancellation laws hold in a ring R if

$$a \neq 0,\ ab = ac \Rightarrow b = 0$$

and $\quad a \neq 0,\ ab = ac \Rightarrow b = 0$ where $a, b, c \in R$.

Theorem:

A ring R is without zero divisors if and only if the cancellation laws hold in R i.e., R is without zero divisors $\Leftrightarrow$ cancellation laws hold in R.

Proof:

Let R has no zero divisors. Let a, b, c be any three elements of R such that $a \neq 0$, $ab = ac$.

We have $ab = ac \Rightarrow ab - ac = 0 \Rightarrow a(b - c) = 0$.

Since R is without zero divisors, therefore we have

$a(b - c) = 0$ and $a \neq 0 \Rightarrow b - c = 0$ *i.e.*, $b = c$.

Thus the left cancellation law holds in R. Similarly, we can prove that the right cancellation law holds in R.

Conversely suppose that the cancellation laws hold in R. If possible let $ab = 0$, $a \neq 0$, $b \neq 0$.

Then we have $ab = a0$, since $a0 = 0$.

Now $a \neq 0$, $ab = a0 \Rightarrow b = 0$ by left cancellation law.

Thus we get a contradiction. Hence R is without zero divisors.

INTEGRAL DOMAIN, FIELDS, DIVISION, RINGS, INTEGRAL, DOMAIN

Definition: *A ring is called an integral domain if it (1) is commutative, (2) has unit element, (3) is without zero divisors.*

A ring known as an integral domain if:

(i) It is commutative.

(ii) It has unit element.

(iii) It is without zero divisor.

An integral domain is a commutative ring with unity, which has no proper divisor or zero.

Example:

Ring I as integers is an example of integral domain. We have proved that I is a commutative ring with unity. Also I does not possess zero divisors. We know that if a, b are integers such that ab = 0, then either a or b must be zero.

Also it can be seen that the algebraic structures (C, +, .), (Q, +, .), (R, +, .) are all integral domains. As an example of a finite integral domain we have the ring ({0, 1, 2, 3, 4}, $+_5$, $\times_5$).

Inversible Elements in a Ring with Unity: *If R is a ring with unity, then a element a* $\in$ *R is called inversible, if there exists b* $\in$ *R such that ab = 1 = ba. Also then we write* $b = a^{-1}$.

Example:

(i) 1 and –1 are the only two inversible elements of the ring of all integers.

(ii) n $\times$ n non-singular matrices with real numbers as elements are the only inversible elements of the ring of all n $\times$ n matrices with elements as real numbers.

ISOMORPHISM OF RINGS

A ring R is said to be isomorphic to another ring S' if there exists a one-one mapping f of R onto R' such that

$$f(a+b) = f(a) + f(b),\ f(ab) = f(a)\, f(b)\ \forall\ a,\ b \in S$$

Also such a mapping f is said to be an isomorphism of R onto S'.

A mapping f : S $\rightarrow$ S' is an isomorphism of S onto S' if f is one-one and if f(a + b) = f(a) + f(b), f(ab) = f(a) f(b) $\forall$ a, b $\in$ S. For f to be an isomorphism of R onto R', it must be one-one onto and it must preserve both addition and multiplication compositions in R and R'. There may be more than one isomorphisms of R onto R' if R is isomorphic to S'.

If a ring R is isomorphic to another ring R', we shall write in symbols S $\cong$ S'. Also R' is said to be an isomorphic image of S.

RELATION OF ISOMORPHISM IN THE SET OF ALL RINGS

Example:

Let R be the ring of integers under ordinary addition and multiplication. Let S' be the set of all even integers. Let us define multiplication in S' to be denoted by '' by the relation*

$$a * b = \frac{ab}{2},$$

where ab is the ordinary multiplication of two integers a and b.

1. *Prove that (S, +, ∗) is a commutative ring where + stands for ordinary addition of integers.*
2. *Prove that S is isomorphic to S′*
3. *What acts as the unit element of S′ ?*

Solution:

Obviously S' is an abelian group with respect to addition.

If a and b are both even integers then $\frac{ab}{2}$ is also an even integer.

Therefore $a * b = \left(\frac{ab}{2}\right) \in S' \ \forall\ a, b, \in S'$.

Thus S' is closed with respect to ∗.

Also if a, b, c are any elements of R', then

$$a * (b * c) = a * \left(\frac{ab}{2}\right) = \frac{a\,(bc/2)}{2} = \frac{(ab/2)\,c}{2} = \left(\frac{ab}{2}\right) * c = (b * c) * c.$$

∴ ∗ is associative.

Further $(a * b) = \frac{ab}{2} = \frac{ba}{2} = (b * a)$.

∴ ∗ is commutative.

Again $a * (b + c) = \frac{a(b+c)}{2} = \frac{ab}{2} + \frac{ac}{2} = (a * b) + (a * c)$.

Similarly $(b + a) * a = (b * a) + (c * a)$.

∴ ∗ is distributive with respect to +.

∴ (S, +, ∗) is a commutative ring.

2. We shall now show that S is isomorphic to S'. Consider the mapping f defined by

   $f : S \to S'$ such that $f(x) = 2x \ \forall\ x \in S$.

The mapping f is obviously one-one onto.

Also $\forall\ x_1, x_2 \in S$ we have

$$f(x_1\ x_2) = 2\,(x_1 + x_2) = 2x_1 + 2x_2 = f(x_1) + f(x_2)$$

and $f(x_1x_2) = 2\,(x_1x_2) = \dfrac{(2x_1)(2x_2)}{2} = (2x_1) * (2x_2) = f(x_1) * f(x_2)$.

∴ The mapping f is an isomorphism of S onto S'.

3. Here 1 is the unit element of S. We have f(1) = 2. Since f is an isomorphism of S onto S', therefore 2 must be the unity element of R'. We have for all $a \in S'$, $2 * a = \dfrac{2a}{2} = a = a * 2$.

∴ 2 is the unit element of S'.

QUOTIENT RINGS OR RINGS OF RESIDUE CLASSES

Suppose R is an arbitrary ring and S is an ideal (two sided ideal) in R. Then S is a subgroup of the additive abelian group of R. We can form the cosets (right as well as left) of S in R. Since R is an abelian additive group, therefore if $a \in R$, then the right coset S + a will be equal to the corresponding left coset a + s. Thus, we shall call S + a as simply a coset of S in R. We remember from our study of cosets in group theory, that if $a, b \in R$, Then

$$S + a = S + b \Rightarrow a - b \in S.$$

The cosets of S in R are called the residue classes of S in R. We denote the set of all residue classes of S in R by the symbol R/S.

Thus $\quad R/S = \{S + a : a \in R\}$.

We shall now impose a ring structure on the set R/S by defining addition and multiplication of residue classes.

Theorem:

If S is an ideal of a ring R, then the set

$$R/S = \{S + a : a \in R\}$$

of all residue classes of S in R forms a ring for the two compositions in R/S defined as follows:

$(S + a) + (S + b) = S + (a + b)$ [*Addition of residue classes*]

$(S + a)(S + b) = S + ab.$ [*Multiplication of residue classes*]

Proof:

Since S + (a + b) and S + ab are also residue classes of S in R, therefore R/S is closed with respect to addition and multiplication of residue classes. First of all, we shall show that both addition and multiplication in R/S are well defined. For this we arc to show that if S + a = S + a' and S + b = S + b', then

$$(S + a) + (S + b) = (S + a') + (S + b')$$

and $\quad (S + a)(S + b) = (S + a')(S + b')$.

We have $\quad S + a = S + a' \Rightarrow a' \in S + a$

and $\quad S + b = S + b' \Rightarrow b' \in S + b$.

Therefore there exist $\alpha, \beta \in S$ such that $a' = \alpha + a$, $b' = \beta + b$.

Now $\quad a' + b' = (\alpha + a) + (\beta + b) = (a + b) + (\alpha + \beta)$.

$\therefore \quad (a' + b') - (a + b) = \alpha + \beta \in S$.

$\therefore \quad S + (a' + b) + (S + b') = (S + a) + (S + b)$.

Thus addition in R/S is well defined.

Again $a'b' = (\alpha + a)(\beta + b) = \alpha b + \alpha\beta + a\beta + ab$

$= ab + \alpha\beta + \alpha b + a\beta$.

$\therefore a'b' - ab = \alpha\beta + \alpha\beta + \alpha\beta \in S$. [Since S is an ideal therefore $\alpha, \beta \in S$ and $a, b \in R$

$\Rightarrow \alpha b \in S, \alpha\beta \in S, \alpha\beta \in S$ and finally $\alpha\beta + \alpha\beta + a\beta \in S$].

Now since $a'b' - ab \in S$, therefore $S + a'b' = S + ab$

$\Rightarrow (S + a')(S + b') = (S + a)(S + b)$.

Hence multiplication in R/S is also well defined.

Associativity of addition in R/S. We have

$(S + a) + [(S + b) + (S + c)] = (S + a) + [S (b + c)]$

$= S + [a + (b + c)] = S + [(a + b) + c] = [S + (a + b)] + (S + c)$

$= [(S + a) + (S + b)] + (S + c)$.

Commutativity of addition in R/S. We have

$(S + a) + (S + b) = S + (a + b) = S + (b + a) = (S + b) + (S + a)$.

Existence of additive identity. We have $S = S + 0 \in R/S$.

If $S + a \in R/S$, then $(S + 0) + (S + a) = S + (0 + a) = S + a$.

$\therefore$ S is the additive identity.

Existence of additive inverse. Let $S + a \in R/S$.

Then $S + (-a) \in R/S$. Also we have

$[S + (-a)] + [S + a] = S + [(-a) + a] = S + 0 = S$.

$\therefore S + (-a)$ or $S - a$ is the additive inverse of $S + a$.

Associativity of multiplication. We have

$[(S + a)(S + b)](S + c) = (S + ab)(S + c) = S + (ab)c$

$= S + a (bc) = (S + a) (S + bc) = (S + a) [(S + b) (S + c)].$

Distributivity of multiplication with respect to addition. We have

$(S + a) [(S + b) + (S + c)] = (S + a) [S + (b + c)]$

$= S + a (b + c) + S + (ab + ac) = (S + ab) + (S + ac)$

$= (S + a) (S + b) + (S + a) (S + c).$

Similarly, we can prove that

$[(S + b) + (S + c)] (S + a) = (S + b) (S + a) + (S + c) (S + a).$

Hence R/S is a ring with respect to the two compositions. The residue class S + 0 or S is the zero element of this ring.

Note: The students should not confuse that by the multiplication (S + a) (S + b) of residue classes we mean the totality of elements obtained on multiplication of residue classes is a new composition which we have defined in the set R/S.

However, the addition (S + a) + (S + b) of residue classes as defined by us coincides with the totality of elements obtained on adding elements of S + a to the elements of S + b as can be easily seen:

$$(S + a) + (S + b) = S + (a + S) + b = S + (S + a) + b$$
$$= (S + S) + (a + b) = S + (a + b).$$

PROPERTIES OF ISOMORPHISM OF RINGS

Theorem:

If f is an isomorphism of a ring S onto a ring S', then

1. *The image of the zero of S is the zero of S'.*
2. *The image of the negative of an element of S is the negative of the image of that element i.e, $f(-a) = -f(a)\ \forall\ a \in S$.*
3. *If S is a commutative ring, then S' is also a commutative ring.*
4. *If S is without zero divisors, then S' is also without zero divisors.*
5. *If S is with unit element, then S' is also with unity element.*
6. *If S is a field, then S' is also a field.*
7. *If S is a skew field, then S' is also a skew field.*

Proof:

1. Let $a \in S$. Then $f(a) \in S'$. Let 0 denote the zero element of S'. To prove that $f(0) = 0$.

   We have $f(a)\ 0' = f(a) = f(a + 0) = f(a) + f(0)$. By cancellation law for addition in R', we get from $f(a) + 0' = f(a) + f(0)$, the result that $0' = f(0)$.

2. We have f (a) + f (–a) = f [a + (–a)] = f (0) = 0'.

∴ f (–a) is the additive inverse of f(a) in S'. Thus

$$f(-a) = -f(a).$$

3. Let f (a) and f (b) be any two elements of S'. Then

$$a, b \in S.$$

we have f (a) f (b) = f (ab) = (ba)

[∵ R is commutative ⇒ ab = ba]

= f (b) f (a).

∴ S' is also commutative.

4. We have f (0) = 0'. Also f is one-one. Therefore 0 is the only element of S whose f-image is 0'.

Let f (a), f (b) be two non-zero elements of S'. Then f (a) ≠ 0', f (b) ≠ 0' ⇒ a ≠ 0, b ≠ 0. Since S is without zero divisors, therefore

$$a \neq 0, b \neq 0 \Rightarrow ab \neq 0, \quad \Rightarrow f(ab) \neq f(0)$$

$$\Rightarrow f(a)\, f(b) \neq 0' \Rightarrow S' \text{ is without zero divisors.}$$

5. Let 1 be the unit element of S. Then f (1) ∈ S'. If f (a) is any element of S', we have

f (1) f (a) = f (1a) = f (a) and f (a) f(1) = f (a1) = f (a).

∴ f (1) is the unit element of S'.

6. If r is a field, then S is commutative, with unity and each non-zero element of S will possess multiplicative inverse. Now as proved in (3) and (5), S' will be commutative and will also have the unit element *i.e.*, f(1).

Let f (a) be any non-zero element of S'. Then

$f(a) \neq 0' \Rightarrow a \neq 0 \Rightarrow a^{-1}$ exists.

Now $f(a^{-1}) \in S'$ and we have

$f(a^{-1})\, f(a) = f(a^{-1}a) = f(1)$ and $f(a)\, f(a^{-1}) = f(aa^{-1}) = f(1)$.

∴ $f(a^{-1})$ is the multiplicative inverse of f (a).

Hence S' is a skew-field.

TRANSFERENCE OF RING STRUCTURE

Subrings

Definition: *Let R be a ring. A non-empty subset S of the set R is said to be a subring of R if S is closed with respect to the operations of addition and multiplication in R and S itself is a ring for these operations.*

If S is a subring of a ring R, it is obvious that S is a subgroup of the additive group of R.

If R is any ring, then {0} and k itself are always subrings of R. These are know as *Improper subrings* of R. Other subrings, if any, of R are called *proper subrings* of R.

Theorem 1:

The necessary and sufficient conditions for a non-empty subset S of a ring R to be a subring of R are

1. $a \in S, b \in S \Rightarrow a - b \in S$
2. $a \in S, b \in S \Rightarrow ab \in S$.

Proof:

The conditions are necessary. Suppose (S, +, ·) is a subring of (R, +,·).

Since S is a group with respect to addition, therefore b = ∈ S

$\Rightarrow -b \in S$.

Now S is closed with respect to addition.

$\therefore a \in S, b \in S \Rightarrow a \in S, -b \in S$

$$\Rightarrow a + (-b) \in S \Rightarrow a - b \in S.$$

Also S is closed with respect to multiplication.

$\therefore a \in S, b \in S \Rightarrow ab \in S$.

Hence the conditions are necessary.

The conditions are sufficient: Suppose S is a non-empty subset of R and the conditions (1) and (2) are satisfied. From (1), we have

$$a \in S, a \in S \Rightarrow a - a \in S$$

$$\Rightarrow 0 \in S \text{ i.e., the zero element} \in S.$$

Now since $0 \in S$, therefore from (1), we have

$0 \in S, a \in S \Rightarrow 0 - a \in S \Rightarrow -a \in S$ *i.e.*, each element of S Possesses additive inverse.

Now if a, b are any elements of S, then $-b \in S$.

From (1), we have

$$a \in S, -b \in S \Rightarrow a(-b) \in S \Rightarrow a + b \in S.$$

$\therefore$ S is closed with respect to addition.

Now, S is a subset of R. Therefore associativity and commutativity of addition must hold in S since they hold in R.

$\therefore$ (S, +) is an abelian group.

From (2) S is closed with respect to multiplication.

Associativity of multiplication and distributivity of multiplication over addition must hold in S since they hold in R.

Hence S is a subring of R.

Cor. : *The necessary and sufficient conditions for a non-empty subset S of a ring R to be a subring of R are*

(1) $S + (-S) = S$ (2) $SS \subseteq S$.

Proof:

The conditions are necessary: Suppose S is a subring of R. Then S is a subgroup of the additive group of R.

Let $a + (-b)$ be any element of $S + (-S)$.

We have

$$a + (-b) \in S + (-S) \Rightarrow a \in S, -b \in -S$$
$$\Rightarrow a \in S, b \in S$$
$$\Rightarrow a - b \in S \qquad [\because \text{ S is a subgroup}]$$

$\therefore \quad S + (-S) \subseteq S$.

Also let a be any element of S. We can write $a = a + 0$.

Now s is a subgroup. Therefore $0 \in S$ or $0 \in -S$.

So $\qquad a + 0 \in S + (-S)$.

$\therefore \quad S \subseteq S + (-S)$.

$\therefore \quad S = S + (-S)$.

Also S must be closed with respect to multiplication.

$\therefore \quad a \in S, b \in S \Rightarrow ab \in S$.

Now ab is an arbitrary element of SS.

$\therefore \quad SS \subseteq S$.

The conditions are sufficient: Suppose S is a non-empty subset of R satisfying the two given conditions.

We have $SS \subseteq S \Rightarrow ab \in S \ \forall \ a, b \in S$.

Therefore S is closed with respect to multiplication.

Also $\quad S + (-S) = S \Rightarrow S + (-S) \subseteq S$

$\Rightarrow a + (-b) \in S$ if $a, b \in S$

$\Rightarrow$ S is a subgroup of the additive group of R.

$\therefore$ S is a subring of R.

Theorem 2(a):

An arbitrary intersection of subrings is a subring.

Proof:

Let R be a ring and let $\{S_t : t \in T\}$ be any family of subrings of R. Here T is an index set and is such that $\forall\ t \in T$, S_t is a subring of R.

Let $S = \bigcap_{t \in T} S_t = \{x \in R : x \in S_t\ \forall\ t \in T\}$

be the intersection of this family of subrings of R. Then to prove that S is also a subring of R.

Obviously $S \neq \emptyset$, since at least the zero element 0 of R is in $S_t\ \forall\ t \in T$.

Now let a, b be any two elements of S. Then

$$a \in \bigcap_{t \in T} S_t \Rightarrow a \in S_t\ \forall\ t \in T$$

and $b \in \bigcap_{t \in T} S_t \Rightarrow b \in S_t\ \forall\ t \in T.$

But $\forall\ t \in T$, S_t is a subring of R. Therefore

$$a, b \in S_t \Rightarrow a - b,\ ab \in S_t\ \forall\ t \in T.$$

Consequently $a - b,\ ab \in \bigcap_{t \in T} S_t$. Thus we have shown that

$$a, b \in \bigcap_{t \in T} S_t \Rightarrow a - b,\ ab \in \bigcap_{t \in T} S_t.$$

Therefore $\bigcap_{t \in T} S_t$ is a subring of R.

Theorem 2(b):

The intersection of two subrings is a subring.

Proof:

Let S_1 and S_2 be two subrings of a ring R. Then $S_1 \cap S_2$ is not empty since at least $0 \in S_1 \cap S_2$.

Now in order to prove that $S_1 \cap S_2$ is a subring, it is sufficient to prove that

1. $a \in S_1 \cap S_2,\ b \in S_1 \cap S_2 \Rightarrow S_2 \Rightarrow a - b \in S_1 \cap S_2$ and
2. $a \in S_1 \cap S_2,\ b \in S_1 \cap S_2 \Rightarrow S_2 \Rightarrow ab \in S_1 \cap S_2$.

We have $a \in S_1 \cap S_2 \Rightarrow a \in S_1,\ a \in S_2$

$$b \in S_1 \cap S_2 \Rightarrow b \in S_1, b \in S_2.$$

Now S_1 and S_2 are both subrings.

$\therefore$ $a \in S_1, b \in S_1 \Rightarrow a - b \in S_1$ and $ab \in S_1$

and $a \in S_2, b \in S_2 \Rightarrow a - b \in S_2$ and $ab \in S_2$.

Now $a - b \in S_1, a - b \in S_2 \in a - b \in S_1 \cap S_2$

and $ab \in S_1, ab \in S_2 \Rightarrow ab \in S_1 \cap S_2$.

Thus $a \in S_1 \cap S_2, b \in S_1 \cap S_2$

$\Rightarrow a - b \in S_1 \cap S_2$ and $ab \in S_1 \cap S_2$.

$\therefore S_1 \cap S_2$ is a subring of R.

HOMOMORPHISM OF RINGS

Definition: *Homomorphism into: A mapping f from a ring R into a ring R' is said to be a homomorphism of R into R' if*

(i) $f(a + b) = f(a) f(a) + f(b) \ \forall \ a, b \in R$

(ii) $f(ab) = f(a) f(b)$ for all $a, b, \in R$.

Homomorphism onto: *A mapping f from a ring R onto a ring R' is said to be a homomorphism of R onto R' if*

(i) $f(a + b) = f(b) \ \forall \ a, b \in R$.

(ii) $f(ab) = f(a) f(b) \ \forall \ a, b, \in R$.

Also then R' is said to be a homomorphic image of R.

Theorem 1:

Let ϕ be a homomorphic mapping of a ring R into a ring R'. Let S' be the homomorphic image of R in R'. Then S' is a subring of R'.

Proof:

Since S' is the image of R in R' under the mapping ϕ, therefore $\phi(R) = S \subseteq R'$.

Let a', b' be any two elements of S'. Since $S' = \phi(R)$, therefore there exist elements $a, b, \in R$ such that $\phi(a) = a'$ $\phi(b) = b'$.

We have $a' - b' = \phi(a) - \phi(b) = \phi(a - b)$.

[$\therefore$ ϕ is a homomorphism]

Now $a - b \in R$ is such that $a' - b' = \phi(a - b)$. Therefore

$$a' - b' \in S'.$$

Further $a'b' = \phi(a)\, \phi(b) = \phi(ab) \in S'$, since $ab \in R$.

Thus $a', b' \in S'$

$\Rightarrow$ $a' - b' \in S'$ and $a'b' \in S'$.

Therefore S' is a subring of R'.

Theorem 2:

If f is a homomorphism of a ring R into a ring R', then

(i) $f(0) = 0'$, *where 0 is the zero element of the ring R and 0' is the zero element of R'.*

(ii) $f(-a) = -f(a)\ \forall\ a \in R$.

Proof:

(i) Let $a \in R$. Then $f(a) \in R'$. We have

$f(a) + 0' = f(a)$ [$\therefore$ 0' is the additive identity of R']

$= f(a + 0) = f(a) + f(0)$.

Now R' is a group with respect to addition. Therefore

$f(a) + 0' = f(a) + f(0)$

$\Rightarrow 0' = f(0)$. [by left cancellation law]

(ii) Let a be any element of R. Then $-a \in R$.

We have $0' = f(0) = f[a + (-a)] = f(a) + f(-a)$.

$\therefore$ $f(-a)$ is the additive inverse of $f(a)$ in the ring R'. Thus

$f(-a) = -f(a)$.

FUNDAMENTAL THEOREM ON HOMOMORPHISM OF RINGS

Theorem:

Every homomorphic image of a ring R is isomorphic to some residue class ring (quotient ring) thereof.

Proof:

Let R' be the homomorphic image of a ring R and f be the corresponding homomorphism. Then f is a homomorphism of R onto R'. Let S be the kernel of this homomorphism. Then S is an ideal of R. Therefore R/S is a ring of residue classes of R relative to S. We shall prove that $R/S \cong R'$.

If $a \in R$, then $S + a \in R/S$ and $f(a) \in R'$. Consider the mapping $\phi: R/S \to R'$ such that

$\phi (S + a) = f (a) \; \forall \; a \in R.$

First we shall show that the mapping ϕ is well defined *i.e.*, if a, b $\in$ R and S + a = S + b, then $\phi (S + a) = \phi (S + b)$.

We have S + a = S + b

$\Rightarrow \quad a - b \in S$

$\Rightarrow \quad f (a - b) = 0'$ [*i.e.*, zero element of R']

$\Rightarrow \quad f [a + (- b)] = 0' \Rightarrow f (a) + f (- b) = 0'$

$\Rightarrow \quad f (a) - f (b) = 0'$

$\Rightarrow \quad f (a) = f (b) \Rightarrow \phi (S + a) = \phi (S + b).$

$\therefore$ ϕ is well-defined.

ϕ is one-one. We have $\phi (S + a) = \phi (S + b)$

$\Rightarrow \quad f (a) = f (b) \Rightarrow f (a) - f (b) = 0'$

$\Rightarrow \quad a - b \in S$ [$\therefore$ S is kernel of f]

$\Rightarrow \quad S + a = S + b.$

$\therefore$ ϕ is one-one.

ϕ is onto R'. Let y be any element of R'. Then y = f (a) *for some* a $\in$ R because f is onto R'. Now S + a $\in$ R/S and we have

$\phi (S + a) = f (a) = y$. Therefore ϕ is onto R'.

Finally we have

$$\phi [(S + a) + (S + b) = \phi [S + (a + b)] = f (a + b)$$
$$= f (a) + f (b) = \phi (S + a) + \phi (S + b).$$

Also $\phi [(S + a) (S + b)] = \phi (S + ab) = f (ab) = f (a) f (b)$

$$= [\phi (S + a)] [\phi (S + b)].$$

$\therefore$ ϕ is an isomorphism of R/S onto R'.

Hence $\quad R/S \cong R'.$

Theorem 2(a):

If f is a homomorphism of a ring R into a ring R' with kernel S, then S is an ideal of R.

Proof:

Let f be a homomorphism of a ring R into a ring R'. Let 0, 0' be the zero elements of R, R' respectively. Let S be the kernel of f. Then S = {x $\in$ R: f (x) = 0'}.

Since f (0) = 0', therefore at least $0 \in S$. Thus S is not empty.

Let a, b, $\in$ S. Then f (a) = 0', f (b) = 0'.

We have $f(a-b) = f[a + (-b)] = f(a) + f(-b)$

$= f(a) - f(b) = 0' - 0' = 0'.$

$\therefore\ a - b \in S.$

Also if r be any element of R, then

$f(ar) = f(a) f(r) = 0' f(r) = 0'$

and $f(ra) = f(r) f(a) = f(r) 0' = 0'.$

$\therefore$ $ar \in S,\ ra \in S.$

Thus $a, b \in S, r \in R$

$\Rightarrow (a - b) \in S,\ ar \in S,\ ra \in S.$

$\therefore$ S is an ideal of R.

Theorem 2(b):

The homomorphism of ϕ of a ring R into a ring R' is an isomorphism of R into R' if and only if I (ϕ) = (0), where I (ϕ) denotes the kernel of ϕ.

Proof:

Let ϕ be a homomorphism of a ring R into a ring R'. Let 0, 0' be the zero elements of R, R' respectively. Let S = I (ϕ) be kernel of ϕ. Then S is an ideal of R and

$S = \{a \in R : \phi(a) = 0'\}.$

Suppose ϕ is an isomorphism of a ring R into R'. Then ϕ is one-one. Let $a \in S$. Then

$\phi(a) = 0'$ [by def. of kernel]

$\Rightarrow \phi(a) = \phi(0)$ [$\because \phi(0) = 0'$]

$\Rightarrow\ a = 0.$ [$\because \phi$ is one-one]

Thus $a \in S \Rightarrow a = 0$. In order words 0 is the only element of R which belongs to S. Therefore S = (0).

Conversely suppose that S = (0). Then to prove that ϕ is an isomorphism of R into *i.e.*, to prove that ϕ is one-one.

If $a, b \in R$, then $\phi(a) = \phi(b)$

$\Rightarrow\ \phi(a) - \phi(b) = 0'$ [$\because \phi(a), \phi(b)$ are in the ring R']

$\Rightarrow\ \phi(a - b) = 0'$ [$\because \phi$ is a homomorphism]

$\Rightarrow\ a - b \in S$ [by def. of kernel]

$\Rightarrow \quad a - b = 0 \quad [\because S = (0)]$

$\Rightarrow \quad a = b.$

$\therefore$ ϕ is one-one. Hence f is an isomorphism of R into R'.

KERNEL OF A RING HOMOMORPHISM

Definition: *If f is a homomorphism of a ring R into a ring R', then the set S of all those elements of R which are mapped onto the zero element of R' is called the kernel of the homomorphism f.*

Thus, if f is a homomorphism of R into R', then S is the kernel of f if $S = \{x \in R: f(x) = 0'$ where 0' is the zero element of R'$\}$.

Theorem 1:

Suppose R is a ring, S an ideal of R. Let f be a mapping from R to R/S defined by f (a) = S + a ∀ a ∈ R. Then f is a homomorphism of R onto R/S.

Proof:

Consider the mapping f: R → R/S such that

$f(a) = S + a \ \forall \ a \in R.$

Let S + x be any element of R/S. Then $x \in R$.

We have f (x) = S + x. Therefore the mapping is onto R/S.

Let a, b ∈ R. Then

$$f(a + b) = S + (a + b) = (S + a) + (S + b) = f(a) + f(b),$$

Also $\quad f(ab) = S + ab = (S + a)(S + b) = f(a) f(b).$

$\therefore$ if is a homomorphism of R onto R/S.

Thus every quotient ring of is a homomorphic image of the ring.

SMALLEST SUBRING CONTAINING A GIVEN SUBSET OF A RING

Suppose R is a ring and M is any subset of R. Further suppose that S is a subring of r such that $M \subseteq S$ and if T is any subring of R containing M then $S \subseteq T$. Then S is called the subring of R generated by the subset M. In short if S is the smallest subring of R containing M, then S is called the subring generated by M. Also we write S = (M).

Theorem:

The intersection of the family of subrings which contain a given subset M of a ring R is the smallest subring containing the subset M.

Proof:

The family of subrings which contain M is not empty since at least R is a subring of R which contains M. Further the intersection of all subrings of R which contain M is also a subring of R and M is contained in this intersection. Also this intersection will be contained in any subring of R which contains M. Therefore this intersection is the smallest subring of R containing M *i.e.*, it is the subring of R generated by M.

Example 1:

The set of integers is a subring of the ring of rational numbers.

If a, b $\in$ I, then a – b $\in$ I and ab $\in$ I.

$\therefore$ I is a subring of the ring of rational numbers.

Example 2:

Let r be the ring of integers. Let m be any fixed integer and let S be any subset of R such that

S = {..., –3m, – 2m, – m, 0, 2m, m,...}. Then S is a subring of R.

Let a = rm and b = sm be any two elements of S. Then r and s are some integers. We have

$$a - b = rm - sm = (r - s)\, m \text{ and } ab = (rm)\,(sm) = (rsm)\, m.$$

Since r – s is some integer and ab = (rsm) is also some. Integer, therefore both a – b and ab are elements of S. Hence S is a subring of R.

Example 3:

Let R be the ring of all 2 × 2 matrices over the field of real numbers. Let M be a subset of R and let the elements of M be matrices of the type

$$\begin{bmatrix} a & 0 \\ b & 0 \end{bmatrix}$$

i.e., matrices in which the elements of second column are all zeros. The M is a subring of R.

Let $A = \begin{bmatrix} a_1 & 0 \\ b_1 & 0 \end{bmatrix}, B = \begin{bmatrix} a_2 & 0 \\ b_2 & 0 \end{bmatrix}$ be any two elements of M.

Then

$$A - B = \begin{bmatrix} a_1 - a_2 & 0 \\ b_1 - b_2 & 0 \end{bmatrix}, \text{ Also } AB = \begin{bmatrix} a_1 & 0 \\ b_1 & 0 \end{bmatrix}, B = \begin{bmatrix} a_2 & 0 \\ b_2 & 0 \end{bmatrix} = \begin{bmatrix} a_1 - a_2 & 0 \\ b_1 - b_2 & 0 \end{bmatrix}.$$

Now A – B and AB are both members of M since the second column of A – B and also of AB consists of zeros only.

$\therefore$ M is subring of R.

EUCLIDEAN RINGS OR EUCLIDEAN DOMAINS

Definition: *Let R be an integral domain i.e., let R be a commutative ring without zero divisors. Then R is said to be a Euclidean ring if to every non-zero element $a \in R$ we can assign a non-negative integer d (a) such that :*

(i) For all $a, b \in R$, both non-zero, $d(ab) \geq d(a)$.

(ii) For any $a, b \in R$ and $b \neq 0$, there exist $q, r \in R$ such that $a = qb + r$ where either $r = 0$ or $d(r) < d(b)$.

The second part of the above definition is known as division *algorithm.* Also we do not assign a value of d (0). Thus d (a) will remain undefined when a = 0. Also d (a) will be called d-value of a and d (a) must be sone non-negative integer for every non-zero element $a \in R$.

Theorem

Every Euclidean ring possesses unity element.

Proof:

Let R be a Euclidean ring. Obviously R is an ideal of R. Therefore there exists an element $u_0 \in R$ such that $R = (u_0)$ *i.e.*, there exists an element $u_0 \in R$ such that every element in R is a multiple of u_0. Since, in particular, $u_0 \in R$ therefore there exists an element $c \in R$ such that $u_0 = u_0c$. We shall show that c is the required unity element. Let now a be any element of R. Since $a \in R$, therefore there exists some $x \in R$ such that $a = u_0x$.

Now	$ac = (u_0x)\,c$	$[\because a = u_0x]$
	$= (u_0c)\,x$	$[\because$ R is a commutative ring$]$
	$= u_0x$	$[\because u_0 = u_0c]$
	$= a.$	$[\because a = u_0x]$

Thus we have $ac = a = ca \ \forall \ a \in r$.

Hence c is the unity element.

PROPERTIES OF EUCLIDEAN RINGS

Theorem 1:

Let R be a Euclidean ring and a and be any two elements R, not both of which are zero. Then a and b have a greatest common divisor d which can be expressed in the form

$$d = \lambda a + \mu b \text{ for some } \lambda, \mu \in R.$$

Proof:

Consider the set

$S = \{sa + tb: s, t \in R).$ ...(1)

We claim that S in an ideal of R. The proof is as follows:

Let $x = s_1a + t_1b$, and $y = s_1a + t_2b$ be any two elements of S.

Then $s_1, t_1, s_2, t_2 \in R$. We have

$x - y = (s_1a + t_1b) - (s_2a + t_2b) = (s_1 - s_2)\ a + (t_1 - t_2)\ b \in S$

since $s_1 - s_2$ and $t_1 - t_2$ are both elements of R.

Thus S is a subgroup of R with respect to addition.

Also if u be any element of R, then

$xu + ux = u\ (s_1a + t_1b) = (us_1)\ a + (ut_1)\ b \in S$ since $us_1, ut_1 \in R$.

Therefore S is an ideal of R. Row every ideal in R is a principal ideal. Therefore there exists an element d in S such that every element in S is a multiple of d.

Since $d \in S$, therefore from (1), we see that there exist elements $\lambda, \mu \in R$ such that $d = \lambda a + \mu b$.

Now R is a ring with unity element 1.

$\therefore$ Putting $s = 1, t = 0$ in (1), we see that $a \in S$. Also putting $s = 0, t = 1$ in (1), we see that $b \in S$.

Now a, b are elements of S. Therefore they are both multiples of d. Hence $d \mid a$ and $d \mid b$.

Now suppose $c \mid a$ and $c \mid b$.

Then $c \mid \lambda a$ and $c \mid \mu b$. Therefore c is also a divisor of $\lambda a + \mu b$ *i.e.*, c is a divisor of d.

Thus d is a greatest common divisor of a and b.

Theorem 2:

Every Euclidean ring is a principal ideal ring.

Proof:

Let R be a Euclidean ring. Let S be an arbitrary ideal of R. If S is the null ideal, then S = (0) *i.e.*, the ideal of R generated by 0. Therefore S is a principal ideal. So let us suppose that S is not a null ideal. Then there exist elements in S not equal to zero. Let b be a non-zero element in S such that d (b) is minimum *i.e.*, there exists no element c in S such that $d\ (c) < d\ (b)$. We shall show that S = (b) *i.e.*, S is nothing but the ideal generated by b.

Let a be any element of S. Then by definition of Euclidean ring there exist elements q and r in R such that

$$a = qb + r \text{ where either } r = 0 \text{ or } d(r) < d(b).$$

Now $\quad q \in R, b \in S \Rightarrow qb \in S$ because S is an ideal.

Further $a \in S, qb \in S \Rightarrow a - qb = r \in S$.

Thus $r \in S$ and we have either $r = 0$ or $d(r) < d(b)$.

If $r \neq 0$, then $d(r) < d(b)$ which contradicts our assumption that no element in S has d-value smaller than d (b). Therefore we must have $r = 0$.

Then $\quad a = qb$.

Thus every element a in S is a multiple of the generating element b. Thus $a \in S \Rightarrow a \in (b)$. Therefore $S \subseteq (b)$.

Again if xb is any element of (b), then $x \in R$.

Now $x \in R, b \in S \Rightarrow xb \in S$. Therefore $(b) \subseteq S$.

Hence $\quad S = (b)$.

Thus, every ideal S in R is a principal ideal. Therefore R is a principal ideal ring.

Theorem 3(a):

The necessary and sufficient condition that the non-zero element in the Euclidean ring R is a unit is that

$$d(a) = d(1).$$

Proof:

Let a be a unit in R. Then to prove that $d(a) = d(1)$.

By the definition of Euclidean ring

$$d(1a) \geq d(1)$$

$$\Rightarrow \quad d(a) \geq d(1). \qquad ...(1)$$

Since a is a unit in R, therefore a^{-1} exists and we have

$$1 = aa^{-1}$$

$$\Rightarrow \quad d(1) = d(aa^{-1}).$$

But $\quad d(aa^{-1}) \geq d(a)$.

$$\therefore \quad d(1) \geq d(a). \qquad ...(2)$$

Conversely, let $d(a) = d(1)$. Then to prove that a is a unit in R. If a is not a unit in R, then by theorem 6, we have

$$d(1a) > d(1).$$

$\Rightarrow \qquad d(a) > d(1).$

Thus we get a contradiction. Hence a must be a unit in R.

DIVISION RING OR SKEW FIELD

Definition: *A ring R with at least two elements is called a division ring or a skew field if it (i) has unity, (ii) is such that each non-zero element possesses multiplicative inverse.*

Thus a commutative division ring is a field.

Every field is also a division ring. But a division ring is a field if it is also commutative. We shall later on give an example of a skew field which is not commutative *i.e.*, which is not a field.

FIELDS

Although, the algebraic structures of rings and integral domains are widely used and play an important part in the applications of mathematics, we still cannot solve the simple equation ax = b, a ≠ 0 in all rings or in all integral domains. Yet this is one of the first equations we learn to solve in elementary algebra and its solvability is basic to numerable questions. Certainly, if we wish to solve a wide range of problems in a system we nee at least all of the laws true for rings and the cancellation laws together with the ability to solve the equation ax = b, a ≠ 0. We summarize the above in a definition and list several theorems without proof which will place this concept in the contact of the previous section.

Definition 1: ***Field :*** *A field is a commutative ring with unity such that each non-zero element has a multiplicative inverse. A field is frequently designated generically by the letter F.*

Definition 2: *A ring R with at least two elements is called a field if it:*

(i) is commutative, (ii) has unity

(iii) is such that each non-zero element possesses multiplicative inverse.

Example:

*[Q; + , ·], [**R**; + , ·], [C; + , ·], and* $[Z_p; +_p, \times_p]$ *(p a prime) are all fields.*

Example:

The ring of rational numbers (Q, +, .) is a field since it is a commutative ring with unity and each non-zero element is inversible.

The rings of real numbers and complex numbers.

DEFINITION OF AN IRREDUCIBLE POLYNOMIAL OVER A FIELD

Let F be a field and f (x) be a non-zero and non-unit polynomial in F [x] i.e., f (x) be a polynomial of positive degree. Then f (x) is said to be irreducible over F (or prime) if it has no proper divisors in F [x]; f (x) is reducible over F if it has a proper divisor in F [x].

Thus, a positive degree polynomial f (x) in F [x] is irreducible over F if whenever f (x) = a (x) b (x) with a (x), b (x) $\in$ F [x] then one of a (x) or b (x) in is a unit in F [x] *i.e.*, has degree 0. Also f (x) is reducible over F if and only if we can find two polynomials a (x) and b (x) in F [x] such that f (x) = a (x) b (x) and none of a (x) and b (x) is a unit in F [x] *i.e.*, has degree 0.

Irreducibility depends on the field. The polynomial $x^2 - 2$ is irreducible over the field of rational numbers while it is reducible over the field of real numbers, since $x^2 - 2 = (x + \sqrt{2})(x - \sqrt{2})$.

The polynomial $x^2 + 1$ is irreducible over the field of real numbers while it is reducible over the field of complex numbers since $x^2 + 1 = (x + i)(x - i)$.

MONIC POLYNOMIALS

Definition: *Let $f(x) = a_0 + a_1x + ... + a_nx^n$, with $a_n \neq 0$, be a polynomial in F [x] over an arbitrary field F. If the leading coefficient a_n of f (x) is equal to 1, the unity element of F, then the polynomial f (x) will be called monic.*

The polynomial $2x - 3x^3$ over the field of real numbers is not monic since its leading coefficient is – 3. But the polynomial $x^3 - 3x + 4$ over the field of real numbers is monic since its leading coefficient is 1.

Greatest Common Divisor of Two Polynomials Over a Field

Definition: *Suppose F is any field. Let f (x) and g (x) be two elements of F [x]. A greatest common divisor of F (x) and g (x) is a non-zero polynomial d (x) such that*

(i) d (x) | f (x) and d (x) g (x)

(ii) If c (x) is a polynomial such that c (x) | f (x) and c (x) | g (x) then c (x) | d (x).

Relatively Prime Polynomials

Two polynomials f (x) and g (x) $\in$ F [x] are said to be relatively prime if their greatest common divisor is 1, the unity element of F.

Divisibility of Polynomials Over a Field

Suppose F is a field. Then F [x] is an integral domain. If $a(x) \neq 0$ and f(x) are elements of F [x], then a(x) is a *divisor (or factor)* of f (x) if and only if there is a polynomial b (x) in F [x] such that f (x) = a (x) b (x). Symbolically we write a (x) | f (x).

UNIT

A *unit* is an element of F [x] which has multiplicative inverse. All the polynomials of zero degree belonging to F [x] are units of F [x]. Thus, the non-zero elements of F are the only units of F [x].

If f [x] and g [x] are polynomials in F [x], then we call f [x] and g (x) associates if f (x) = c g (x) for some $0 \neq c \in F$. It can be easily proved that two non-zero polynomials f (x) and g (x) in F [x] are associates if and only if f (x) | g (x) and g (x) | f (x).

If f (x) is any non-zero polynomial in F [x], then f (x) is always divisible by its associates and by all units of F [x]. These divisors of f (x) are called its *improper divisor.* All other divisors of f (x), if there are any, R called its *proper divisors.*

Division Algorithm for Polynomials Over a Field

Theorem:

Let f (x), g (x) ≠ 0 be any two polynomials of the polynomial domain F [x], over the field F. Then there exist uniquely two polynomials q (x) and r (x) in F [x] such that

$$f(x) = q(x)\, g(x) + r(x)$$

where either r (x) = 0 or deg r (x) < deg g (x).

Proof:

Suppose

$$f(x) = a_0 + a_1x + a_2x^2 + \ldots + a_mx^m,\ a_m \neq 0$$

and $$g(x) = b_0 + b_1x + b_2x^2 + \ldots + b_nx^n,\ b_n \neq 0.$$

If degree m of f (x) is smaller than the degree n of g (x) or if f (x) =0, then we are nothing to prove. Because we can always write f (x) = 0 . g (x). So in this case q (x) = 0, r (x) = f (x) and we have either f (x) = 0 or deg f (x) < deg g (x).

Now let us assume that $m \geq n$. In this case we shall prove the theorem by induction on m. *i.e.*, degree of f (x).

If m = 0, then $m \geq n \Rightarrow n = 0$. Therefore f (x) and g (x) are both non-zero constant polynomials, $f(x) = a_0$, $a_0 \neq 0$, and $g(x) = b_0$, $b_0 \neq 0$. We have in this case

$$f(x) = a_0 = \left(a_0 b_0^{-1}\right) b_0 + 0 = \left(a_0 b_0^{-1}\right) g(x) + 0.$$

Thus, the theorem is true when m = 0 or when the degree of f (x) is less than 1.

We shall now assume that the theorem is true when f (x) is a polynomial of degree less than m and then we shall show that it is also true if f (x) is of degree m and then the proof will be complete by induction.

Let $\quad f_1(x) = f(x) - (a_m \; b_n^{-1})\, x^{m-n}\, g(x).$ ...(1)

Obviously deg $f_1(x) < m$. Therefore by our assumed hypothesis, there exist polynomials s (x) and r (x) such that

$$f_1(x) = s(x)\, g(x) + r(x),$$

where $\quad r(x) = 0$ or deg $r(x) <$ deg $g(x)$.

Now putting the value of $f_1(x)$ in (1)

$$s(x)\, g(x) + r(x) + f(x) - \left(a_m\, b_n^{-1}\right)\; x^{m-n}\, g(x)$$

$$\Rightarrow \qquad f(x) = \left[\left(a_m\, b_n^{-1}\right)\; x^{m-n} + s(x)\right] g(x) + r(x).$$

If we write q (x) in place of $\left(a_m\, b_n^{-1}\right)\; x^{m-n} + s(x)$, we get

$$f(x) = q(x)\, g(x) + r(x)$$

where $\quad r(x) = 0$ or deg $r(x) <$ deg $g(x)$.

This proves the existence of polynomials q (x) and r (x). Now to show that q (x) and r (x) are unique. Let us assume that

$$f(x) = q_1(x)\, g(x) + r_1(x) = q_2(x)\, g(x) + r_2(x).$$

Then $\quad q_1(x)\, g(x) + r_1(x) = q_2(x)\, g(x) + r_2(x)$

$\Rightarrow \qquad [q_1(x) - q_2(x)\, g(x) = r_2(x) - r_1(x).$ ...(2)

If $[q_1(x) - q_2(x)] \neq 0$, then $[q_1(x) - q_2(x)]\, g(x)$ cannot be equal to the zero polynomial because g (x) ' 0 and F [x] is without zero divisors. Also then the degree of $[q_1(x) - q_2(x)]\, g(x)$ is at least n, the degree of g (x). But $r_2(x) - r_1(x)$ is either equal to the zero polynomial or else its less than n because the degrees of $r_2(x)$ and $r_1(x)$ are both less than n. Hence the equality (2) among two polynomials holds only if $q_1(x) - q_2(x) = 0$ and $r_2(x) - r_1(x) = 0$

i.e., only if $q_1(x) = q_2(x)$ and $r_1(x) = r_2(x)$.

∴ the polynomials q (x) and r (x) are unique.

Definition: *In the division algorithm, the polynomial q (x) is called the quotient on dividing f (x) by g (x) and the polynomial r (x) is called the remainder.*

Theorem:

A polynomial domain F [x] over a field F is a principal ideal ring.

Proof:

Obviously F [x] is a commutative ring with unity and without zero divisors. Therefore F [x] is a principal ideal ring if every ideal in F [x] is a principal ideal.

Let S be an arbitrary ideal of F [x]. If S is the null ideal, then S = (0) *i.e.*, the ideal of F [x] generated by 0. Therefore S is a principal ideal, so let us suppose that S is not a null ideal. Then there exist non-zero polynomials f (x) in S. Let g (x) be a polynomial of lowest degree m belonging to S. We shall show that S is the principal ideal generated by g (x).

Let f (x) be any arbitrary member of S. By division algorithm there exist two polynomials $q(x) \in F[x]$, $r(x) \in F[x]$, such that

$f(x) = q(x)\, g(x) + r(x)$, where $r(x) = 0$ or $\deg r(x) < \deg g(x)$.

Since S is an ideal, therefore

$$q(x) \in F[x],\ g(x) \in S \Rightarrow q(x)\, g(x) \in S.$$

Also $\quad f(x) \in S,\ q(x)\, g(x) \in S \Rightarrow f(x) - q(x)\, g(x) \in S.$

But $\quad f(x) - q(x)\, g(x) = r(x)$. Therefore $r(x) \in S$.

Now either $r(x) = 0$ or $\deg r(x) < \deg g(x)$. But we have assumed that g (x) is a polynomial of lowest degree belonging to S. Hence deg r (x) cannot be less than deg (x). Therefore we must have $r(x) = 0$. Then $f(x) = q(x)\, g(x)$. Thus $g(x) \in S$ is such that $f(x) \in S \Rightarrow f(x) = q(x)\, g(x)$ for some $q(x) \in F[x]$. Therefore S is a principal ideal of F [x] generated by g (x). Hence F [x] is a principal ideal ring.

Note: A polynomial ring over an arbitrary field is a principal ideal ring. But a polynomial ring over an arbitrary ring is not a principal ideal ring as is obvious from the following example.

Example:

Show that the polynomial ring **I** *[x] over the ring of integers is not a principal ideal ring.*

Solution:

To prove this statement we shall show that the ideal (2, x) of the ring I [x] generated by two elements 2 and x of **I** [x] is not a principal ideal. Let x) be a principal ideal in **I** [x]. Then there will exist a non-zero element **I** [x] such that (2, x) = (g (x)).

Since $2 \in (g(x))$ and $x \in (g(x))$ therefore there will exist elements $\phi(x)$ and $\psi(x)$ belonging to **I** [x] such that

$$2 = \phi(x)\, g(x) \qquad ...(1)$$

$$x = \psi(x)\, g(x). \qquad ...(2)$$

From (1), we get $2x = [\phi(x)\, g(x)]\, x$ and from (2), we get

$$2x = 2\, \psi(x)\, g(x).$$

$\therefore 2\, \psi(x)\, g(x) = x\, \phi(x)\, g(x)$ [$\because$ **I** [x] is a commutative ring].

$\therefore 2\, \psi(x) = x\, \phi(x)$ since $g(x) \neq 0$, and **I** [x] is without zero divisors.

Now $2\, \psi(x) = x\, \phi(x)$ implies that the coefficients of $\phi(x)$ must be even integers. Therefore $\phi(x) = 2h(x)$ where $h(x)$ is some polynomial in **I** [x]. Putting this value of $\phi(x)$ in (1) we get

$$2 = 2h(x)\, g(x)$$

$$\Rightarrow \qquad 1 = h(x)\, g(x).$$

Now $1 = h(x)\, g(x) \Rightarrow 1 \in (g(x))$. Therefore element of **I** [x] will belong to $(g(x))$. Thus, we have **I** [x] $= (g(x)) = (2, x)$. Therefore each element of **I** [x] will belong to $(2, x)$. We shall show that $1 \notin (2, x)$ and this contradiction will mean that $(2, x)$ is not a principal ideal in **I** [x].

Now $1 \in (2, x) \Rightarrow$ we can write

$$1 = 2p(x) + xq(x)$$

where $p(x)$ and $q(x)$ are some elements in **I** [x].

Let $p(x) = a_0 + a_1 x + a_2x^2 + ...$ and $q(x) = b_0 + b_1x + b_2x^2 + ...$

Then $1 = 2(a_0 + a_1x + a_2x^2 + ...) + x(b_0 + b_1x + b_2x^2 + ...)$

$\Rightarrow 1 = 2a_0 + (2a_1 + b_0)x + (2a_2 + b_1)x^2 + ...$

This equality implies $1 = 2a_0$ where $a_0 \in$ **I**.

But for no integer a_0 we can have $1 = 2a_0$. Hence $1 \notin (2, x)$.

$\therefore (2, x)$ is not a principal ideal in **I** [x].

EUCLIDEAN ALGORITHM FOR POLYNOMIALS OVER A

Theorem:

Field : *Let F be a field and f (x) and g (x) be any two polynomial in F [x], not both of which are zero. Then f (x) and g (x) have a greatest common divisor d (x) which can be expressed in the form*

$$d(x) = m(x)\, f(x) + n(x)\, g(x)$$

for polynomials m (x) and n (x) in F [x].

Proof:

Consider the set

$S = \{s(x) f(x) + t(x) g(x) : s(x), t(x) \in F[x]\}$. ...(1)

We claim that S is an ideal of F [x]. The proof is as follows.

Let $s_1(x) f(x) + t_1(x) g(x)$ and $s_2(x) f(x) + t_2(x) + g(x)$ be any two elements of S.

Then $[s_1(x) f(x) + t_1(x) g(x)] - [s_2(x) f(x) + t_2(x) g(x)]$

$= [s_1(x) - s_2(x)] f(x) + [t_1(x) - t_2(x)] g(x) \in S$

since $s_1(x) - s_2(x)$ and $t_1(x) - t_2(x)$ are both members of F [x].

Also if a (x) be any member of F [x], then

$$\alpha(x) [s_1(x) f(x) + t_1(x) g(x)]$$
$$= [a(x) s_1(x)] f(x) + [a(x) t_1(x)] g(x) \in S.$$

Therefore S is an ideal of F [x]. Now every ideal in F [x] is a principal ideal. Therefore there exists an element d (x) in S such that every element in S is a multiple of d (x).

Since d (x) ∈ S, therefore from (1) we see that there exist elements m (x), n (x) ∈ F [x] such that

$$d(x) = m(x) f(x) + n(x) g(x).$$

Now F [x] is a ring with unity element 1.

∴ Putting $s(x) = 1, t(x) = 0$ in (1), we see that f (x) ∈ S. Also putting $s(x) = 0, t(x) = 1$ in (1), we see that g (x) ∈ S.

Now f (x), g (x) are elements of S. Therefore they are both multiples of d (x). Hence d (x) | f (x) and d (x) | g (x).

Now suppose c (x) | f (x) and c (x) | g (x).

Then c (x) | [m (x) f (x)] and c (x) | [n (x) g (x)]. Therefore c (x) is also a divisor of m (x) f (x) + n (x) g (x) *i.e.*, c (x) is a divisor of d (x).

Thus d (x) is a greatest common divisor of f (x) and g (x).

Note: If d (x) is a greatest common divisor of f (x) and g (x) then any associate of d (x) *i.e.*, k d (x) where ≠ k ∈ F will also be a greatest common divisor of f (x) and g (x). In particular if $0 \neq b$ is the leading coefficient of the polynomial d (x), then the monic polynomial b^{-1} d (x) will also be a greatest common divisor of f (x) and g (x). *Often while defining greatest common divisor of two polynomials over a field we include one more condition in our definition that the greatest common divisor should be a monic polynomial.* The advantage of this extra condition is that now we shall get a unique greatest common divisor as shown below:

Supposed d_1 (x) and d_2 (x) are two monic polynomials and each is a greatest common divisor of f (x) and g (x). Then d_1 (x) | d_2 (x) and d_2 (x) | d_1 (x). Therefore d_1 (x_1) and d_2 (x) are associates and we have $d_1 (x) = ud_2 (x)$ for some $0 \neq u \in F$. Since d_1 (x) and d_2 (x) are both monic, therefore, u = 1.

UNIQUE FACTORIZATION DOMAIN

Definition: *An integral domain, R, with unity element 1 is a unique factorization domain if*

(a) any non-zero element in R is either a unit or can be written as the product of a finite number of irreducible (prime) elements of R;

(b) the decomposition in part (a) is unique upto the order and associates of the irreducible elements.

Thus, if R is a unique factorization domain and if $a \neq 0$ is a non-unit in R, then a can be expressed as a product of a finite of prime elements of R. Also if

$$a = p_1 p_2 p_3 \dots p_n = p_1' p_2' p_3' \dots p_m'$$

where the p_i and p_j' are prime elements of R, then m = n and each p_i, $1 \leq i \leq n$ is an associate of some p_j' $1 \leq j \leq m$ and conversely each p_k' is an associate of some p_l.

THE UNIQUE FACTORIZATION THEOREM FOR POLYNOMIALS OVER A FIELD

We shall now prove that every polynomial over a field can be factored uniquely into irreducible factors. Before stating the main factorization theorem, we shall give two preliminary theorems that are needed for its proof.

Theorem 1:

If f (x) is an irreducible polynomial in F [x] for a field F and f (x) | g (x) h (x) where g (x), h (x) $\in$ F [x] then f (x) divides at least one of g (x) or h (x).

Proof:

Suppose that f (x) does not divide g (x). Since f (x) is prime therefore f (x) does not divide g (x) implies that f (x) and g (x) are relatively prime. Therefore the greatest common divisor of f (x) and g (x) is 1. Hence by theorem 1, we get that f (x) | h (x).

Corollary: *If f (x) an irreducible polynomial in F [x] for a field F, and if f (x) divides the product g_1 (x) g_2 (x) ... g_n (x) of polynomials in F [x], then f (x) divides g_i (x) for some i, $1 \leq i \leq n$.*

This result follows immediately by repeated application of theorem.

Theorem 2:

Let f (x), g (x) and h (x) be polynomials in F [x] for a field F. If f (x) | g (x) h (x) and the greatest common divisor of f (x) and g (x) is 1, then f (x) | h (x).

Proof:

If the greatest common divisor of f (x) and g (x) is 1, then by theorem of there exist polynomials m (x) and n (x) $\in$ F [x] such that 1 = m(x) f(x) + n (x) g (x). Multiplying both members of this equations by h (x), we get

$$h(x) = m(x)\,f(x)\,h(x) + n(x)\,g(x)\,h(x) \qquad \ldots(1)$$

But f (x) | g (x) h (x), so there exists a polynomial q (x) $\in$ F [x] such that g (x) h (x) = q (x) f (x).

Substituting this value of g (x) h (x) in (1), we get

$$h(x) = m(x)\,f(x)\,h(x) + n(x)\,q(x)\,f(x)$$
$$= f(x)\,[m(x)\,h(x) + n(x)\,q(x)],$$

which shows that f (x) is a divisor of h (x).

Hence the theorem.

THE UNIQUE FACTORIZATION THEOREM FOR POLYNOMIALS OVER A FIELD

Theorem:

Let f (x) be a non-zero polynomial in F [x], where F is a field. Then either f (x) is a unit n F [x] or f (x) = a (x) p_1 (x) p_2 (x) ... p_m (x), where each p_i (x), $1 \le i \le m$, is an irreducible monic polynomial in F [x] and a $\in$ F is the leading coefficient of f (x). Further the factors p_1 (x), p_2 (x), ..., p_m (x) are unique except for the order in which they appear.

Proof:

We shall prove the theorem in two parts. First we shall prove that f (x) can be factored as required, and then we shall show that the factors are unique.

Let f (x) be a non-zero element of F [x]. Then either f (x) is a unit in F [x] *i.e.*, deg f (x) is 0 or deg f (x) > 0. If deg f (x) > 0, and the leading coefficient of f (x) is a we are to prove that f (x) can be expressed as a product of a and a finite number of irreducible monic polynomials in F [x].

The proof will be any induction on the degree of f (x).

Suppose f (x) is of degree one. Let f (x) = b + ax for a, b $\in$ F and a $\neq$ 0. We can write f (x) = a (a^{-1} b + x). Therefore the theorem holds in the case where f (x) has degree one since a^{-1} b + x is irreducible and monic.

Now assume, as the induction hypothesis, that every polynomial of degree less than n can be factored as stated in the theorem. Consider an arbitrary polynomial f (x) of degree n having a as its leading coefficient. We can write f (x) = af_1 (x), where f_1 (x) = a^{-1} f (x) and f_1 (x) is monic. If f (x) is irreducible, then f_1 (x) is also irreducible and the theorem holds. If f (x) is reducible, then it can be factored as f (x) = g (x) h (x) where neither g (x) nor h (x) is a unit in F [x]. Now the degree of f (x) is equal to the sum of the degrees of g (x) and h (x). Also g (x) and h (x) are not units in F [x], so each of them must be of degree one or larger. Hence both g (x) and h (x) have degrees less than n. Therefore by our induction hypothesis we can write

$$g(x) = c\alpha_1(x)\,\alpha_2(x) \ldots \alpha_s(x),\; h(x) = d\,\beta_1(x)\,\beta_2(x) \ldots \beta_t(x)$$

where each α_i (x) and each β_j (x) is monic and irreductivle and where c and d are leading coefficients of g (x) and h (x) respectively.

Thus $f(x) = cd\alpha_1(x)\,\alpha_2(x)\,\alpha_3(x) \ldots \alpha_s(x)\,\beta_1(x)\,\beta_2(x) \ldots \beta_t(x)$.

Since the leading coefficient of f (x) is a, therefore we must have a = cd because each α (x) and each β (x) is monic. Therefore

$$f(x) = a\,\alpha_1(x)\,\alpha_2(x) \ldots \alpha_s(x)\,\beta_1(x)\,\beta_2(x) \ldots \beta_i(x).$$

The factorization of f (x) satisfies the requirements of the theorem. Hence the theorem holds for all polynomials of degree n, and by the principle of induction, for all polynomials of arbitrary degree.

In order to prove that the factors are unique, let us suppose that $f(x) = a\,p_1(x)\,p_2(x) \ldots p_m(x) = a\,q_1(x)\,q_2(x) \ldots q_n(x)$ where each p (x) and each q (x) is irreducible and monic. Then we shall prove that n = m and each p (x) is equal to some q (x) and each q (x) is equal to some p (x). From these two decompositions of f (x), we have

$$p_1(x)\,p_2(x) \ldots p_m(x) = q_1(x)\,q_2(x) \ldots q_n(x).$$

Now $\quad p_1(x) \mid p_1(x)\,p_2(x) \ldots p_m(x)$. Therefore

$$p_1(x) \mid q_1(x)\,q_2(x) \ldots q_n(x).$$

By Corollary to theorem of this article p_1 (x) must divide at least one of q_1 (x), q_2 (x), ..., q_n (x). Since F [x] is a commutative ring, therefore without loss of generality we may suppose that p_1 (x) divides q_1 (x). But p_1 (x) and q_1 (x) are both irreducible polynomials in F [x] and $p_1(x) \mid q_1(x)$. Therefore p_1 (x) and q_1 (x) must be associates and we have $q_1(x) = up_1(x)$ where u is a unit in F [x] *i.e.*, u is a non-zero element of F. Since q_1 (x) and p_1 (x) are monic therefore u must be equal to 1 and we have $p_1(x) = q_1(x)$. Thus we have

$$p_1(x)\, p_2(x) \dots p_m(x) = p_1(x)\, q_2(x) \dots q_n(x).$$

Cancelling 0 ¹ $p_1(x)$ from both sides, we get

$$p_2(x)\, p_3(x) \dots p_m(x) = q_2(x)\, q_3(x) \dots q_n(x) \qquad \dots(1)$$

Now we can repeat the above argument on the relation (1) with $p_2(x)$. If $n > m$, then after m steps the left hand side becomes 1 while the right hand side reduces to a product of a certain number of q (x) (the excess of n over m). But the q (x) are irreducible polynomials so they are not units of F [x] *i.e.*, they are not polynomials of zero degree.

So their product will be a polynomial of degree ≥ 1. So it cannot be equal to 1. Therefore n cannot be greater than m. Then $n \leq m$. Similarly interchanging the roles of p (x) and q (x), we get $m \leq n$. Hence $m = n$.

Also in the above process we have shown that every p (x) is equal to some q (x) and conversely every q (x) is equal to some p (x). Hence the theorem has been completely established.

Thus, we can say that *the ring of polynomials over a field is a unique factorization domain.*

FIELD EXTENSION

From high-section algebra we realize that to solve a polynomial equation means to find its roots (or, equivalently, to fine the zeros of the polynomials). The previous section we know that the zeros may not lie in the given ground field. Hence, to solve a polynomial really involves two steps; first, find the zeros, and second, find the field in which the zeros lie. For economy's sake we would like this field to be the smallest field which contains all the zeros of the given polynomial.

Example:

*Let $f(x) = x^2 - 2 \in \mathbf{Q}[x]$. It is important to remember that we are considering $x^2 - 2$ over **Q**, no other field. We would like to find all zeros of f (x) and the smallest field, call it S, which contains them. The zeros are $x = \pm\sqrt{2}$, neither of which is an element of **Q**. The set we are looking for must.*

(1) contain **Q** as a subfield.

(2) contain all zeros of $f(x) = x^2 - 2$ and

(3) be a field.

Certainly $\sqrt{2}$ must be an element of S, and if S is to be a field, the sum, product, difference and quotient of elements in S must be in S. So $\sqrt{2}$, $(\sqrt{2})^2$, $(\sqrt{2})^3$, $\sqrt{2} + \sqrt{2}$, $\sqrt{2} - \sqrt{2}$ and $\sqrt{2} \div \sqrt{2}$ must all be elements of S. Further, since S contains BQ as a subset, any element of **Q** combined with $\sqrt{2}$ under any field operation must be an element of S. Hence, every element of the form $a \div b\sqrt{2}$, where a and b can be any elements in **Q**, is an element of S. We leave to the reader to show that S is a field. We note that the second zero of $x^2 - 2$, namely $-\sqrt{2}$, is an element of S. To see this, simply take a = 0 and b = –1. The filed S is frequently designated as $\mathbf{Q}(\sqrt{2})$, and it is referred to as an *extension field of* **Q**. Note that the polynomial $x^2 - 2 = (x + \sqrt{2})(x - \sqrt{2})$ factors into linear factors or splits in $\mathbf{Q}(\sqrt{2})[x]$; that is, all coefficients of both factors are elements of the field $\mathbf{Q}(\sqrt{2})$.

Theorem 1:

As field (skew-field) has no divisors of zero.

Proof:

Let D be a skew-field. Then D is a ring with unit element 1 and each non-zero element of D possesses multiplicative inverse.

Let a, b be elements of D with $a \neq 0$ such that $ab = 0$.

Since $a \neq 0$, a^{-1} exists and we have

$$ab = 0 \Rightarrow a^{-1}(ab) = a^{-1}0$$

$$\Rightarrow (a^{-1}a)b = 0 \Rightarrow 1b = 0 \Rightarrow b = 0.$$

Similarly, let $ab = 0$ with $b \neq 0$.

Since $b \neq 0$, b^{-1} exists and we have

$$ab = 0 \Rightarrow (ab)b^{-1}$$

$$\Rightarrow a(bb^{-1}) = 0 \Rightarrow a1 = 0 \Rightarrow a = 0.$$

Therefore a skew-field has no zero divisors.

Theorem 2(a):

A finite commutative ring without zero divisors is a field.

or

Every finite integral domain is a field

Proof:

Let us consider D be a finite commutative ring without zero divisors having n elements $a_1, a_2 ..., a_n$. In order to prove that D is a field, we must produce an element $1 \in D$ such that $1a = a \ \forall \ a \in D$. Also we should show that for every element $a \neq 0 \in D$ there exists an element $b \in D$ such that $ba = 1$.

Let $a \neq 0 \in D$. Consider the n products $aa_1, aa_2 ..., aa_n$.

All these are elements of D. Also they are distinct. For suppose that $aa_i = aa_j$ for $i \neq j$.

Then $\quad a (a_i - a_j) = 0.$ ...(i)

Since D is without zero divisors and $a \neq 0$, therefore (1) implies

$$a_i - a_j = 0 \Rightarrow a_i = a_j, \text{ contradicting } i \neq j.$$

$\therefore$ $aa_1, aa_2 ..., aa_n$ are all the n distinct elements of D placed in some order. So one of these elements will be equal to a. Thus there exists an element, say, $1 \in D$ such that

$$a1 = a = 1a. \qquad [\because \ D \text{ is commutative}]$$

We shall show that this element 1 is the multiplicative identity of D. Let y be any arbitrary element of D. Then from the above discussion for some $x \in D$, we shall have $ax = y = xa$.

Now $\quad 1y = 1 (ax)$ $\quad [\because \ ax = y]$

$= (1a) x$

$= ax$ $\quad [\because \ 1a = a]$

$= y$ $\quad [\because \ ax = y]$

$= y1$ $\quad [\because \ D \text{ is commutative}]$

Thus, $1y = y = y1, \ \forall \ y \in D$. Therefore 1 is the unit element *i.e.*, the multiplicative identity of the ring D.

Now $1 \in D$. Therefore from the above discussion one of the n products $aa_1, aa_2 ..., aa_n$ will be equal to 1. Thus there exists an element, say, $b \in D$ such that

$$ab = 1 = ba.$$

$\therefore$ b is the multiplicative inverse of the non-zero element $a \in D$. Thus every non-zero element of D is inversible.

Hence D is a field.

Theorem 2(b):

Every field is an integral domain.

Proof:

Let a, b be elements of F with $a \neq 0$ such that $ab = 0$.

Since $a \neq 0$, a^{-1} exists and we have

$$ab = 0 \Rightarrow a^{-1}(ab) = a^{-1}0$$
$$\Rightarrow (a^{-1}a)b) = 0$$
$$\Rightarrow 1b = 0 \qquad [\because a^{-1}a = 1]$$
$$\Rightarrow b = 0. \qquad [\because 1b = b]$$

Similarly, let ab = and $b \neq 0$.

Since $b \neq 0$, b^{-1} exists and we have

$$ab = 0 \Rightarrow (ab)\,b^{-1} = 0\,b^{-1}$$
$$\Rightarrow a\,(bb^{-1}) = 0 \Rightarrow al = 0 \Rightarrow a = 0.$$

Thus in a field $ab = 0 \Rightarrow a = 0 \Rightarrow b = 0$. Therefore a field has no zero divisors. Therefore every field is an integral domain.

But the converse is a not true i.e., every integral domain is not a field. For example the ring of integers is an integral domain and it is not a field. The only inversible elements of the ring of integers are 1 and –1.

Note: A field has no zero divisors. Therefore in a field the product of two non-zero elements will again be a non-zero element. Also the unit element $1 \neq 0$ and each non-zero element possesses multiplicative inverse which is again a non-zero element. The multiplication is commutative as well as associative. Therefore *the non-zero elements of a field form an abelian group with respect to multiplication.*

Existence of Identity for the Operation * on I

The integer $e \in I$ will be the identity for the operation * on the set I if

$e * a = a \;\forall\; a \in I.$ [Note that $e * a = a * c$]

Now $e * a = a \Rightarrow e + a - 1 = a \Rightarrow 0 \Rightarrow e = 1$.

Therefore the integer 1 is identity for the operation * on the set I and will be the zero element of the ring c I, *, o).

Existence of Inverse of Each Element of I for the Operation *

Let $a \in I$. Then $b \in I$ will be the inverse of a for the operation * if $a * b = 1$.

Now $a * b = 1 \Rightarrow a + b - 1 = 1 \Rightarrow b = 2 - a$.

Thus if $a \in I$, then $2 - a \in I$ and is the inverse of a for the operation *.

$\therefore$ (I, *) is an abelian group.

I is closed for the operation o. Let $a, b \in I$.

Then $a \circ b = a + b - ab$ which is also an integer. Therefore I is closed with respect to the operation o.]

Commutativity and associativity of the operation o on the set I. Let a, b, c ∈ I. Then

$$a \circ b = a + b - ab = b + a - ba = b \circ a$$

Also $(a \circ b) \circ c = (a + b - ab) \circ c = (a + b - ab) + c - (a + b - ab)\, c$

$$= a + b + c - ab - ac - bc + abc$$

and $a \circ (b \circ c) = a \circ (b + c - bc) = a + (b + c - bc) - a\,(b + c - bc)$

$$= a + b + c - bc - ab - ac + abc.$$

$\therefore a \circ (b \circ c) = (a \circ b) \circ c.$

Thus, the operation o on the set I is commutative as well as associative.

Distributivity of o Over ∗

Let a, b, c ∈ I. Then

$a \circ (b * c) = a \circ (b + c - 1) = a\,(b + c - 1) - a\,(b + c - 1)$

$$= a + b\,c - 1 - ab - ac + a = 2a + b + c - ab - ac - 1.$$

Also $(a \circ b) * (a \circ b) = (a + b - ab) * (a + c - ca)$

$= a + b - ab + a + c - ac - 1 = 2a + b + c - ab - ac - 1.$

$\therefore a \circ (b * c) = (a \circ b) * (a \circ c).$

Similarly we can show that the other distributive law

$$(b * c) \circ a = (b \circ a) * (c \circ a)$$

also holds good.

Existence of Identity for the Operation o on I

The integer u ∈ I will be the identity for the operation o on the set I if

$u \circ a = a \;\forall\; a \in I$. [Note $u \circ a = a \circ u$]

Now $u \circ a = a \;\forall\; a \in I$

$\Rightarrow\; u + a - ua = a \;\forall\; a \in I$

$\Rightarrow\; u\,(1 - a) = 0 \;\forall\; a \in I$

$\Rightarrow\; u = 0.$

Therefore the integer 0 is the identity for the operation o on the set I and so will be the unity element of this ring with unity. We observe that ∀ a ∈ I, we have

$$0 \circ a = 0 + a - 0a = a = a \circ 0.$$

Hence the algebraic structure (I, ∗ o) is a commutative ring with identity. The operation ∗ is the addition of this ring and the operation o is the multiplication of this ring. The integer 1 is the zero element of this ring and the integer 0 is the unity element of this ring with unity.

Example of a Finite Field: The ring ({0, 1, 2,}, $+_3$, $\times_3$) is an example of a finite field because the number of distinct elements in this ring is 3 and so it is a finite ring. This ring is a commutative ring with unity element the integer 1. The zero elements 1 and 2 of this ring possess multiplicative inverse. The multiplicative inverse of 1 is 1 because $1 \times_3 1 = 1$ and the multiplicative inverse of 2 is 2 because $2 \times_3 2 = 1$.

Hence ({0, 1, 2,}, $+_3$, $\times_3$) is a field and it is a finite field.

Difference Between an Integral Domain and a Skew Field: A skew field need not be an integral domain because in a skew field multiplication need not be cumulative while in an integral domain multiplication must be commutative. Also an integral domain need not be a skew field. For example, the ring of integers is an integral domain but it is not a skew field.

Difference Between a Field and a Skew Field: Every field is a skew-field but a skew-field need not be a field. A skew field will be a field only if the operation of multiplication in it is commutative. (4) is not a field because the operation of multiplication of matrices in this skew field is not commutative.

Difference Between a Field and an Integral Domain: Every field is an integral domain but every integral domain need not be a field. For example the field of rational numbers (Q, +, .) is an integral domain because it does no possess zero divisors. The product of two non-zero rational numbers is never zero. On the other had, the ring of integers (I, +, ·) is an integral domain but it is not a field. The only inversible elements of this ring are 1 and –1 while in a field every non-zero element must be inversible.

However, every finite integral domain is always a field. For example, the ring ({0, 1, 2}, $+_3$, ×3) ισ an integral domain and it is also a field.

SUBFIELDS

Definition: *Let F be a field. A non-empty subset K of the set F is said to be a subfield of F if K is closed with respect to the operations of addition and multiplication in F and K itself is a field for these operations.*

Conditions for a Subfield

Theorem:

The necessary and sufficient conditions for a non-empty subset K of a field F to be a subfield of F are

1. $a \in K, b \in K \Rightarrow a - b \in K$,
2. $a \in K, 0 \neq b \in K \Rightarrow ab^{-1} \in K$.

Proof:

The conditions are necessary: Suppose K is a subfield of the field F. Now K is a group with respect to addition. Therefore $b \in K \Rightarrow -b \in K$. Also K is closed with a respect to addition.

$$\therefore \quad a \in K, b \in K \Rightarrow a + (-b) \in K \Rightarrow a - b \in K.$$

Now each non-zero element of K possesses multiplicative inverse. Therefore $0 \neq b \in K \Rightarrow b^{-1} \in K$.

But K is closed with respect to multiplication.

$$\therefore \quad a \in K, 0 \neq b \in K \Rightarrow ab^{-1} \in K.$$

Hence the conditions are necessary.

The conditions are sufficient: Suppose K is a non-empty subset of F and the conditions (1) and (2) are satisfied.

As we have proved in subrings, we can prove that with the help of condition (1), (K, +) is an abelian group. (Give the same proof here].

Now let a be any non-zero element of K. Then from (2) we have $a \in K$ $0 \neq a \in K \Rightarrow aa^{-1} \in K \Rightarrow 1 \in K$

Now $1 \in K$, therefore again from (2), we have

$$1 \in K \quad 0 \neq a \in K \Rightarrow 1a^{-1} \in K \Rightarrow a^{-1} \in K.$$

$\therefore$ Each non-zero element of K possesses multiplicative inverse.

Now let $a \in K$ and $0 \neq a \in K$. Then $b^{-1} \in K$.

From (2), we have

$$a \in K, 0 \neq b^{-1} \in K \Rightarrow a\,(b^{-1})^{-1} \in K \Rightarrow ab \in K$$

Also if $b = 0$, then $ab = 0$ and $0 \in K$.

$$\therefore \quad ab \in K \ \forall\ a, b, \in K.$$

Associativity of multiplication and distributivity of multiplication over addition must hold in K since they hold in F.

Example:

If R is a ring, show that

$$Z(R) = \{x \in R : xy = yx \ \forall\, y \in R\}$$

is a subring of R. Further show that Z(R) is a field if R is a division ring.

Solution:

We have $0y = 0 = y0 \ \forall\, y \in R$. Therefore $0 \in Z(R)$ and so $Z(R) \neq \varnothing$.

Now let $z_1, z_2 \in Z(R)$. Then

$$z_1y = yz_1 \text{ and } z_2y = yz_2 \ \forall y \in R.$$

Now $\forall y \in R$, we have

$$(z_1 - z_2)\,y = z_1y - z_2y = yz_1 - yz_2 = y\,(z_1 - z_2)$$

and $(z_1z_2)\,y = z_1\,(z_2y) = z_1\,(yz_2)\,z_2 = (z_1y)\,z_2 = (yz_1)\,z_2 = y\,(z_1z_2)$.

$\therefore$ by definition of Z(R), both $z_1 - z_2$ and $z_1z_2 \in Z(R)$.

Thus $z_1, z_2 \in Z(R) \Rightarrow z_1 - z_2 \in Z(R)$ and $z_1z_2 \in Z(R)$.

$\therefore$ Z(R) is a subring of R.

Now suppose R is a division ring *i.e.*, R is a ring with unity and every non-zero element of R possesses multiplicative inverse. Then to prove that Z(R) is a field.

Z(R) is a commutative ring. Let $z_1, z_2 \in Z(R)$.

Now $z_1 \in Z(R) \Rightarrow z_1y = yz_1 \ \forall y \in R$. Since $z_2 \in R$, therefore $z_1z_2 = z_2z_1$.

Thus $z_1z_2 = z_2z_1 \ \forall z_1, z_2 \in Z(R)$ and so Z(R) is a commutative ring.

The Ring Z(R) Possesses Multiplicative Identity: If 1 denotes the unity element of the division ring R, then $1y = y = y1 \ \forall y \in R$. Therefore $1 \in Z(R)$ and is also the unity element of Z(R). Thus Z(R) is also a ring with unity.

Each Non-zero Element of the Ring Z(R) Possesses Multiplicative Inverse: Let $0 \neq z \in Z(R)$ and let z^{-1} denote the multiplicative inverse of z in the division ring R. We shall show that $z^{-1} \in Z(R)$.

We have $z \in Z(R) \Rightarrow zy = yz \ \forall y \in R$

$\Rightarrow \ z^{-1}(zy)\,z^{-1} = z^{-1}(yz)\,z^{-1} \ \forall y \in R$

$\Rightarrow \ (z^{-1}z)\,yz^{-1} - (z^{-1}y)\,zz^{-1} \ \forall y \in R$

$\Rightarrow \ 1\,(yz^{-1}) = (z^{-1}y)\,1 \qquad \forall y \in R$

$\Rightarrow \ yz^{-1} = z^{-1}y \ \forall y \in R.$

$\therefore$ by the definition of Z(R), $z^{-1} \in Z(R)$.

Since $zz^{-1} = 1 = z^{-1}z$. where 1 is the unity element of the ring Z(R), therefore z^{-1} is also the multiplicative inverse of z in the ring Z(R).

Thus each non-zero element of the ring Z(R) is inversible.

Since Z(R) is a commutative ring with unity and each non-zero element of Z(R) is inversible, therefore Z(R) is a field.

Now $S_1 \cup S_2$ is not a subring of the ring of integers because $S_1 \cup S_2$ is not closed for addition.

We have $2 \in S_1 \cup S_2$ is because $2 \in S_1$ and $3 \in S_1 \cup S_2$ because $3 \in S_2$. But $2 + 3 = 5 \notin S_1 \cup S_2$ because neither $5 \in S_1$ nor $5 \in S_2$. Thus $S_1 \cup S_2$ is not closed for addition. Hence $S_1 \cup S_2$ is not a subring of the ring of integers.

However $S_1 \cup S_3 = S_1$ because $S_3 \subset S_1$. Thus $S_1 \cup S_2$ is a subring of the ring of integers.

From the above example it is obvious that the union of two subrings may or may not be a subring.

CHARACTERISTIC OF A RING

Definition: *Let R be a ring with zero element 0 and suppose there exists a positive integer n such that na = a + a +... upto n terms = 0 for every a ∈ R. The smallest such positive integer n is called the characteristic of the ring. If there exists no such positive integer, then R is said to be for characteristic zero or infinite.*

If any element of a ring R is of order zero when regarded as an element of the additive group (R, +), Then R will be of zero characteristic.

The ring of integers is of characteristic zero. The ring of rational numbers is also of characteristic zero.

If $I_6 = \{0, 1, 2, 3, 4, 5,\}$, then the ring $(I, +_6, \times_6)$ *i.e.*, the ring of integers modulo 6 has characteristic 6 since 6x = 0 for every x in the ring. Obviously, no integer smaller than 6 satisfies this property. For instance, 5 cannot be the characteristic, since 5(2) = 4 in I_6 and $4 \neq 0$.

Theorem 1:

The characteristic of a ring with unity is 0 or n > 0 according as the unity element 1 regarded as a member of the additive group of the ring has the order zero or n.

Proof:

Let R be a ring with unity element 1. If 1 has order zero, then the characteristic of the ring is zero.

Suppose 1 is of finite order n so that

$$1 + 1 + 1 + \ldots \text{ upto n terms} = 0 \text{ i.e., } n1 = 0.$$

Let a be any element of R. Then, we have

$$\begin{aligned} na &= a + a + a + \ldots \text{ upto n terms} \\ &= 1a + 1a + 1a + \ldots \text{ upto n terms} \\ &= (1 + 1 + 1 + \ldots \text{ upto n terms})\, a \qquad \text{[by dist. law]} \\ &= (n1)\, a = 0a = 0. \end{aligned}$$

∴ order of a is ≤ n.

Hence the characteristic of the ring is n.

Theorem 2(a):

The characteristic of an integral domain is 0 or n > 0 according as the order of any non-zero element regarded as a member of the additive group of the integral domain is either 0 or n.

Proof:

Let D be an integral domain.

If a non-zero element a of D is of order zero, then the characteristic of D is zero.

Let the order of the non-zero element a be finite and equal to n. Then na = 0.

Suppose b is any other non-zero element of D.

We have na = 0

$\Rightarrow$ (na) b = 0

$\Rightarrow$ (a + a + a + ... upto n terms) b = 0

$\Rightarrow$ (ab + ab + ab + ... upto n terms) = 0

$\Rightarrow$ a (b + b + b + ... upto n terms) = 0

$\Rightarrow$ a (nb) = 0.

But D is without zero divisors. Therefore a $\neq$ 0 and a (nb) = 0 $\Rightarrow$ nb=0.

But the order of a is n $\Rightarrow$ n is the least positive integer such that na =0. Also we have n0 = 0. Thus n is the least positive integer such that nx = 0 $\forall$ x $\in$ D, hence D is of characteristic n.

Theorem 2(b):

Each non-zero element of an integral domain D, regarded as a member of the additive group of D, is of the same order.

Proof:

Let D be an integral domain, Suppose a is a non-zero element of D and o (a) is finite and say, equal to n.

Suppose b is any other non-zero element of D and o (b) = m.

We have o (a) = n $\Rightarrow$ na = 0

$\Rightarrow$ nb = 0

$\Rightarrow$ o (b) $\leq$ n $\Rightarrow$ m $\leq$ n.

Similarly o (b) = m $\Rightarrow$ mb = 0 $\Rightarrow$ a (mb) = 0

$\Rightarrow$ a (b + b + ... upto m times) = 0

$\Rightarrow$ (ab + ab + ab + ... upto m times) = 0

$\Rightarrow$ (a + a + a + ... upto m times) b = 0

$\Rightarrow$ (ma) b = 0

$\Rightarrow$ ma = 0 $\quad$ [$\because$ b $\neq$ 0 and D is without zero divisors]

$\Rightarrow$ o (a) $\leq$ m $\Rightarrow$ n $\leq$ m.

Now m $\leq$ n, n $\leq$ m $\Rightarrow$ m = n. Hence o (a) = o (b).

Also if o (a) is zero, then o (b) cannot be finite. Because o (b) = m $\Rightarrow$ ma = 0 i.e, the order of a is finite. Hence o (b) must also be zero. Hence the theorem.

Theorem 2(c):

The characteristic of an integral domain is either 0 or a prime number.

Proof:

Suppose D is an integral domain. Let $0 \neq a \in D$. If o (a) is zero, then the characteristic of D is 0. If o (a) is finite, let o (a) = p. Then the characteristic of D will be p. We are to prove that p must be prime.

Suppose p is not prime. Let $p = p_1p_2$ where $p_1 \neq 1$, $p_2 \neq 1$ and $p_1 < p$ also $p_2 < p$.

Since D is an integral domain, therefore the product to two non-zero elements of d cannot be equal to 0.

$\therefore \quad aa \neq 0$ *i.e.*, $a^2 \neq 0$.

Now in an integral domain two non-zero elements are of the same order.

$\therefore \quad$ o (a) = p $\Rightarrow$ o (a^2) = p $\Rightarrow$ $pa^2 = 0$

$\Rightarrow (p_1p_2)\, a^2 = 0 \qquad [\because \; p = [p_1p_2]$

$\Rightarrow (a^2 + a^2 + a^2 + \ldots \text{ upto } p_1p_2 \text{ terms}) = 0$

$\Rightarrow (p_1a)\,(p_1a) = 0$

$\Rightarrow$ either $p_1a = 0$ or $p_2a = 0$ [$\because$ D is without zero divisors]

But $p_1 < p$ and $p_2 < p$. Also p is the least positive integer such that pa = 0. Hence p must be prime.

CHARACTERISTIC OF FIELD

Every field is an integral domain. Therefore the characteristic of a field F is 0 or n > 0 according as any non-zero element (in particular the unit element 1) of F is of order 0 or n.

Thus in order to find the characteristic of a field F, we should find the order of the unit element 1 of F when regarded as a member of the additive group of F. If the order of 1 is zero then f is of characteristic 0. If the order of 1 is finite, say, n then the characteristic of F is n.

Then characteristic of the field of real numbers is 0.

The characteristic of the finite field $(I_7, +_7, \times_7)$ is 7 where $I_7 = \{0, 1, 2, 3, 4, 5, 6\}$.

ORDERED INTEGRAL DOMAINS

Definition: *An integral domain $(D, +, \cdot)$ is said to be ordered if D contains a subset D_+ such that*

1. *D_+ is closed with respect to addition and multiplication as defined on D.*
2. *$\forall a \in D$, one and only one of $a = 0$, $a \in D_+$,... $a \in D_+$ holds (principle of Trichotomy).*

The elements of D_+ are called the positive elements of D, all other non-zero elements of D are called negative elements of D.

Ordered Field: *A field $(F, +, \cdot)$ is to be ordered if it is ordered as an integral domain.*

The integral domain $(I_+, +, \cdot)$ of all integers is ordered.

The set I_+ of all positive integers is the set of the positive elements of this integral domain. We know that the sum and product of two positive integers is again a positive integer *i.e.*, I_+ is closed with respect to addition and multiplication. If $a \in I$, then either is a zero or positive or negative *i.e.*, either $a = 0$ or $a \in I_+$ or $-a \in I_+$.

The field of rational numbers is an ordered field. The field of real numbers is also an ordered field. But the field of complex numbers is not an ordered field.

Theorem 1:

The field $(I_p, +_p, \times_p)$ where p is a prime and $I_p = \{0, 1, ..., p - 1\}$ is not ordered.

Proof:

Suppose I_p is an ordered field and P is the set of positive elements of the field. The zero element of this field is 0.

Now $1 \neq 0$. The additive inverse of 1 is $p - 1$. According to the definition of ordered field either $1 \in P$ or its additive inverse $p - 1 \in P$.

But P is closed with respect to $+_p$.

Therefore $1 \in P \Rightarrow 1 +_p 1 +_p 1 +_p$ ...upto $p - 1$ times $\in P$

$$\Rightarrow p - 1 \in P.$$

This contradicts the principle of trichotomy. Similarly assuming that $p - 1 \in P$ we can show that its additive inverse 1 also belongs to P. This again contradicts the principle of trichotomy. Hence the given field is not an ordered field.

Theorem 2:

Let D be an integral domain with unity element 1. If D is an ordered integral domain show that 1 is a positive element of D

Proof:

Let D be an ordered integral domain with unity element 1. Let D_+ denote the set of positive elements of D.

Suppose $1 \notin D_+$.

Now $1 \neq 0$. Since $1 \notin D_+$, therefore by the definition of an ordered integral domain,

$$-1 \in D_+$$

$$\Rightarrow (-1)(-1) \in D_+$$

[$\because D_+$ is closed with respect to multiplication]

$$\Rightarrow 1 \in D_+ \text{ which is a contradiction.}$$

Hence $1 \in D_+$ *i.e.*, 1 is a positive element of D.

Theorem 3:

The field $(C, +, \cdot)$ of complex numbers is not ordered.

Proof:

Suppose C is an ordered field and C_+ is the set of positive elements of this field. The additive identity *i.e.*, the zero element is $0 + i0$.

Now $i \neq 0$.

By the principle of trichotomy either $i \in D_+$ or $-i \in C_+$.

Now C_+ is closed with respect to multiplication.

$\therefore i \in C_+, -1 \in C_+ \Rightarrow i(-1) = -i \in C_+$. Thus if $i \in C_I$, its additive inverse $-i$ also belongs to C_+. This contradicts the principle of trichotomy *i.e.*, the condition (2) of the definition of an ordered integral domain.

Similarly if we assume that $-1 \in C_+$, we can show that its additive inverse i also belongs to C_+. This again contradicts the principle of trichotomy.

Hence the field of complex numbers is not an ordered field.

ORDER RELATIONS IN AN ORDERED INTEGRAL DOMAIN

Definition: *Let D be an ordered integral domain and D_+ be the set of positive elements of D. Then we define less than' (<) 'greater than' (>) relations in D as follows:*

For all a, b, ∈ D, we have

1. $a > b$ when $a - b \in D_+$,
2. $a < b$ when $b - a \in D_+$.

Obviously a > iff b < a.

Theorem:

The order relation in an ordered integral domain is transitive i.e., $a > b$, $b > c \Rightarrow a > c$.

Proof:

Let D be an ordered integral domain and let D_+ be the set of positive elements of D.

We have $a > b \Rightarrow a - b \in D_+$ [by def. of >]

and $b > c \Rightarrow b - c \in D_+$.

Now D_+ is closed with respect to addition.

$$\therefore a - b \in D_+, b - c \in D_+ \Rightarrow (a - b) + (b - c) \in D_+$$
$$\Rightarrow a - c \in D_+ \Rightarrow a > c.$$

ORDERED FIELD

Definition: A field (F, + , ·) is said to be ordered if F contains a subset F_+ such that

1. F_+ is closed with respect to addition and multiplication as defined on F.
2. $\forall a \in F$, one and only on of $a = 0$, $a \in F_+$, $-a \in F_+$ holds *(Principle of Trichotomy).*

The elements of F_+ are called the positive elements of F, all other non-zero elements of F are called negative elements of F.

An Example of an Ordered Field: The field of real numbers (R, +, ·) is an ordered field. The set R_+ of all positive real numbers is the set of the positive elements of this field. We know that the sum and product of two positive real numbers is again a positive real number *i.e.*, R_+ is closed with respect to addition and multiplication. If a $\in$ R, then either a is zero or positive or negative *i.e.*, either a = 0 or a $\in R_+$ or –a $\in R_+$.

The field of rational numbers (Q, +, ·) is also an ordered field. But the field of complex numbers (C, + , ·) is not an ordered field.

Example:

Show that every finite integral domain is of finite characteristic.

Solution:

Let (R, +, ·) be a finite integral domain.

Let a be any non-zero element of R. Then we know that the characteristic of the integral domain (R, +, ·) is equal to the order of the element a regarded as a member of the additive group of the integral domain *i.e.*, regarded as a member of the additive group (R, +).

But from our study of the order of an element in the group theory we know that the order of every element of a finite group is finite. Therefore the order of a as a member of the additive group (R, +) is finite.

Hence very finite integral domain is of finite characteristic.

IDEALS

Definition:

(a) **Left Ideal:** *A non-empty subset S of a ring R is said to be a left ideal of R if:*

1. *S in a subgroup of R with respect to addition.*
2. *rs $\in$ S $\forall$ r $\in$ R and $\forall$ s $\in$ S.*

(b) **Right Ideal:** *A non-empty subset S of a ring R is said to be a right ideal of R if:*

1. *s is a subgroup of R under addition,*
2. *sr $\in$ R $\forall$ r $\in$ R and $\forall$ s $\in$ R*

(c) **Ideal:** A non-empty subset S of a ring R is said to be an *ideal* (also a *two sided ideal*) if and only if it is both a left ideal and a right ideal. *Thus, f a non-empty subset S of a ring R is said to be an ideal of R if:*

1. *S is a subgroup of R under addition i.e., S is a subgroup of the additive group of R.*
2. *rs $\in$ S and sr $\in$ S for every r $\in$ R and for every s $\in$ S.*

If S is an ideal of a ring R, then S is also a subring of R. The obvious reason is that S is a subgroup of R under addition and from condition (2), we have xs ∈ S ∀ x s ∈ S because x ∈ S ⇒ x ∈ R. Thus s is closed with respect to multiplication. Therefore S is a subring of *R. Thus every ideal of a ring R is also a subring of R.* But every subring is not an ideal. An ideal requires a stronger closure property than the subring. If S is a subgroup of R under addition, then S will be a subring if S is closed with respect to multiplication *i.e.*, the product of two elements of S with any element of R is in S.

If R is commutative ring, then very left ideal will also be a right ideal. Therefore in a commutative ring every left (right) ideal is an ideal.

Notes:

1. If we are to prove that a non-empty subset S of a ring R is an ideal of R, then it is sufficient to prove that

   (i) a ∈ S, b ∈ S ⇒ a – b ∈ S, and

   (ii) rs ∈ S and sr ∈ S ∀ r ∈ R and ∀ r ∈ S.

   Obviously (i) is a sufficient condition for S to be a subgroup of R under addition.

   If R is a commutative ring, then the condition (ii) will be come more simple. Then it will become

   rs ∈ S ∀ r ∈ R and ∀ s ∈ S.

2. Every ring R always possesses two improper ideals: one R itself and the other consisting of 0 only. These are respectively know as the unit ideal and the null ideal.

   Any other ideals of R are called proper ideals. A ring having no proper ideals is called a simple ring.

Theorem 1:

An arbitrary intersection of left ideals of a ring is a left ideal of the ring.

Proof:

Let R be a ring and let $\{S_t : t \in T\}$ be any family of left ideals of R. Here T is an index set and is such that ∀ t ∈ T, S_t is a left ideal of R. Let

$$S = \bigcap_{t \in T} S_t = \{x \in R : x \in S_t \ \forall\, t \in T\}$$

be the intersection of this family of left ideals of R. Then to prove that S is also a left ideal of R.

Obviously S ≠ ∅, since at least 0 is in S_t ∀ t ∈ T.

Now let a, b be any two elements of S. Then

$a, b, \in S \Rightarrow a, b, \in S_t \; \forall \; t \in T$

$\Rightarrow a - b \in S_t \; \forall \; t \in T$

$[\because \forall \; t \in T, S_t$ is a left ideal of R]

$\Rightarrow a - b \in \bigcap_{t \in T} S_t \Rightarrow a - b \in S.$

Now let a be any element of S and R be any element of R.

We have $a \; a \in S \Rightarrow a \; a \in \bigcap_{t \in T} S_t \; \forall \; t \in T$

$\Rightarrow \quad rs \in S_t \; \forall \; t \in T$ $\quad [\because \forall \; t \in T, S_t$ is a left ideal of R]

$\Rightarrow \quad rs \in \bigcap_{t \in T} S_t \Rightarrow ra \in S.$

Thus $a, b, \in S \Rightarrow a - b \in S$ and $r \in R, a \in S \Rightarrow ra \in S.$

$\therefore$ S is a left ideal of R.

Theorem 2:

The intersection of any two left ideals of a ring is again a left ideal of the ring.

Proof:

Let I_1 and I_2 be two left ideals of a ring R. Then I_1, I_2 are subgroups of R under addition. Therefore $I_1 \cap I_2$ is also a subgroup of R under addition.

Now to show that $I_1 \cap I_2$ is a left ideal of R we are only to show that $r \in R, s \in I_1 \cap I_2 \Rightarrow rs \in I_1 \cap I_2$.

We have $s \in I_1 \cap I_2 \Rightarrow s \in I_1, s \in I_2$.

But I_1 and I_2 are left ideals of R. Therefore

$r \in R, s \in I_1 \Rightarrow rs \in I_1$ and $r \in R, s \in I_2 \Rightarrow \in I_2$.

Now $rs \in I_1, rs \in I_2 \Rightarrow rs \in I_1 \cap I_2$.

$\therefore I_1 \cap I_2$ is also a left ideal of R.

Note: A similar result can be proved for right ideals as well as for ideals.

SMALLEST LEFT IDEAL CONTAINING A GIVEN SUBSET

Definition: *Let M be a non-empty subset of a ring. Then a left ideal I of R is called the smallest left ideal of R containing M, if I contains M and if 1 is contained in every left ideal of R containing M.*

The smallest left ideal of R containing M is called the left ideal generated by M and will be denoted by (M)

It can be easily seen that the intersection of the family of left ideals containing M is the left ideal generated by M.

Remark: *A similar definition can be given for the right ideal generated by M as well as for the ideal generated by M. For this purpose simply replace the word 'left ideal' by 'right ideal' or by 'ideal'.*

Sum of Two Left Ideals

Theorem:

The left ideal generated by the union $I_1 \cup I_2$ of the left ideals is the set $I_1 + I_2$ consisting of the elements of R obtained on adding any element of I_1 to any element of I_2.

Proof:

Let $a_1 + a_2, b_1 + b_2 \in I_1 + I_2$.

Then $a_1, b_1 \in I_1$ and $a_2, b_2 \in I_2$.

Since I_1, I_2 are left ideals of R, therefore they are subgroups of the additive group of R. Therefore

$a_1, b_1 \in I_1 \Rightarrow a_1 - b_1 \in I_1$ and $a_2, b_2 \in I_2 \Rightarrow a_2\ b_2 \in I_2$.

Consequently $(a_1 + a_2) - (b_1 + b_2) = (a_1 - b_1) + (a_2 - b_2) \in I_1 + I_2$.

Therefore $I_1 + I_2$ is a subgroup of the additive group of R.

Now let $r \in R$ and $a_1 + a_2 \in I_1 + I_2$. Then $a_1 \in I_1, a_2 \in I_2$.

We have $r(a_1 + a_2) = ra_1 + ra_2 \in I_1 + I_2$.

[$\because$ I_1 is a left ideal implies $ra_1 \in I_1$ and similarly $ra_2 \in I_2$]

$\therefore I_1 + I_2$ is a left ideal of R.

Since $0 \in I_2$, therefore $a_1 \in I_1$ can be written as $a_1 + 0$. Thus

$$a_1 \in I_1 \Rightarrow \in I_1 + I_2.$$

$$\therefore \quad I_1 \subseteq I_1 + I_2.$$

Similarly $I_2 \subset I_1 + I_2$.

$$\therefore \quad I_1 \cup I_2 \subseteq I_1 + I_2.$$

Thus $I_1 + I_2$ is a left ideal containing $I_1 \cup I_2$.

Also if any left ideal contains $I_1 \cup I_2$, then it must contain $I_1 + I_2$.

$\therefore I_1 + I_2$ is the smallest left ideal containing $I_1 \cup I_2$.

$\therefore I_1 + I_2 =$ the left ideal generated by $I_1 \cup I_2 = (I_1 \cup I_2)$.

Note: A similar result can be proved for right ideals as well as for ideals.

Distinction Between Subrings and Ideals in a Ring: Let R be any ring and S be any non-empty subset of R.

So far as the operation of addition is concerned, whether S is a subring or an ideal of R, S must be a subgroup of the additive group of R *i.e.*, $a \in S, b \in S \Rightarrow a - b \in S$.

But so far as the operation of multiplication is concerned, for S to be a subring of R, the product of any two elements of S must remain in S *i.e.*, $a \in S, b \in S \Rightarrow ab \in S$. While on the other hand for S to be an ideal of R the product of any element of R and any element of S must remain in S *i.e.*, $r \in r, s \in S \Rightarrow rs \in S$ and $sr \in S$.

Thus, an ideal requires a stronger closure property for multiplication then a subring. *In fact every ideal is a subring while a subring may or may not be an ideal.* For example the set of integers is only a subring but not an ideal of the ring of rational numbers. On the other hand the set of even integers is a subring as well as an ideal of the ring of integers.

Second Part of the Question: Let M be the ring of all 2×2 matrices with elements as integers for addition and multiplication of matrices as the two ring operations.

Let S be the subset of M consisting of matrices of the form

$$\begin{bmatrix} a & 0 \\ b & c \end{bmatrix},$$ where a, b, c are any integers.

To show that S is a subring of M.

Let $A = \begin{bmatrix} a_1 & 0 \\ b_1 & c_1 \end{bmatrix}$, $B \begin{bmatrix} a_2 & 0 \\ b_2 & c_2 \end{bmatrix}$ be any two elements of S.

Then $A - B = \begin{bmatrix} a_1 - a_2 & 0 \\ b_1 - b_2 & c_1 - c_2 \end{bmatrix}$ which is obviously an element of S.

Also $AB = \begin{bmatrix} a_1 - a_2 & 0 \\ b_1 a_2 + c_1 b_2 & c_1 c_2 \end{bmatrix}$ which is also an element of S.

Hence S is a subring of the ring M.

The subring S of M is not an integral domain because it possesses zero divisors. The zero element of the subring S is the mull matrix $\begin{bmatrix} 0 & 0 \\ 0 & 0 \end{bmatrix}$.

Now $A = \begin{bmatrix} 1 & 0 \\ 0 & 0 \end{bmatrix}$ and $B = \begin{bmatrix} 0 & 0 \\ 1 & 0 \end{bmatrix}$ are two non-zero elements of S but

$AB = \begin{bmatrix} 0 & 0 \\ 0 & 0 \end{bmatrix}$ = the zero element of S. Thus both A and B are zero divisors.

Hence S is not an integral domain.

SOLVED EXAMPLES

Example 1:

Prove that in a field

1. $\frac{a}{b} = \frac{c}{d} \Leftrightarrow ad = bc.$
2. $\frac{a}{b} - \frac{c}{d} = \frac{ad - bc}{bd}$
3. $(-a)^{-1} = -(a^{-1}).$
4. $\frac{(-a)}{(-b)} = \frac{a}{b}.$

Solution:

Let a, b, c, d ∈ F where F is a field. Here $b \neq 0$, $d \neq 0$.

1. We have $\frac{a}{b} = \frac{c}{d} \Leftrightarrow ab^{-1} = cd^{-1} \Leftrightarrow (ab^{-1})$ $(bd = (cd^{-1})$ (bd)

$\Leftrightarrow (ad)\,(b^{-1}\,b) = (bc)\,(d^{-1}\,d)$ [∵ in a field multiplication is commutative as well as associative]

$\Leftrightarrow (ad)\,1 = (bc)\,1 \Leftrightarrow ad = bc.$

2. We have $\frac{a}{b} - \frac{c}{d} = (ab^{-1}) - (cd^{-1})$

$= (bd)^{-1}\,(bd)\,[(ab^{-1}) - (cd^{-1})]$

$= (bd)^{-1}\,[(bd)\,(ab^{-1}) - (bd)\,(cd^{-1})]$

$= (bd)^{-1}\,(ad - bc) = \frac{ad - bc}{bd}.$

3. We have $(-a)\,[-(a^{-1})] = aa^{-1}$ [∵ in a field $(-a)\,(-b) = ab$] $= 1.$

∴ $(-a)^{-1} = -(a^{-1}).$ [∵ in a field $ab = 1 \Leftrightarrow a^{-1} = b$]

4. We have $\frac{(-a)}{(-b)} = (-a)\,(-b)^{-1} = (-a)\,[-\,(b^{-1})]$

[$\because$ by part (3) of this question $(-b)^{-1} = -\,(b^{-1})$]

$= ab^{-1}$ [$\because$ in a field $(-a)\,(-b) = ab$]

$= \frac{a}{b}$.

Example 2:

The set I of all integers is a ring with respect to addition and multiplication of integers as the two ring compositions. The ring is called the ring of integers.

Solution:

As in groups, we should first prove that I is an abelian group with respect to addition of integers

1. The product of two integers is also an integer. Therefore I is closed with respect to multiplication of integers.
2. Multiplication of integers is an associative composition.
3. Multiplication of integers is distributive with respect to addition of integers *i.e.*, if a, b, c are any elements of I, then

$$a\,(b + c) = ab + ac \text{ and } (b + c)\,a = ba + ca.$$

Therefore I is a ring with respect to addition and multiplication of integers. The integer 0 is the zero element of the ring. Also the multiplicative identity exists and it is the integer 1. We have $1a = a = a1 \; \forall \; a \in I$. Thus the ring of integers is a ring with unity. The integer 1 is the unit element of this ring.

Since the multiplication of integers is a commutative composition. Therefore it is also a commutative ring.

Example 3(a):

The set $R = \{0, 1, 2, 3, 4, 5\}$ is a commutative ring with respect to '$+_6$' and '$\times_6$' as the two ring compositions.

Solution:

We have already prove in group, we should first prove that R is an abelian group with respect to '$+_6$'.

We form a composition table for the given composition $\times_6$.

$\times_6$	0	1	2	3	4	5
0	0	0	0	0	0	0
1	0	1	2	3	4	5
2	0	2	4	0	2	4
3	0	3	0	3	0	3
4	0	4	2	0	4	2
5	0	5	4	3	2	1

From the composition table we find that R is closed with respect to the composition '$\times_6$'.

Also we know that '$\times_6$' is an associative composition in R *i.e.*, $a \times_6 (b \times_6 c) = (a \times_6 b) \times_6 c \ \forall \ a, b, c \in R$. Hence it held associate low

Further '$\times_6$' is distributive in R with respect to '$\times_6$'. If a, b, c are any elements of R, then, we have

$a \times_6 (b \times_6 c) = a \times_6 (b + c)$ $\qquad [\because \ b + c \equiv {}_6c \pmod 6]$

= least non-negative remainder when a (b + c) is divided by 6

= least non-negative remainder when a ab + ac is divided by 6

$= (ab) +_6 (ac)$

$= (a \times_6 b) +_6 (ac)$ $\qquad [\because \ a \times {}_6b \equiv ab \pmod 6]$

$= (a \times_6 b) +_6 (a \times_6 c)$ $\qquad [\because \ a \times {}_6c \equiv ac \pmod 6]$

Similarly, we can prove that $(b \times_6 c) \times_6 a$

$= (b \times_6 a) +_6 (c \times_6 a)$.

$\therefore$ R is a ring with respect to the given compositions. Since '$\times_6$' is a commutative composition in R as is clear from the composition table also, therefore R is commutative ring. Also 1 is the identity element for the composition '$\times_6$'. Therefore R is a ring with unity. The integer 0 is the zero element of the ring.

Example 3(b):

The set R = {0, 1, 2, 3, 4, 5} is a commutative ring with respect to '$+_6$' and '$\times_6$' as the two ring compositions.

Solution:

As we have proved in groups, we should first prove that R is an abelian group with respect to '$+_6$'.

The composition table for R is as unclear.

$\times_6$	0	1	2	3	4	5
0	0	0	0	0	0	0
1	0	1	2	3	4	5
2	0	2	4	0	2	4
3	0	3	0	3	0	3
4	0	4	2	0	4	2
5	0	5	4	3	2	1

From the composition table we have R is closed with respect to the composition '$\times_6$'.

Also we know that '$\times_6$' is an associative composition in R *i.e.*, $a \times_6 (b \times_6 c)$ $(a \times_6 b) \times_6 c$ $\forall$ a, b, c $\in$ R.

Further '$\times_6$' is distributive in R with respect to '$\times_6$'. If a, b, c are any elements of R, then

$a \times_6 (b +_6 c) = a \times_6 (b + c)$ $\quad$ [$\because$ $b + c \equiv b +_6 c \pmod 6$]

= least non-negative remainder when a (b + c) is divided by 6

= least non-negative remainder when ab + ac is divided by 6

$= (ab +_6 (ac)$

$= (a \times_6 b) +_6 (ac)$ $\quad$ [$\because$ $a \times_6 b \equiv ab \pmod 6$]

$= (a \times_6 b) +_6 (a \times_6 c)$ $\quad$ [$\because$ $a \times_6 c \equiv ac \pmod 6$]

Similarly, we can prove that $(b \times_6 c) \times_6 a = (b \times_6 a) +_6 (c \times_6 a)$.

$\therefore$ R is a ring with respect to the given compositions. Since '$\times_6$' is a commutative composition in R as is clear from the composition table also, therefore R is a commutative ring. Also 1 is a ring with unity. The integer 0 is the zero element of this ring.

Note: We see that in this ring R neither 2 nor 3 is equal to the zero element of the ring. But $2 \times_6 3 = 0$ (zero element of the ring). Thus, in a ring it is possible that the product of two non-zero elements is equal to the zero element. Also the number of elements in R is finite. Therefore this is an example of a *finite ring.*

Example 3(c):

The set 2I of all even integers is a commutative ring without unity, addition and multiplication of integers being the two ring compositions.

Solution:

Do yourself.

Example 4:

The set M of all $n \times n$ matrices with their elements as real numbers (rational numbers, complex numbers, integers) is a non-commutative ring with unity, with respect to addition and multiplication of matrices as the two ring compositions.

Solution:

We know that the sum and product of two $n \times n$ matrices with their elements as real numbers are again $n \times n$ matrices with their elements as real numbers. Therefore M is closed with respect to addicted and multiplication of matrices.

Here we observe that

1. $A + (B + C) = (A + B) + C \ \forall\ A, B, C \in M$, since the addition of matrices is an associate the composition
2. $A + B = B + A \ \forall\ A, \in M$ since the addition of matrices is commutative.
3. If O is the null matrix of the type $n \times n$, then $O \in M$ and we have $O + A = A \ \forall\ A \in M$.
4. To each matrix $A \in M$ there exists a matrix $-A \in M$ such that $(-A) + A = O$ (null matrix).
5. $(AB)\,C = A\,(BC), \ \forall\ A, B, C \in M$, since multiplication of matrices is associative.
6. $A\,(B + C) = AB + AC$,

   and $(B + C)\,A = BA + CA \ \forall\ A, B, C \in M$, since matrix multiplication is distributive with respect to matrix addition.

Hence, M is a ring with respect to the two given compositions. The null matrix O of the type $n \times n$ is the zero element of this ring *i.e.*, $O = 0$.

Since the multiplication of matrices is not always commutative, therefore the ring is a non-commutative ring. $[n > 1]$

Now if I be the unit of the type $n \times n$, then $I \in M$ and we have $IA = A$ $AI \ \forall\ A \in M$. Therefore the matrix 1 is the multiplicative identity. Thus the ring is with unity and the matrix I is the unity element of the ring *i.e.*, $I = 1$.

Example 5:

If a ring R has a left identity as well as right identity, then the two are equal.

Solution:

Let R be a ring which possesses left identity e and right identity e' *i.e.*, ea = a $\forall$ a $\in$ R and ae' = a $\forall$ a $\in$ R.

To prove that e = e'.

Since e is left identity and e $\in$ R, therefore

$$ee' = e'. \tag{1}$$

Again e' is right identity and e $\in$ R.

$$\therefore \quad ee' = e. \tag{2}$$

Since ee' is a unique element of R, therefore from (1) and (2) we can lead that eqn.

Example 6(a):

Prove that the set {0, 1, 2} (mod 3) is a field with respect to addition and multiplication.

Solution:

Let R = {0, 1, 2,}. To prove that (R, $+_3$, $\times_3$) is a field. First we shall show that (R, $+_3$) is an abelian group. The composition table for R for the operation $+_3$ is as given below.

$+_3$	0	1	2
0	0	1	2
1	1	2	0
2	2	0	1

From the composition table we see that R is closed for $+_3$ and the operation $+_3$ on the set R is commutative. The operation $+_3$ on the set R is also associative as we know it.

The element 0 $\in$ R is identity for $+_3$. The inverses of 0, 1, 2, for $+_3$ are 0, 2, 1 respectively. Therefore (R, $+_3$) is an abelian group.

Now we prepare composition table for R for the operation $\times_3$.

$+_3$	0	1	2
0	0	0	0
1	0	1	2
2	0	2	1

From the composition table we see that R is closed for $\times_3$ and the operation $\times_3$ on the set R is commutative.

The operation $\times_3$ on the set R is associative. For if a, b, c $\in$ R, then

$$a \times_3 (b \times_3 c) = (a \times_3 b) \times_3 c \text{ because } a (bc) = ab) c.$$

Also the operation $\times_3$ distributes over the operation $+_3$ For if a, b, c, $\in$ R, then

$$a \times_3 (b +_3 c) = (a \times_3 b) +_3 (a \times_3 c)$$

and $$(b +_3 c) \times_3 a = (b \times_3 a) +_3 (c \times_3 a).$$

The element $1 \in$ R is identity for the operation $\times_3$ because from the composition table we see that

$$1 \times_3 a = a\ a \times_3 1, \forall a \in R.$$

Also each non-zero element of R possesses multiplicative inverse. From the composition table we see that the inverses of the non-zero elements 1 and 2 of R for $\times_3$ are 1 and 2 respectively.

Thus, (R, $+_3$, $\times_3$) is a commutative ring with unity element 1 and each non-zero element of R is inversible. Hence (R, $+_3$, $\times_3$) is a field.

Example 6(b):

Prove that the set of integers is on integral domain with respect to addition and multiplication.

Solution:

Let I = {..., – 3, – 2, – 1, 0, 1, 2, 3,...} be the set of integers. To prove that the algebraic structure (I, +, .) is an integers is also an integer. Therefore I is closed for addition of integers.

Also addition of integers is commutative as well as associative. The integer 0 is identity for addition of integers. Also if a is any integer in I, then the integer –a is its additive inverse. Thus (I, +) is an abelian group.

Now the product of two integers is also an integer. Therefore I is closed for multiplication of integers.

Also multiplication of integers is associative as well as commutative and it distributes over addition of integers. The integer 1 is identity for multiplication. Further the product of two integers can be zero only if at least one of them is zero.

Hence the algebraic structure (I, +, .) is a commutative ring with unity and without zero divisors and so it is an integral domain.

Example 6(c):

Define a ring and an integral domain. Give an example of a ring which is not an integral domain.

Solution:

For the definition of a ring refer § 1 and for the definition of an integral domain refer § 6.

For an example of ring which is not an integral domain consider the ring $(R, +_6, \times_6)$ where $R = \{0, 1, 2, 3, 4, 5\}$.

The above ring $(R, +_6, \times_6)$ is a commutative ring and possesses unity element which is the integer 1. But this ring is not an integral domain because it possesses zero divisors.

We observe that $2 \in R$, $3 \in R$, $2 \neq 0$, $3 \neq 0$, but $2 \times_6 3 = 0$. Thus both 2 and 3 are zero divisors. Hence the ring

$$(\{0, 1, 2, 3, 4, 5\}, +_6, \times_6)$$

is not an integral domain.

Example 7:

Define a ring and furnish an example of (1) a non-commutative ring with unity, (2) a commutative ring without unity.

Solution:

1. *Example of a non-commutative ring with unity:* Let M be the set of all 2×2 matrices with elements as real numbers. Then M is a ring for addition and multiplication of matrices as the two ring operations.

This ring is a non-commutative ring because the operation of multiplication of matrices on the set M is not commutative.

For example, if we take $A = \begin{bmatrix} 1 & 2 \\ 3 & 5 \end{bmatrix}$, $B = \begin{bmatrix} 1 & 2 \\ 0 & 1 \end{bmatrix}$ as two members of M, we find that

$$AB = \begin{bmatrix} 2 & 4 \\ 3 & 5 \end{bmatrix} \begin{bmatrix} 1 & 2 \\ 0 & 1 \end{bmatrix} = \begin{bmatrix} 2 & 8 \\ 3 & 11 \end{bmatrix} \text{ and } BA = \begin{bmatrix} 1 & 2 \\ 0 & 1 \end{bmatrix} \begin{bmatrix} 2 & 4 \\ 3 & 5 \end{bmatrix} = \begin{bmatrix} 8 & 14 \\ 3 & 5 \end{bmatrix}.$$

Thus, $AB \neq BA$ and so the ring is a non-commutative ring. But this ring is a ring with unity.

The unit matrix $I = \begin{bmatrix} 1 & 0 \\ 0 & 1 \end{bmatrix} \in M$ and is identity for multiplication of matrices because we have

$$IA = A = AI \ \forall \ A \in M.$$

Therefore the unit matrix I is the unity element of this ring.

Hence the ring of 2×2 matrices with elements as real numbers is a non-commutative ring with unity.

2. *Example of a commutative ring without unity:* The ring of even (2I, +, ·) *i.e.*, the ring

$$(\{..., -6, -4, -2, 0, 2, 4, 6,...\}, +, \cdot)$$

is a commutative ring without unity.

Since the multiplication of integers is a commutative operation, therefore the ring of even integers is a commutative ring. But this ring is without unity because it does not possess multiplicative identity. In the set of even integers there exists no even integer e such that

$$ea = a = ae, \ \forall \ a \in 2I.$$

Hence (2I, +, .) is a commutative ring without unity.

Example 8:

Explain with examples the difference between a field, a skew field and an integral domain.

Solution:

First give the definitions of a field, a skew field and an integral domain. For these definitions.

Example 9:

If in a ring with unity any element a has the multiplicative inverse, then a cannot be a divisor of zero.

Solution:

Let R be a ring with unity 1.

Let $a \in R$ be such that a has multiplicative inverse. To prove that a cannot be a divisor of zero.

Let $a \in R$ be such that $ab = 0 \Rightarrow ba = 0$. Then a cannot be a divisor of zero if we prove that $ab = 0 \Rightarrow ba = 0$ is possible only if $b = 0$

We have $ab = 0 \Rightarrow a^{-1}(ab) = a^{-1}0$ $[\because a^{-1} \text{ exists}]$

$$\Rightarrow (a^{-1}a)\, b = 0 \Rightarrow 1b = 0 \Rightarrow b = 0.$$

Again $ba = 0 \Rightarrow (ba)\, a^{-1} = 0a^{-1} \Rightarrow b(aa^{-1}) = 0 \Rightarrow b1 = 0 \Rightarrow b = 0.$

Hence a cannot be a zero divisor.

Example 10:

Prove that a ring R is commutative if and only if $(a + b)^2 = a^2 + 2ab + b^2 \ \forall\, a, b \in R$.

Solution:

Let R be a commutative ring *i.e.*, $ab = ba \ \forall\, a, b \in R$. Then to prove that $(a + b)^2 = a^2 + 2ab + b^2 \ \forall\, a, b \in R$. Let $a, b \in R$.

We have $(a + b)^2 = (a + b)(a + b)$ $\quad [\because\ a^2 = aa]$

$= aa + ab + ba + bb$, by dist, laws

$= a^2 + ab + ab + b^2$ $\quad [\because\ ba = ab]$

$= a^2 + 2ab + b^2$.

Hence if R is commutative ring, then

$(a + b)^2 = a^2 + 2ab + b^2 \ \forall\, a, b \in R$.

Conversely suppose that $(a + b)^2 = a^2 + 2ab + b^2 \ \forall\, a, b \in R$.

Then to prove that R is a commutative ring.

Let $a, b \in R$. Then

$(a + b)^2 = a^2 + 2ab + b^2$

$\Rightarrow$ $(a + b)(a + b) = a^2 + 2ab + b^2$

$\Rightarrow$ $a^2 + ab + ba + b^2 = a^2 + 2ab + b^2$, by dist. laws

$\Rightarrow$ $ab + ba = 2ab$, by left and right cancellation laws for addition in R

$\Rightarrow$ $ab + ba = ab = ab$ $\quad [\because\ ab + ab = 2ab]$

$\Rightarrow$ $ba = ab$, by left cancellation law for addition in R.

Thus $ab = ba \ \forall\, a, b \in R$ and so R is a commutative ring.

Example 11:

If R is a ring such that $a^2 = a \ \forall\, a \in R$ *prove that* (1) $a + a = 0 \ \forall\, a \in R$ *i.e., each element of R is its own additive inverse.*

(2) $a + b = 0 \Rightarrow a = b$. (3) R is a commutative ring.

Solution:

1. $a \in R \Rightarrow a + a \in R$.

Now $(a + a)^2 = (a + a)$ $\quad$ [given]

$\Rightarrow (a + a)(a + a) = a + a$

$\Rightarrow (a + a)\, a + (a + a)\, a = a + a$ $\quad$ [left dist. law]

$\Rightarrow (a^2 + a) + (a^2 + a^2) = a + a$ $\quad$ [right dist. law]

$\Rightarrow$ (a + a) + (a + a) = a + a [$\because$ $a^2 = a$]

$\Rightarrow$ (a + a) + (a + a) = (a + a) + 0. [$\because$ a + 0 = a]

$\Rightarrow$ (a + a) = 0 [by left cancellation law for addition in R]

2. We have just proved that a + a = 0.

$\therefore$ a + b 0 $\Rightarrow$ a + b = a + a $\Rightarrow$ b = a, by left cancellation law for addition in R.

3. We have

$(a + b)^2 = (a + b)$

$\Rightarrow$ (a + b) (a + b) = (a + b)

$\Rightarrow$ (a + b) a + (a + b) b = a + b [left dist. law]

$\Rightarrow$ $(a^2 + ba) + (ab + b^2) = a + b$ [right dist. law]

$\Rightarrow$ (a + ba) + (ab + b) = a + b [$\because$ $a^2 = a$, $b^2 = b$]

$\Rightarrow$ (a + b) + (ba + ab) = (a + b) + 0 [by commutativity and associativity of addition]

$\Rightarrow$ ba + ab = 0 [by left cancellation law for addition in R]

$\Rightarrow$ ab = ba. [by part (2) of this question]

$\therefore$ R is commutative ring.

Note: An element a of a ring R is said to be idempotent if $a^2 = a$. *A ring R is called a* **Boolean Ring** *if all of its elements are idempotent i.e, if* $a^2 = a \ \forall a \in R$.

Example 12(a):

If R is a ring with unity element 1, then

$(-1)\, a = -\, a = a\, (-1) \ \forall a \in R$ *and* $(-1)(-1) = 1$.

Solution:

Let R be a ring with unity element 1. Let a be any element of R.

We have 0a = 0

$\Rightarrow$ (–1 + 1) a = 0 [$\because$ –1 + 1 = 0]

$\Rightarrow$ (–1) a + 1a = 0 [by a dist. law]

$\Rightarrow$ (–1) a + a = 0 [$\because$ 1a = a]

$\Rightarrow$ (–1) a = – a.

Again a0 = 0

$\Rightarrow$ a (–1 + 1) = 0

$\Rightarrow$ a (–1) + a1 = 0

$\Rightarrow$ a (–1) + a = 0

$\Rightarrow$ a (–1) = – a.

Hence (–1) a = – a = a (–1) $\forall$ a $\in$ R.

Now (–1) (–1) = – (–1) [$\because$ (–1) = – a, as proved above]

= 1. [$\because$ in a ring, – (–a) = a]

Example 12(b):

The set N of all 2 × 2 matrices of the form

$$\begin{bmatrix} a & 0 \\ b & 0 \end{bmatrix}$$

for a, b integers is a left ideal but not a right ideal in the ring R of all 2 × 2 matrices with elements as integers. Here N is the subset of R consisting of those elements whose second column contains only zeros.

Solution:

Let $A = \begin{bmatrix} a & 0 \\ b & 0 \end{bmatrix}$, $B = \begin{bmatrix} c & 0 \\ d & 0 \end{bmatrix}$ be any tow elements of N.

Then $A - B = \begin{bmatrix} a & 0 \\ b & 0 \end{bmatrix} - \begin{bmatrix} c & 0 \\ d & 0 \end{bmatrix} = \begin{bmatrix} a-c & 0 \\ b-d & 0 \end{bmatrix} \in N.$

$\therefore$ N is a subgroup of R under addition.

Now let $U = \begin{bmatrix} w & x \\ y & z \end{bmatrix}$ be any element of R and $A = \begin{bmatrix} a & 0 \\ b & 0 \end{bmatrix}$ be any element of N.

Then $UA = \begin{bmatrix} w & x \\ y & z \end{bmatrix} \begin{bmatrix} a & 0 \\ b & 0 \end{bmatrix} = \begin{bmatrix} wa+xb & 0 \\ ya+zb & 0 \end{bmatrix} \in N.$

Therefore N is a left ideal of R. It is not a right ideal, since

$$\begin{bmatrix} 1 & 0 \\ 1 & 0 \end{bmatrix} \in N, \begin{bmatrix} 1 & 2 \\ 0 & 1 \end{bmatrix} \in R,$$

and the product $\begin{bmatrix} 1 & 0 \\ 1 & 0 \end{bmatrix} \begin{bmatrix} 1 & 2 \\ 0 & 1 \end{bmatrix} = \begin{bmatrix} 1 & 2 \\ 1 & 2 \end{bmatrix}$ which is not an element of N.

Example 12(c):

Define a field. Prove that every field is an integral domain, but there exist some integral domains which are not fields.

Solution:

For the definition of a field refer to prove that every field is an integral domain refer

For an example of a ring which is an integral domain but is not a field, consider the ring of integers (I, +, .) where I is the set of integers *i.e.*,

$$I = \{..., -3, -2, -1, 1, 0, 2, 3, ...\}.$$

The ring of integers (I, +, .) is an integral domain because it is a commutative ring with unity 1 and does not possess zero divisors. But this ring is not a field.

The only elements of the ring of integers which possess multiplicative inverse are 1 and –1 while in a field every non-zero element must possess multiplicative inverse.

Hence, the ring of integers is an integral domain but it is not a field.

Example 12(d):

Prove that the set of integers R,

$$R = \{0, 1, 2, 3, 4\}$$

forms a field under addition modulo 5 and multiplication modulo 5.

Solution:

Do yourself.

Hint : Hence also 0 is identity for the operation $+_5$. Also for the operation $+_5$, the inverses of 0, 1, 2, 3, 4 are 0, 4, 3, 2, 1 respectively.

1 is multiplicative identity *i.e.*, identity for the operation $\times_5$. Also for the operation $\times_5$ the inverses of the non-zero elements 1, 2, 3, 4, of R are 1, 3, 2, 3 respectively.

Example 13:

Give an example of a skew field which is not a field.

Solution:

Let M be the set of all 2 × 2 matrices of the form

$$\begin{bmatrix} a + ib & c + id \\ -c + id & a - ib \end{bmatrix},$$

where a, b, c, d are arbitrary real numbers.

First we shall prove that M is a ring with respect to addition and multiplication of matrices.

Let $$A = \begin{bmatrix} a+ib & c+id \\ -c+id & a-ib \end{bmatrix} \text{ and } B = \begin{bmatrix} p+iq & r+is \\ -r+is & p-iq \end{bmatrix}$$

be any two elements of M. We have

$$A + B = \begin{bmatrix} (a+p)+i(b+q) & (c+r)+i(d+s) \\ -(c+r)+i(d+s) & (a+p)-i(b+q) \end{bmatrix},$$

which is obviously a matrix of the given form. So $A + B \in M$.

Also AB

$$= \begin{bmatrix} (a+ib)(p+iq)+(c+id)(-r+is) & (a+ib)(r+is)+(c+id)(p-iq) \\ (-c+id)(p+iq)+(a-ib)(-r+is) & (-c+id)(r+is)+(a+ib)(p-iq) \end{bmatrix}$$

$$= \begin{bmatrix} (ap-bq-cr-ds) & (ar-bs+cp+dq) \\ \quad +i(aq+bp+cs-dr) & \quad +i(as+br-cq+dp) \\ -(ap+dq+ar-bs) & (-cr-ds+ap-bq) \\ \quad +i(dp-cq+as+br) & \quad -i(cs-dr+aq+bp) \end{bmatrix}$$

which is obviously an element of M. Thus M is closed with respect to addition and multiplication of matrices. Further matrix addition is commutative as well as associative. The zero matrix $\begin{bmatrix} 0+i0 & 0+i0 \\ -0+io & 0-i0 \end{bmatrix}$ is the additive identity and so it is the zero element of M. If $A = \begin{bmatrix} a+ib & c+id \\ -c+id & a-ib \end{bmatrix} \in M$, then obviously $-A = \begin{bmatrix} -a-ib & -c+id \\ c-id & -a+ib \end{bmatrix} \in M$. Thus each element of M possesses additive inverse. Further matrix multiplication is associative and it is distributive with respect to addition. Hence M is a ring with respect to addition and multiplication of matrices.

Existence of multiplicative identity: The matrix

$\begin{bmatrix} 1+i0 & 0+0i \\ -0+i0 & 1-i0 \end{bmatrix} = \begin{bmatrix} 1 & 0 \\ 0 & 1 \end{bmatrix}$ is obviously an element of M. It is the unit matrix and so it is the multiplicative identity. Thus M is a ring with unity.

Existence of multiplicative inverse of each non-zero element of M.

Let $A = \begin{bmatrix} a + ib & c + id \\ -c - id & a - ib \end{bmatrix}$ be any non-zero element of M *i.e.*, a, b, c, d are not all equal to zero. We have | A | *i.e.*, det. A = $a^2 + b^2 + c^2 + d^2 \neq 0$. Therefore the matrix A is non-singular and is therefore inversible. We must show that $A^{-1} \in M$. Let | A | = m. Then m is a real number and m > 0. We have

$$A^{-1} \; 0 \; \frac{1}{|A|} \text{Adj. } A = \frac{1}{m}\begin{bmatrix} a - ib & -c - id \\ c - id & a + ib \end{bmatrix}$$

which is obviously an element of M. Therefore each non-zero element of M possesses multiplicative inverse.

∴ M is a skew field (or a division ring).

Now we shall show that M is not a field *i.e.*, multiplication in M is not commutative. We have $A = \begin{bmatrix} 3 + 4i & 5 + 6i \\ -5 + 6i & 3 - 4i \end{bmatrix} \in M$,

and $B = \begin{bmatrix} 1 + i0 & 1 + i0 \\ -1 + i0 & 1 - i0 \end{bmatrix} = \begin{bmatrix} 1 & 1 \\ -1 & 1 \end{bmatrix} \in M$.

Also $AB = \begin{bmatrix} -2 - 2i & 8 + 10i \\ -8 + 10i & -2 + 2i \end{bmatrix}$, and $BA = \begin{bmatrix} -2 - 10i & 8 + 2i \\ -8 + 2i & -2 + 10i \end{bmatrix}$.

We see that, AB ≠ BA. Therefore multiplication in M is not commutative and so M is not a field.

Hence M is a skew-field which is not a field.

Example 14:

Prove that in general the set of all numbers of the form

$$a + (\sqrt{p})\, b$$

where a, b are rational numbers and p is a prime is a field with respect to ordinary addition and multiplication.

Solution:

We can easily verify that the given system is a commutative ring with unit element $1 + \sqrt{p}0$. Also $0 + \sqrt{p}0$ is the zero element of this ring. Further let $a + \sqrt{p}b$ be any non-zero element of the given set.

$$\text{Then } \frac{1}{a + \sqrt{pb}} = \frac{a - \sqrt{pb}}{a^2 - pb^2} = \frac{a}{a^2 - pb^2} + \left(-\frac{b}{a^2 - pb^2}\right)\sqrt{p}.$$

Now if p is prime and if at least one of the rational numbers a and b is not zero, then we cannot have $a^2 = pb^2$.

$\therefore \dfrac{a}{a^2 - pb^2}$ and $-\dfrac{b}{a^2 - pb^2}$ are both rational numbers and at least one them is not zero.

$\therefore \left(\dfrac{a}{a^2 - pb^2}\right) + \left(\dfrac{b}{a^2 - pb^2}\right)\sqrt{p}$ is a non-zero element of the given set

and it is the multiplicative inverse of the non-zero element $a + b\sqrt{p}$. Hence the given system is a field.

Example 15(a):

Define the characteristic of a ring and prove that if R is a finite ring then the characteristic of R is finite and ≠ 0.

Solution:

Let (R, +, ·) be a finite ring having n elements.

From our study of group theory we know that if a is any element of a finite group G of order n, then $a^n = e$, where e is the identity of the group.

Now the identity of the additive group (R, +) of the ring (R, +, ·) is 0. Therefore if a sin ay element of R, then we have na = 0.

Thus we have $na = 0 \ \forall \ a \in R$.

Therefore the characteristic of the ring R is ≤ n and cannot be infinite or zero.

Hence if R is a finite ring then the characteristic of R is finite and ≠ 0.

Example 15(b):

Let x, y be commutative elements of a ring R of characteristic two. Show that $(x + y)^2 = x^2 + y^2 = (x - y)^2$

Solution:

Since the ring R is of characteristic two, therefore

$$a + a = 0 \ \forall \ a \in R.$$

Now let $x, y \in R$ and $xy = yx$.

Then $(x + y)^2 = (x + y)(x + y)$

$= x^2 + xy + yx + y^2$, by dist, laws

$= x^2 + xy + xy + y^2$ $\quad [\because \ xy = yx]$

$= x^2 + 0 + y^2$ [$\because$ xy $\in$ R and R is of characteristic two implies xy + xy = 0]

$= x^2 + y^2$.

Again $(x - y)^2 = (x - y)(x - y) = x^2 - xy - yx + y^2$

$= x^2 - xy - xy + y^2 = x^2 - (xy + xy) + y^2$

$= x^2 - 0 + y^2 = x^2 + y^2$.

Example 15(c):

Let R be an non-zero ring such that for all $a \in R$, $a^2 = a$. Prove, that R is a commutative ring of characteristic 2.

Solution:

Do your self.

Hint: We have proved there in part (1) of the question that

$$a + a = 0 \ \forall \ a \in R.$$

Therefore the ring R is of characteristic two.

Again in parts (2) and (3) of that question we have proved that the ring R is commutative.

Example 16:

Show that the set of even integers forms a subring of the ring of integers.

Solution:

Let (I, +, .) be the ring of integers.

Let S = 2I = {..., –6, –4, –2, 0, 2, 4, 6, ...} be the set of even integers. Then $S \subset I$.

To prove that S is a subring of the ring of integers.

Let a = 2r and b 2s be any two elements of S where r and s are some integers.

Then $a - b = 2r - 2s = 2(r - s)$ which is an even integer and so $a - b \in S$.

Also ab = (2r) (2s) = 2 (2rs) which is also an even integer and so $ab \in S$.

Thus, $a, b \in S \Rightarrow a - b \in S$ and $ab \subset S$. Hence the set of even integers S is a subring of the ring of integers.

Example 17:

Prove or disprove that any subring of a non-commutative ring is non-commutative.

Solution:

A subring of a non-commutative ring may be commutative as is obvious from the following example.

Let M be the ring of all 2 × 2 matrices with elements as integers for addition and multiplication of matrices as the tow ring operations. Then M is a non-commutative ring because the operation of multiplication of matrices on the set M is not commutative.

Let S be the subset of M consisting of matrices of the type

$$\begin{bmatrix} a & 0 \\ 0 & 0 \end{bmatrix}, \text{ where a is any integer.}$$

Let $A = \begin{bmatrix} a_1 & 0 \\ 0 & 0 \end{bmatrix}, B = \begin{bmatrix} a_2 & 0 \\ 0 & 0 \end{bmatrix}$ be any two members of the set S.

$$\text{The } A - B = \begin{bmatrix} a_1 - a_2 & 0 \\ 0 & 0 \end{bmatrix} \in S$$

$$\text{and} \quad AB = \begin{bmatrix} a_1 a_2 & 0 \\ 0 & 0 \end{bmatrix} \in S$$

∴ S is a subring of the ring M.

Now the subring S of the ring M is a commutative ring. For let

$$A = \begin{bmatrix} a_1 & 0 \\ 0 & 0 \end{bmatrix} \text{ and } B = \begin{bmatrix} a_2 & 0 \\ 0 & 0 \end{bmatrix}$$

be any two members of S. Then

$$AB = \begin{bmatrix} a_1 a_2 & 0 \\ 0 & 0 \end{bmatrix} \text{ and } BA = \begin{bmatrix} a_2 a_1 & 0 \\ 0 & 0 \end{bmatrix}.$$

Since $a_1 a_2 = a_2 a_1$, therefore AB = BA.

Thus, multiplication of matrices is a commutative operation on the set S. Hence the subring S of the non-commutative ring M is a commutative ring.

Example 18:

Show that the set of all 2-rowed matrices of the form

$$\begin{bmatrix} a & 0 \\ b & c \end{bmatrix}$$

where a, b, c are integers is a subring of the ring M of all 2-rowed matrices with integral entries.

Solution:

Let M be the set of all 2 × 2 matrices with elements as integers. Then M is a ring for addition and multiplication of matrices as the two ring operations.

Let S be the subset of M consisting of matrices of the type

$$\begin{bmatrix} a & 0 \\ b & c \end{bmatrix}, \text{ where a, b, c are integers.}$$

To prove that S is a subring of the ring M.

Let $A = \begin{bmatrix} a_1 & 0 \\ b_1 & c_1 \end{bmatrix}, B = \begin{bmatrix} a_2 & 0 \\ b_2 & c_2 \end{bmatrix}$ be any two members of the set S.

The $A - B = \begin{bmatrix} a_1 - a_2 & 0 \\ b_1 - b_2 & c_1 - c_2 \end{bmatrix}$ which is obviously a member of the set S.

Also $AB = \begin{bmatrix} a_1a_2 & 0 \\ b_1a_2 + c_1b_2 & c_1c_2 \end{bmatrix}$ which is also a member of the set S.

Thus A, B $\in$ S $\Rightarrow$ A – B $\in$ S and AB $\in$ S

Hence S is a subring of the ring M

Example 19:

Prove that the totality R of all ordered pairs (a, b) of real numbers is a commutative ring with zero divisors under the addition and multiplication of ordered pairs defined as

$$(a, b) + (c, d) = (a + c, b + d)$$

$$(a, b) + (c, d) = (ac, bd)$$

$$\forall (a, b) + (c, d) \in R.$$

Solution:

We see that R is closed with respect to the two compositions since a + c, b + d, ac, bd are all real numbers. Now let (a, b), (c, d), (e, f) be any elements of R. Then we observe:

Associativity of Addition: We have

[(a, b) + (c, d)] + (e, f) = (a + c, b + d) + (e, f)

$= ([a + c] + e, [b + d] + f)$

$= (a + [c + e], b + [d + f])$

[$\because$ addition of real numbers is associative]

$= (a, b) + (c + e, d + f)$

$= (a, b) + [(c, d) + (e, f)].$

$\therefore$ addition in R is associative.

Commutativity of Addition: We have

$(a, b) + (c, d) = (a + c, b + d) = (c + a, d + b) = (c, d) + (a, b).$

Existence of Additive Identity: We have $(0, 0) \in R$, Also $(0, 0) + (a, b) = (0 + a, 0 + b) = (a, b)$.

Existence of Additive Inverse: If $(a, b) \in R$, then $(-a, -b) \in R$ and we have $(-a, -b) + (a, b) = (-a + a, -b + b) = (0, 0)$.

$\therefore (-a, -b)$ is the additive inverse of (a, b)

Associativity of Multiplication: We have

$[(a, b) (c, d)] (e, f) = (ac, bd) (e, f) = ([ac] e, [bd] f)$

$= (a [ce], b [df]).$

[$\because$ multiplication of real numbers is associative]

$= (a, b) (ce, df) = (a, b) [(c, d) (e, f)].$

Distributive Laws: We have

$(a, b) [(c, d) + (e, f)] = (a, b) (c + e, d + f)$

$= (a [c + e], b [d + f])$

$= (ac + ae, bd + bf) = (ac, bd) + (ae, bf)$

$= (a, b) (c, d) + (a, b) (e, f).$

Similarly we can show that the other distributive law also holds good.

$\therefore$ R is a ring with respect to the given compositions.

Commutativity of Multiplication: We have

$(a, b) (c, d) = (ac, bd) = (ca, db) = (c, d) (a, b).$

$\therefore$ R is s commutative ring.

Existence of Multiplicative Identity: We have $(1, 1) \in R$. If $(a, b) \in R$, then $(1, 1) (a, b) = (1a, 1b) = (a, b) = (a, b) (1, 1)$.

$\therefore (1, 1)$ is the multiplicative identity and is therefore the unit element of the ring. So r is a ring with unity also.

The zero element of this ring is the ordered pair $(0, 0)$.

Now in order to show that R is a ring with zero divisors we should show that there exist two non-zero elements of R whose product is equal to the zero element of R. Obviously neither (3, 0) nor (0, 5) is equal to the zero element of R. But $(3, 0)\ (0, 5) = (3 \times 0, 0 \times 5) = (0, 0)$ which is the zero element of R.

∴ R is a ring with zero divisors.

Example 20(a):

Give an example each of:

1. *a commutative ring without unity,*
2. *a non-commutative ring,*
3. *a ring without zero divisors,*
4. *division ring.*

Solution:

1. A commutative ring without unity.
2. A non-commutative ring
3. A ring without zero divisors. The ring of integers (I, +, .) is a ring without zero divisors. We know that the product of two non-zero integers is never zero. Thus if $a, b, \in I$, then

   $$ab = 0 \Rightarrow a = 0 \Rightarrow b = 0.$$

   Hence the ring of integers is a ring without zero divisors.
4. Division ring. Let M be the set of all 2 × 2 matrices of the form

   $$\begin{bmatrix} a + ib & c + id \\ -c + id & a - ib \end{bmatrix},$$

where a, b, c, d are arbitrary real numbers.

Then M is a ring for addition and multiplication of matrices as the two ring operations. This ring possesses unity element.

The unit matrix $I = \begin{bmatrix} 1 & 0 \\ 0 & 1 \end{bmatrix} \in M$ and we have

$$IA = A = AI, \forall A \in M.$$

Also in this ring every non-zero element possesses multiplicative inverse. But the operation of multiplication of matrices on the set M is not commutative. Hence the above ring M is a division ring or a skew field.

Example 20(b):

Prove that the set of rational number (real numbers or complex numbers) is a field with respect to addition and multiplication.

Solution:

We shall give proof in the case of the set of rational numbers and a similar proof can be given in these of the set of real numbers or the set of complex numbers.

Let Q denote the set of rational numbers. Remember that a rational number is of the form a/b, where a and b are integers and b ≠ 0.

To prove that the algebraic structure (Q, +, .) is a field

(Q, +) is an abelian group: We know that the sum of two rational numbers is also a rational number. Therefore Q is closed for addition of rational numbers.

Also addition of rational numbers is commutative as well as associative. The rational number 0 is identity for addition of rational numbers. Also if a/b is its additive inverse. Thus (Q, +) is an abelian group.

Now the product of two rational numbers is also a rational number. Therefore Q is closed for multiplication of rational numbers.

Also multiplication of rational numbers is commutative as well as associative and it distributes over addition of rational numbers. The rational number 1 is identity for multiplication. Thus (Q, +, .) is a commutative ring with unity 1.

If a/b is any non-zero rational number, then the integer a ≠ 0 and so b/a is also a rational number.

We have (a/b) (b/a) = 1, so that b/a is the multiplicative inverse of a/b. Thus each non-zero rational number possesses multiplicative inverse. Hence (Q, +, .) is a field.

Remark: *To give the proof in the case of the set of complex numbers denote the set of the complex numbers by C.*

Remember that a complex number is of the form a + ib where a and b are any real numbers.

If a + ib and c + id are any two complex numbers, then

(a + ib) + (c + id) = (a + c) + i (b + c) which is also a complex number

and (a + ib) (c + id) = (a + c) + i (b + d) which is also a complex number.

The complex number 0 *i.e.*, 0 + i0 is identity for addition of complex numbers. Also if a + ib is any complex number, then the complex number –a – ib *i.e.*, (–a) + i (–b) is its additive inverse.

The complex number 1 *i.e.*, 1 + i0 is identity for multiplication.

Finally every non-zero complex number possesses multiplicative inverse as shown below.

Let a + ib be any non-zero complex number *i.e.*, a + ib ≠ 0 + i0. Then a and b are real numbers and at least one of them is not zero.

Suppose the complex number x + iy is the multiplicative inverse of a + ib. Then

$$(x + iy)(a + ib) = 1 + i0$$

$$\Rightarrow (xa - yb) + i(xb + ya) = 1 + i0$$

$$\Rightarrow xa - yb = 1,\ xb + ya = 0.$$

Solving the equations xa – yb = 1, xb + ya = 0, we get

$$x = \frac{a}{a^2 + b^2},\ y = \frac{-b}{a^2 + b^2}.$$

Since at least one of a and b is not zero, therefore the real number $a^2 + b^2 \neq 0$ and so both x and y are some real numbers.

Therefore the complex number $\frac{a}{a^2 + b^2} + i\left(\frac{-b}{a^2 + b^2}\right)$ is the multiplicative inverse of the non-zero complex number a + ib.

Example 21:

Show that the set M of all 2 × 2 matrices of the form

$$\begin{bmatrix} 0 & a \\ 0 & b \end{bmatrix}$$

a, b, integers is a left ideal but not a right ideal in the ring of all 2×2 matrices with elements as integers.

Solution:

Let R be the ring of all 2 × 2 matrices with elements as integers for addition and multiplication of matrices as the two ring operations. Let M be the subset of R consisting of matrices of the form

$\begin{bmatrix} 0 & a \\ 0 & b \end{bmatrix}$, where a, b are any integers. First to show that M is a left ideal of the ring R.

Let $A = \begin{bmatrix} 0 & a_1 \\ 0 & b_1 \end{bmatrix}$ and $B = \begin{bmatrix} 0 & a_2 \\ 0 & b_2 \end{bmatrix}$ be any two elements of M.

Then $\quad A - B = \begin{bmatrix} 0 & a_1 - a_2 \\ 0 & b_1 - b_2 \end{bmatrix} \in M.$

$\therefore$ M is a subgroup of the additive group of the ring R.

Now let $U = \begin{bmatrix} w & x \\ y & z \end{bmatrix}$ be any element of R and $A = \begin{bmatrix} 0 & a \\ 0 & b \end{bmatrix}$ be any element of M.

$$\text{Then } UA = \begin{bmatrix} w & x \\ y & z \end{bmatrix}\begin{bmatrix} 0 & a \\ 0 & b \end{bmatrix} = \begin{bmatrix} 0 & wa+xb \\ 0 & ya+zb \end{bmatrix} \in M$$

Therefore M is a left ideal of R.

But M is not a right ideal of R. Since

$$\begin{bmatrix} 0 & 1 \\ 0 & 1 \end{bmatrix} \in M, \begin{bmatrix} 2 & 4 \\ 3 & 5 \end{bmatrix} \in R,$$

and the product $\begin{bmatrix} 0 & 1 \\ 0 & 1 \end{bmatrix}\begin{bmatrix} 2 & 4 \\ 3 & 5 \end{bmatrix} = \begin{bmatrix} 3 & 5 \\ 3 & 5 \end{bmatrix}$ which is not an element of M. Hence M is not a right ideal of R.

Example 22:

If U is an ideal of a ring R with unity and 1 $\in$ U prove that U = R.

Solution:

We have $U \subseteq R$ since U is an ideal of R. Let x be any element of R. Since u is an ideal of R, therefore

$$1 \in U, x \in R \Rightarrow 1x \in U \Rightarrow x \in U.$$

$$\therefore \quad R \subseteq U.$$

$$\therefore \quad U = R.$$

Example 23(a):

Show that S is an ideal of S + T where S is any ideal of ring R, and T any subring of R.

Solution:

Since S is an ideal of R therefore S is subring of R. Also Ti is a subring of R. First we shall show that S + T is a subring of R. Let $a + \alpha$, $b + \beta \in S + T$, where a, b, $\in$ S and α, β, □$\in$ T.

Since S is a subring, therefore $a - b \in S$. Similarly $\alpha - \beta \in T$.

$\therefore (a + \alpha) - (b + \beta) = (a - b) + (\alpha - \beta) \in S + T$.

Also $(a + \alpha)(b + \beta) = ab + a\beta + \alpha b + \alpha\beta = (ab + a\beta + \alpha b) + \alpha\beta$.

Now S is subring. Therefore $a, b \in S \Rightarrow ab \in S$.

Also S is an ideal, therefore $a, b, \in S$ and $\alpha, \beta \in R \Rightarrow a\beta, \alpha b \in S$. Therefore $ab + a\beta + \alpha b \in S$.

Further t is a subring implies $\alpha\beta \in T$ if $\alpha, \beta \in T$.

$\therefore (a + \alpha)(b + \beta) = (ab + a\beta + \alpha b) + \alpha\beta \in S + T$.

$\therefore S + T$ is a subring of R.

Since $0 \in T$, therefore $a \in S$ can be written as

$$a = a + 0 \in S + T.$$

$$\therefore \quad S \subseteq S + T.$$

Thus, $S \subseteq S + T$ and $S + T$ is a subring of R. Since S is an ideal of R, therefore S is also an ideal of $S + T$.

Example 23(b):

Show that in an integral domain all non-zero elements generate additive cholic groups of the same order which is equal to the characteristic of the integral domain.

Solution:

Let d be an integral domain.

First we shall prove that each non-zero element of D, regarded as a member of the additive group of D, is of the same order.

Now we know that the order of a cyclic group is equal to the order of its generator. Hence in an integral domain all non-zero elements generate additive cyclic groups of the same order.

In the end we shall prove that the characteristic of the integral domain D is equal to the order of any non-zero element of D regarded as a member of the additive group of D.

Example 23(c):

Give without proof, an example of an integral domain which contains only five elements. Is this an ordered integral domain? Give reason.

Solution:

The integral domain $(\{0, 1, 2, 3, 4\}, +_5, \times_5\}, +_5, \times_5)$ contains only five elements

This integral domain is not an ordered integral domain as shown below.

Suppose the above integral domain is an ordered integral domain and P is the set of positive elements of this integral domain. The zero element of this integral domain is 0.

Now 1 is an element of this integral domain and $1 \neq 0$. The additive inverse of 1 is 4 because $1 +_5 4 = 0 = 4 +_5 1$.

According to the definition of an ordered integral domain either $1 \in P$ or its additive inverse $4 \in P$.

But P is closed with respect to $+_5$.

$\therefore 1 \in P \Rightarrow 1 +_5 1 +_5 1 +_5 1 \in P \Rightarrow 4 \in P$.

This contradicts the principle of trichotomy.

Similarly $4 \in P \Rightarrow 4 +_5 4 +_5 4 +_5 4 \in P \Rightarrow 1 \in P$. This again contradicts the principle of trichotomy.

Hence, the integral domain ({0, 1, 2, 3, 4}, + 5, × 5) is not an ordered integral domain.

Example 23(d):

If U, V are ideals of a ring R, let

$$U + V = \{u + v : u \in U, v \in V\}.$$

Prove that U + V is also an ideal of R.

Solution:

Let $u_1 + v_1 \in U + V$ and $u_2 + v_2 \in U + V$. Then

$u_1 + u_2 \in U$ and $v_1 + v_2 \in V$.

We have $(u_1 + v_1) - (u_2 + v_2) = (u_1 - u_2) + (v_1 + v_2)$.

Since U is an ideal of R, therefore

$u_1, u_2 \in U \Rightarrow u_1 - u_2 \in U$.

Similarly V is also an ideal of R, therefore

$v_1, v_2 \in V \Rightarrow v_1 - v_2 \in V$.

$\therefore (u_1 - u_2) + (v_1 - v_2) \in U + V$.

$\therefore (u_1 + v_1) - (u_2 + v_2) \in U + V$.

$\therefore$ U + V is a subgroup of the additive group of R.

Now let r be any element of R and u + v be any element of U + V where $u \in U$, $v \in V$.

Then $\quad r(u + v) = ru + rv$

$\in U + V$ since $r \in R$, $u \in U$ and U is an ideal

$\Rightarrow \quad ru \in U$ and similarly $rv \in V$

Similarly $(u + v)\, r = ur + vr \in U + V$ since $ur \in U$, $vr \in V$.

Hence U + V is an ideal of R.

Example 24:

Show that the set of matrices $\begin{bmatrix} a & b \\ 0 & c \end{bmatrix}$ *is a subring of the* 2×2 *matrices with integral elements.*

Solution:

Let R be the ring of 2×2 matrices and let M be the subset of R and let the elements of M be matrices of the type

$$\begin{bmatrix} a & b \\ 0 & c \end{bmatrix}.$$

Let $A = \begin{bmatrix} a_1 & b_1 \\ 0 & c_1 \end{bmatrix}$, $B = \begin{bmatrix} a_2 & b_2 \\ 0 & c_2 \end{bmatrix}$ be any two elements of M.

Then $A - B = \begin{bmatrix} a_1 - a_2 & b_1 - b_2 \\ 0 & c_1 - c_2 \end{bmatrix}$ which is obviously an element of M.

Also $AB = \begin{bmatrix} a_1 & b_1 \\ 0 & c_1 \end{bmatrix} = \begin{bmatrix} a_2 & b_2 \\ 0 & c_2 \end{bmatrix} = \begin{bmatrix} a_1a_2 & a_1b_2 + b_1c_2 \\ 0 & c_1c_2 \end{bmatrix}$

which is obviously an element of M.

$\therefore$ M is a subring of R.

Example 25:

Verify the following for being true or false:

1. *The set of all positive rationals is a subring of the ring of all rational numbers.*
2. *A subring of any field is a field.*
3. *Any subring of the ring of integers, Z, is an ideal of Z.*

Solution:

1. Let $(Q, +, \cdot)$ be the ring of all rational numbers and Q_+ be the set of all positive rational numbers.

Then Q_+ is not a subring of the ring $(Q, +, \cdot)$. Obviously the rational number 0 which is the zero element of the ring of all rational numbers does not belong to Q_+. Therefore Q_+ is not a subring of the ring (Q, +, ·). Hence the given statement is false.

2. The statement that a subring of any field is a field is false. For example consider the field of rational numbers (Q, +, ·). The set of integers I is a subring of the field of rational numbers because a ∈ I, b ∈ I ⇒ a – b ∈ I and ab ∈ I.

But the subring (I, + , ·) of the field (Q, + , ·) is not a field. The only element in I which possess multiplicative inverse are 1 and –1 while in a field every non-zero element must possess multiplicative inverse. Hence (I, + , ·) is not a field and so the given statement is false.

3. The statement that any subring of the ring of integers, Z, is an ideal of Z is true.

If S is the zero subring of Z, then obviously S is an ideal of Z.

If S is any subring other than the zero subring of the ring of integers Z and m is the smallest positive integer belonging to S, then we S = {xm : x is an integer}.

Now show that S is an ideal of the ring Z.

Hence any subring of the ring of integers, Z, is also an ideal of Z.

Example 26:

Prove that the only idempotent elements of an integral domain with unity are 0 and 1.

Solution:

Let R be an integral domain with unity 1.

Let a ∈ R and a be idempotent *i.e.*, $a^2 = a$.

We have $a^2 = a$

$\Rightarrow \quad aa = a1$

$\Rightarrow \quad aa - a1 = 0$

$\Rightarrow \quad a(a - 1) = 0.$

But in an integral domain the product of two elements is 0 only if at least one of them is zero.

$\therefore \quad a(a - 1) = 0$

$\Rightarrow \quad a = 0$

$\Rightarrow \quad a - 1 = 0$

$\Rightarrow \quad a = 0 \Rightarrow a = 1.$

Hence the only idempotent elements of an integral domain with unity are 0 and 1.

Example 27:

Is every field also a division ring ? Does the set of all integers under usual addition and multiplication form a field ? Give some example of a field which is finite.

Solution:

First give the definitions of a field and a division ring. For these definitions refer.

From these two definitions we conclude that every field is also a division ring.

The set of integers I under usual addition and multiplication is a ring. The ring of integers (I, +, .) is a commutative ring with unity element the integer 1. But this ring is not a field. In this ring the only inversible elements are 1 and –1 while in a field very non-zero element must be inversible.

Example 28:

A Gaussian integer is a complex number a + ib, where a and b are integers. Show that the set J[i] of Gaussian integers forms a ring under ordinary addition and multiplication of complex numbers. Is it an integral domain? Is it a field?

Solution:

Let a + ib and c + id be any two Gaussian integers.

Then $\quad (a + ib) + (c + id) = (a + c) + i(b + d)$

and $\quad (a + ib) + (c + id) = (ac - bd) + i(ad + bc).$

These are again Gaussian integers. Therefore J[i] is closed with respect to ordinary addition and multiplication of complex numbers.

Further in complex numbers both addition and multiplication are associative as well as commutative compositions. Also multiplication distributes with respect to addition. The Gaussian integer 0 + i0 is the additive identity. The additive inverse of a + ib is (–a) + i (–b). The Gaussian integer 1 + i0 is the multiplicative identity.

Therefore the set of Gaussian integers is a commutative ring with unity for the given compositions.

Also this ring is free from zero divisors since the product of two non-zero complex numbers cannot be zero. Therefore J[i] is an integral domain.

But this is not field since the multiplicative inverse of a + ib will be $\frac{a}{a^2+b^2} + i\left(-\frac{b}{a^2+b^2}\right)$ which is not always a Gaussian integer as $\frac{a}{a^2+b^2}$ and $-\frac{b}{a^2+b^2}$ are not necessarily integers.

Example 29:

If two operations $*$ *and* o on the set I of integers are defined as follows.

$$a * b = a + b - 1,\ a \circ b = a + b - ab,$$

prove that the system (I, $*$, o*) is a commutative ring with identity.*

Solution:

First we shall show that the algebraic structure (I, $*$) is an abelian group.

In is closed for the operation $*$. Let a, b, $\in$ I.

Then a $*$ b = a + b – 1 which is also a member of I. Therefore I is closed with respect to the operation *.

Commutativity and associativity of the operation * on these I.

Let a, b, c $\in$ I. Then

$$a * b = a + b - 1 = b + a - 1 = b * a$$

and $(a * b) * c = (a + b - 1) * c = (a + b - 1) + c - 1 = a + (b + c - 1) - 1$

$$= a * (b + c - 1) = a * (b * c).$$

Thus the operation $*$ on the set I is commutative as well as associative.

Example 30:

Add and multiply the following polynomials over the ring of integers:

$$f(x) = 2x^0 + 5x + 3x^2 - 4x^2,\ g(x) = 3x^0 + 4x - x^3 + 5x^4.$$

Solution:

By our definition of the sum of two polynomials, we have

$$f(x) + g(x) = (2 + 3)\,x^0 + (5 + 4)\,x + (3 + 0)\,x^3 + (-4 - 1)\,x^3 + (0 + 5)\,x^4$$

$$= 5x^0 + 9x + 3x^2 - 5x^3 + 5x^4.$$

Also $f(x)\,g(x) = (2x^0 + 5x + 3x^2 - 4x^3)(3x^0 + 4x - x^3 + 5x^4)$

$$= 6x^0 + (8 + 15)\,x + (20 + 9)\,x^2 + (-2 + 12 - 12)\,x^3 + (10 - 5 - 16)\,x^4 + (25 - 3)\,x^5 + (15 + 4)\,x^6 - 20x^7$$

$= 6x^0 + 23x + 29x^2 - 2x^3 - 11x^4 + 22x^5 + 19\, x^6 - 20x^7.$

Example 31(a):

Show that id a ring R has no zero divisors, then the ring R [x] has also no zero divisors.

Solution:

It is given that a ring R has no zero divisors and we have to show that the ring R [x] has also no zero divisors.

Let $f(x) = a_0 + a_1x + a_2x^2 + ... + a_mx^m,\ a_m \neq 0$

and $g(x) = b_0 + b_1x + b_2x^2 + ... + b_nx^n,\ b_n \neq 0$

be two non-zero elements of R [x].

Then f (x) g (x) cannot be the zero polynomial *i.e.*, the zero element of R [x].

Then reason is that at least on coefficient of f (x) g (x) namely a_mb_n of x^{m+n} is $\neq 0$ because a_m, b_n are non-zero elements of R and R is without zero divisors.

Thus, in R [x] the product of no two non-zero elements can be the zero element. Hence the ring R [x] has no zero divisors.

Example 31(b):

Consider the following polynomials over the ring

$(I_8, +_8, \times_8)$:

$f(x) = 2 + 6x + 4x^2$, $g(x) = 2x + 4x^2$, $h(x) = 2\ 4x$ *and find*

1. deg [f (x) + g (x)]
2. deg [f (x) g (x)]
3. deg [h (x) h (x)].

Solution:

1. We have $f(x) + g(x) = (2 + 6x + 4x^2) + (0 + 2x + 4x^2)$

   $= (2 +_8 0) + (6 +_8 2)\, x + (4 +_8 4)\, x^2 = 2 + 0x + 0x^2 = 2.$

   Thus f (x) + g (x) is a non-zero constant polynomial and so deg [f (x) + g (x)] = 0.
2. We have $f(x)\, g(x) = (2 + 6x + 4x^2)\,(2x + 4x^2)$

   $= (2 \times_8 2)\, x + [(2 \times_8 4) +_8 (6 \times_8 2)]\, x^2$

   $+ [(6 \times_8 4) +_8 (4 \times_8 2)]\, x^3 + (4 \times_8 4)\, x^4$

$= 4x + (0 +_8 4)\ x^2 + (0 +_8 0)\ x^3 + 0x^4.$

$= 4x + 4x^2 + 0x^2 + 0x^4 = 4x + 4x^2.$

$\therefore \quad \deg [f(x)\ g(x)] = 2.$

3. We have $h(x)\ h(x) = (2 + 4x)\ (2 + 4x)$

$= (2 \times_8 2) + [\ (2 \times_8 4) +_8 (4 \times_8 2)]\ x + (4 \times_8 4)\ x^2$

$= 4 + (0 +_8 0)\ x + 0x^2 = 4 + 0x + 0x^2 = 4.$

Thus $h(x)\ h(x)$ is a non-zero constant polynomial and so $\deg [h(x)\ h(x)] = 0$.

Example 32:

Prove that the relation of divisibility in an integral domain is reflexive and transitive.

Solution:

Let D be an integral domain with unity element 1.

If $a \in D$, then a is said to divide be $\in D$, if there exists an element $c \in D$ such that $b = ca$. If a is a divisor of b, then symbolically we write $a \mid b$.

1. The relation of divisibility on D is reflexive. Let a be any element of D.

   We can write $a = 1\ a$, where $1 \in D$.

   $\therefore \qquad a \mid a.$

   Thus $a \mid a$, $\forall\ a \in D$. Hence the relation of divisibility on D is reflexive.

2. The relation of divisibility on D is transitive *i.e.*, $a \mid b$ and $b \mid c \Rightarrow a \mid c$.

   We have $\quad a \mid b \Rightarrow b = ap$ for some $p \in D$

   and $\quad b \mid c \Rightarrow c = bq$ for some $q \in D$.

   Now $\quad c = bq$ and $b = aq \Rightarrow c = (ap)\ q$

   $\Rightarrow \quad c = a\ (pq) \Rightarrow a \mid c$ since $pq \in D$.

   Hence the relation of divisibility on D is transitive.

Example 33:

Add and multiply the following polynomials over the ring $(I_6, +_6 \times_6)$:

$f(x) = 2x^0 + 5x + 3x^2,\ g(x) = 1\ x^0 + 4x + 2x^3.$

Solution:

$f(x) + g(x) = (2 +_6 1)\, x^0 + (5 +_6 4)\, x + (3 +_6 0)\, x^2 + (0 +_6 2)\, x^3$

$= 3x^0 + 3x + 3x^2 + 2x^3.$

Also $f(x)\, g(x) = (2x^0 + 5x + 3x^2)\,(1x^0 + 4x + 2x^3)$

$= (2 \times_6 1)\, x^0 + [(2 \times_6 4) +_6 (5 \times_6 1)]\, x + [(5 \times_6 4) +_6 (3 \times_6 1)]\, x^2$

$+ [(2 \times_6 2) +_6 (3 \times_6 4)]\, x^3 + (5 \times_6 2)\, x^4 + (3 \times_6 2)\, x^5$

$= 2x^0 + (2 +_6 5)\, x + (2 +_6 3)\, x^2 + (4 +_6 0)\, x^3 = 4x^4 + 0x^5$

$= 2x^0 + 1x + 5x^2 + 4x^2 + 4x^4.$

Note: Here degree of f (x) = 2, degree of g (x) = 3 and degree of f (x) g (x) = 4. The point to note is that degree f (x) g (x) may be less than the sum of the degrees of f (x) and g (x).

Example 34:

If U is an ideal of a ring R, then prove that R/U is a ring and is a homomorphic image of R.

Solution:

We have $R/U = \{U + a : a \in R\}$.

Here $U + a = \{u + a : u \in U\}$ and is called residue class of U in R generated by a.

The operations of addition and multiplication in R/U are defined as follows:

$(U + a) + (U + b) = U + (a + b)$ [Addition of residue classes]

$(U + a)\,(U + b) = U + ab$ [Multiplication of residue classes]

To prove that R/U is a ring for these two compositions see the proof of theorem of this chapter.

Now to show that the ring R/U is a homomorphic image of the ring.

Consider the mapping $f: R \to R/U$ defined as

$$f(a) = U + a \ \forall\, a \in R.$$

The mapping f is onto. Let U + x be any element of R/U. Then $x \in R$.

We have $f(x) = U + x$. Therefore the mapping f is onto R/U.

The mapping f is a homomorphism. Let $a, b \in R$. Then

$$f(a + b) = U + (a + b) = (U + a) + (U + b) = f(a) + f(b).$$

Also f (ab) = U + ab = (U + a) (U + b) = f (a) f (b).

$\therefore$ f is a homomorphism of R onto R/U.

Hence the ring R/U is a homomorphic image of the ring R under. The mapping f.

Example 35(a):

If U is a left ideal of a ring R, let

λ (U) = {x $\in$ R: xu = 0 $\forall$ u $\in$ U}.

Prove that λ (U) is a two sided of R.

Solution:

First we see that λ (U) $\neq \varnothing$ because 0 $\in$ R is such that

$$0u = 0 \ \forall \ u \in U.$$

Now let x_1, x_2 be any two elements of λ (U). Then

$x_1u = 0 \ \forall \ u \in U$ and $x_2u = 0 \ \forall \ u \in U$.

We have $(x_1 - x_2)\, u = x_1u - x_2u = 0 - 0 = 0$ for all $u \in U$.

$\therefore \quad x_1 - x_2 \in \lambda\,(U)$.

Now let x be any element of λ (U) and r be any element of R. Then

$xu = \forall \ u \in U$ [by def. of λ (U)

$\Rightarrow \ r\,(xu) = r0 \ \forall \ u \in U$

$\Rightarrow \ (rx)\,u =$ for all $u \in U \Rightarrow rx \in \lambda\,(U)$.

Further U is a left ideal of R. Therefore $ru \in U \ \forall \ u \in U$.

Since $x \in \lambda$ (U), therefore by def. of λ (U), we have

$x \in \lambda\,(U),\ ru \in U \Rightarrow x\,(ru) = 0$ for all $u \in U$

$\Rightarrow (xr)\,u = 0$ for all $u \in U$

$\Rightarrow xr \in \lambda\,(U)$.

Thus $x \in \lambda\,(U),\ r \in R \Rightarrow xr,\ rx \in \lambda\,(U)$

Hence λ (U) is a two sided ideal of R.

Example 35(b):

Let R be the ring of all real-valued continuous functions on the closed interval [0, 1] of the real line, the compositions being the usual pointwise addition and multiplication of functions. Let M be a subset of R defined by:

$$M = \{f \in R: f\left(\frac{1}{2}\right) = 0\}.$$

Prove that M is a maximal ideal of R.

Solution:

First of all we observe that M is non-empty because the real valued function e (x) on [0, 1] defined by

$$e(x) = 0 \ \forall \ x \in [0, 1]$$

belongs to M.

Now let f (x), g (x) be any two elements of M. Then

$$f\left(\frac{1}{2}\right) = 0, \ g\left(\frac{1}{2}\right) = 0, \text{ by definition of M.}$$

Let $h(x) = f(x) - g(x)$. Then

$$h\left(\frac{1}{2}\right) = f\left(\frac{1}{2}\right) - g\left(\frac{1}{2}\right) = 0 - 0 = 0.$$

Therefore $h(x) \in M$.

Thus $f(x), g(x) \in M$

$\Rightarrow$ $h(x) = f(x) - g(x) \in M$.

Further let f (x) be any element of M and r (x) be any element of R. Then $f\left(\frac{1}{2}\right) = 0$, by definition of M.

Let $t(x) = r(x)\, f(x) = f(x)\, r(x)$. [$\therefore$ R is a commutative ring]

Then $t\left(\frac{1}{2}\right) = r\left(\frac{1}{2}\right) f\left(\frac{1}{2}\right) = r\left(\frac{1}{2}\right) . 0 = 0$. Therefore (x) Î M

Thus $r(x) \in R,\ f(x) \in M \Rightarrow r(x)\, f(x) \in M$.

Hence M is an ideal of R.

Clearly $M \in R$ because $i(x) \in R$ defined by $i(x) = 1 \ \forall \ x \in [0, 1]$ does not belong to M.

The ring R is with unity and the element i (x) is its unity element.

Let N be an ideal of R properly containing M *i.e.*, $M \subseteq N$ and $M \neq N$. Then M will be a maximal ideal of R if N = R, which will be so if the unity i (x) of R belongs to N. Since M is a proper subset of N, therefore there exists $\lambda(x) \in N$ such that $\lambda(z) \notin M$. This means $\lambda\left(\frac{1}{2}\right) \neq 0$. Put $\lambda\left(\frac{1}{2}\right) = c$ where $c \neq 0$.

Let us define $\beta(x) \in R$ by $\beta(x) = c \ \forall \ x \in [0, 1]$. Now consider $\mu(x) \in R$ given by $\mu(x) = \lambda(x) - \beta(x)$.

We have m $\left(\frac{1}{2}\right) = \lambda\left(\frac{1}{2}\right) - \beta\left(\frac{1}{2}\right)$.

Therefore $\mu(x) \in M$ and so $\mu(x)$ also belongs to N because N is a superset of M. Now N is an ideal of R and $\lambda(x)$, $\mu(x)$ are in N. Therefore $\lambda(x) - \mu(x) = \beta(x)$ is also an element of N.

Now define $\gamma(x) \in R$ by $\gamma(x) = 1/c \; \forall \; x \in [0, 1]$. Since N is an ideal of R, therefore $\gamma(x) \in R$ and $\beta(x) \in N \Rightarrow \gamma(x)\,\beta(x) \in N$.

We shall show that $\gamma(x)\,\beta(x) = i(x)$.

For every $x \in [0, 1]$, we have

$$\gamma(x)\,\beta(x) = (1/c) = 1.$$

Therefore $\gamma(x)\,\beta(x) = i(x)$, by definition of $i(x)$.

Thus the unity element $i(x)$ of R belongs to N and consequently $N = R$.

Hence M is a maximal ideal of R.

Example 35(c):

In the ring I of integers, we have $3 \mid 6$ since we have $6 = 3 \times 2$ and $2 \in R$ I.

However in the ring of integers 3 is not a divisor of 7.

Solution:

Do yourself.

Example 36:

For any given element a of a ring R let

$$Ra = \{xa : x \in R\}.$$

Prove that Ra is a left ideal of R.

Solution:

Let x_1a, x_2a be any two elements of R a where $x_1, x_2 \in R$. We have $x_1a - x_2a = (x_1 - x_2)\,a \in R$ Ra, since

$$x_1, x_2 \in R \Rightarrow x_1 - x_2 \in R.$$

Thus $x_1a, x_2a \in Ra \Rightarrow x_1a - x_2a \in Ra$.

Now let xa be any element of Ra where $x \in R$ and r be any element of R. We have $r(xa) = (rx)\,a \in Ra$, since

$$r \in R, x \in R \Rightarrow rx \in R.$$

Thus $r \in R$, $xa \in Ra \Rightarrow r(xa) \in Ra$.

$\therefore$ ra is a left ideal of R.

Example 37:

Show that an arbitrary intersection of ideals of a ring is an ideal of the ring.

Solution:

Let R be a ring and let $\{S_t : t \in T\}$ be any family of ideals of R. Here T is an index set and is such that $\forall\, t \in T$, S_t is an ideal of R.

Let $\quad S = \cap\, S_t = \{x \in R : x \in S_t\ \forall\, t \in T\}\ t \in T$

be the intersection of this family of ideals of R. Then to prove that S is also an ideal of R.

Obviously $S \neq \varnothing$, since at least 0 is in $S_t\ \forall\, t \in T$

Now let a, b be any two elements of S. Then

$a, b, \in S \Rightarrow a, b \in S_t\ \forall\, t \in T$

$\Rightarrow a - b\ t \in S_t\ \forall\, t \in T \quad [\because\ \forall\, t \in T, S_t \text{ is an ideal of } R]$

$\Rightarrow a - b \in \bigcap_{t \in T} S_t \Rightarrow a - b \in S.$

$\therefore$ S is a subgroup of the additive group of R.

Now let s be any element of S and r be any element of R.

We have $s \in S \Rightarrow s \in \bigcap_{t \in T} S_t$

$\Rightarrow s \in S_t\ \forall\, t \in T$

$\Rightarrow rs \in S_t$ and $sr \in S_t\ \forall\, t \in T$

$[\because\ \forall\, t \in T, S_t \text{ is an ideal of } R]$

$\Rightarrow rs \in \bigcap_{t \in T} S_t$ and $sr \in \bigcap_{t \in T} S_t$

$\Rightarrow rs \in S$ and $sr \in S$.

Thus $a, b, \in S \Rightarrow a - b \in S$

and $r \in R$, $s \in S \Rightarrow rs \in S$, $sr \in S$.

Hence S is an ideal of R.

Example 38:

If U, V are ideals of a ring R let UV be the set of all those elements of R which can be written as finite sums of elements of the form uv where $u \in U$ and $v \in V$. Prove that UV is an ideal of R.

Also show that $UV \subseteq U \cap V$

Solution:

U and V are ideals of a ring R, Let

$UV = \{u_1v_1 + u_2v_2 + ... + u_nv_n : u_1, u_2, ..., u_n \in U, v_1\ v_2, ..., v_n \in V$ and n is any positive integer$\}$.

To prove the UV is also an ideal of R.

Let $\alpha\ u_1v_1 + u_2v_2 + ... + u_nv_n$, $\beta = u_1'v_1' + u_2'\ v_2' + ... + u_m'v_m'$ be and two elements of UV, where $u_1, u_2, ..., u_n, u_1, u_2, ..., u_m' \in U$ and $v_1, v_2, ..., v_n, v_1', v_2', ..., v_m' \in V$. Also m and n are any positive integers.

We have $\alpha - \beta = u_1v_1 + u_2v_2 + ... + u_nv_n - u_1'v_1' - u_2'v'_m = u_1v_1 + u_2v_2 + ... + u_nv_n + (-u_1')\ v_1' + (-u_2')\ v_2' + ... + (-u_m')\ v_m'$.

This is obviously an element of UV because U is an ideal and therefore $u_1' \in U \Rightarrow (-u_1') \in U$, etc.

Again let $r \in R$ and $\alpha \in UV$. Then

$r\alpha = r\ (u_1v_1 + u_2v_2 + ... + u_nv_n) = (ru_1)\ v_1 + (ru_2)\ v_2 + ... + (ru_n)\ v_n$.

This is an element of UV because U is an ideal and therefore $r \in R$, $u_1 \in U \Rightarrow ru_1 \in U$, etc.

Also $\alpha r = (u_1v_1 + u_2v_2 + ... + u_nv_n)\ r = u_1\ (v_1r) + u_2\ (v_2r) + ... + u_n\ (v_nr)$.

This is an element of UV because V is an ideal and therefore $r \in R$, $v_1 \in V \Rightarrow v_1r \in V$, etc.

Hence UV is an ideal of R.

Now to show that $UV \subseteq U \cap V$.

Let $\alpha = u_1v_1 + ... + u_nv_n$ be any element of UV where

$$u_1, ..., u_n \in U, \text{ and } v_1, ..., v_n \in V.$$

Now $v_1 \in V \Rightarrow v_1 \in R$. Also U is an ideal. Therefore

$$v_1 \in R, u_1 \in U \Rightarrow u_1v_1 \in U.$$

Similarly $u_1 \in U \Rightarrow u_1 \in R$. But V is an ideal. Therefore

$$u_1 \in R, v_1 \in V \Rightarrow u_1v_1 \in V.$$

Thus $\quad u_1v_1 \in U, u_1v_1 \in V \Rightarrow u_1v_1 \in U \cap V$.

Similarly $u_2\ v_2, ..., u_nv_n \in U \cap V$.

Since $\quad U \cap V$ is also an ideal of R, therefore

$$u_1v_1, ..., u_nv_n \in U \cap V \Rightarrow \alpha = u_1v_1 + ... + \in U \cap V.$$

Thus $\alpha \in UV \Rightarrow \alpha \in U \cap V$. Therefore $UV \subseteq U \cap V$.

Example 39:

If U is an ideal of a ring R, let

$[R: U] = \{x \in R: rx \in U \ \forall \ r \in R\}$.

Prove that [R: U] is an ideal of R and that it contains U.

Solution:

First we see that [R: U] is not empty because $0 \in R$ is such that $r0 = 0 \in U$ for all $r \in R$.

Now let x_1, x_2 be any two elements of [R: U]. Then

$rx_1 \in U \ \forall \ r \in R$, and $rx_2 \in U \ \forall \ r \in R$,

Since U is an ideal, therefore

$$rx_1 \in U, rx_2 \in U \Rightarrow rx_1 - rx_2 \in U$$
$$\Rightarrow r(x_1 - x_2) \in U \ \forall \ r \in R$$
$$\Rightarrow x_1 - x_2 \in [R: U], \text{ by def. of } [R: U].$$

Now let x be any element of [R: U] and s be any element of R. Then $rx \in U \ \forall \ r \in R$

$\Rightarrow (rx) s \in U \ \forall \ r \in R$ [$\because$ U is an ideal and so $s \in R$, $rx \in U \Rightarrow (rx) s \in U$]

$\Rightarrow r (xs) \in U$ for all $r \in R \Rightarrow xs \in [R: U]$.

Also $rx \in U \ \forall \ r \in R \Rightarrow sx \in U$ [$\because$ $s \in R$]

$\Rightarrow (sx) r \in U \ \forall \ r \in R$

[$\because$ U is an ideal and so $sx \in U$, $r \in R \Rightarrow (sx) r \in U$]

$\Rightarrow sx \in [R: U]$.

Thus $x \in [R: U]$, $s \in R \Rightarrow xs \in [R: U]$, $sx \in [R: U]$.

Therefore [R: U] is an ideal or R.

Now to show that $U \subseteq [R: U]$. We have

$y \in U \Rightarrow yr \in U \ \forall \ r \in R$ [$\because$ U is an ideal]

$\Rightarrow y \in [R: U]$.

$\therefore \ U \subseteq [R: U]$.

Example 40(a):

The ring of Gaussian integers is a Euclidean ring.

Solution:

Let (G, +,.) be the ring of Gaussian integers where

$$G = \{x + iy: x,y \in \mathbf{I}\}.$$

Let the d function on the non-zero elements of G be defined as

$$d\,(x + iy) = x^2 + y^2 \;\forall\; 0 + i0 \neq x + iy \in G.$$

Now if x + iy is a non-zero element of G, then $(x^2 + y^2)$ is a non-negative integer. Thus we have assigned a non-negative integer to every non-zero element of G.

If x + iy and m + in are two non-zero elements of G, then

$$d\,[(x + iy)\,(m + in)] = d\,[(xm - ny) + i\,(my + xn)]$$

$$= (xm - ny)^2 + (my + xn)^2 = x^2m^2 + n^2y^2 + m^2y^2 + x^2n^2$$

$$= (x^2 + y^2)\,(m^2 + n^2)$$

$$\geq x^2 + y^2. \qquad [\because m^2 + n^2 \geq 1]$$

Thus $d\,[x + iy)\,(m + in)] \geq d\,(x + iy)$.

Now to show the existence of division algorithm in G.

Let $\alpha \in G$ and let β be a non-zero element of G. Let $\alpha = x + iy$ and $\beta + m + in$. Define a complex number λ be the equation

$$\lambda = \frac{\alpha}{\beta} = \frac{x+iy}{m+in} = \frac{(x+iy)\,(m-in)}{m^2+n^2} = p+iq$$

where p, q are rational numbers.

Here λ is not necessarily a Gaussian integer.

Also division by β is possible since $\beta \neq 0$.

Let p' and q' be the nearest integers to p and q respectively.

Then obviously $|p - p'| \leq \frac{1}{2}$, $|q - q'| \leq \frac{1}{2}$.

Let $\lambda' = p' + iq'$. Then λ' is a Gaussian integer.

Now $\lambda = \frac{\alpha}{\beta} \Rightarrow \alpha = \lambda\beta$

$$\Rightarrow \alpha = \lambda'\beta + \lambda\beta - \lambda'\beta.$$

Thus $\alpha = \lambda'\beta + (\lambda - \lambda')b$. ...(1)

Since α, β, γ' are Gaussian integers, therefore from (1) it implies that $(\lambda - \lambda')\,\beta$ is also a Gaussian integer.

Now if p and q are integers then p = p', q = q'.

So $\lambda - \lambda' = (p - p') + i\,(q - q') = 0 + i0$. Thus $(\lambda - \lambda')\,\beta = 0 + i0$.

If p and q are not both integers, then $(\lambda - \lambda')\,\beta$ is a non-zero Gaussian integer and we have

$$d[(\lambda - \lambda')\beta] = d[\{(p - p') + i(q - q')\}(m + in)]$$

$$= [(p - p')^2 + (q - q')^2](m^2 + n^2) = [(p - p')^2 + (q - q')^2]\, d(\beta)$$

$$\leq \left[\frac{1}{4} + \frac{1}{4}\right] d(\beta) \qquad [\because (p - p')^2 \leq \frac{1}{4}, (q - q')^2 \leq \frac{1}{4}]$$

$$= \frac{1}{2} d(\beta) < d(\beta).$$

Thus $\alpha = \lambda'\beta\ (\lambda - \lambda')\ \beta$ where λ' and $(\lambda - \lambda')\ \beta$ are Gaussian integers either $(\lambda - \lambda')\ b = 0$

$\Rightarrow \qquad d\,[(\lambda - \lambda')\ b] < d\,(\beta).$

Hence the ring of Gaussian integers is a Euclidean ring.

Example 40(b):

The ring of polynomials over field is a Euclidean ring.

Solution:

Let F [x] be the ring of polynomials over a field F. Let the d function on the non-zero polynomials in F [x] be defined as

$$d\,[f(x) = \deg f(x),\ \forall\ 0 \neq f(x) \in F[x].$$

Now if $0 \neq f(x) \in F[x]$, then deg f (x) is a non-negative integer.

Thus we have assigned a non-negative integer to every non-zero element f (x) in F [x].

Further if f (x), g (x) $\in$ F [x] and are both non-zero polynomials, then

$$\deg [f(x)] = \deg f(x) + \deg g(x)$$

$$\Rightarrow \deg [f(x)\, g(x) \geq \deg f(x) \qquad [\because \deg g(x) \geq 0]$$

$$\Rightarrow d\,[f(x)\, g(x) \geq d\,[f(x)].$$

Finally we know that if $f(x) \in F[x]$ and $0 \neq g(x) \in F[x]$, then there exist two polynomials q (x) and r (x) in F [x] such that

$$f(x) = q(x)\, g(x) + r(x)$$

where either $r(x) = 0$ or $\deg r(x) < \deg g(x)$

i.e. where either $r(x) = 0$ or $d\,[r(x)] < d\,[g(x)]$.

Hence the ring of polynomials over a field is a Euclidean ring.

Example 41:

The ring of integers is a Euclidean ring.

Solution:

Let (**I**, +,.) be the ring of integers where

I = {..., – 3, – 2, –1, 0, 1, 2, 3,....}.

Let the d function on the non-zero elements of **I** be defined as

$d(a) = |a| \ \forall\ 0 \neq a \in \mathbf{I}$.

Now if $0 \neq a \in \mathbf{I}$, then $|a|$ is a non-negative integer. Thus we have assigned a non-negative integer to every non-negative element $a \in \mathbf{I}$.

[$d(-5) = |-5| = 5$, $d(-1) = |-1| = 1$, $d(4) = |4| = 4$ etc.]

Further if $a, b \in \mathbf{I}$ and are both non-zero, then

$|ab| = |a||b|$

$\Rightarrow |ab| \geq |a|$ [$\because |b| \geq 1$ if $0 \neq b \in \mathbf{I}$]

Finally we known that if $a \in \mathbf{I}$ and $0 \neq b \in \mathbf{I}$, then there exist two integers q and r such that

$$a = qb + r \text{ where } 0 \leq r < |b|$$

i.e., where either r = 0 or $\leq r < |b|$

i.e., where either r = 0 or d (r) < d (b).

It should be noted that d (b) = | b | and if r is a positive integer then r = | r | = d (r).

Hence the ring of integers is a euclidean ring.

Example 42:

Resolve $x^4 + 4$ into factors over the field

$(\{0, 1, 2, 3, 4\}, +_5, \times_5)$.

Solution:

Let F be the field $(\{0, 1, 2, 3, 4\}, +_5, \times_5\}$.

Let $f(x) = x^4 + 4 \in F[x]$.

We have

$$f(0) = 0^4 +_5 4 = 4 \neq 0,\ f(1) = 1^4 +_5 4 = 1 +_5 4 = 0,$$

$$f(2) = 2^4 +_5 4 = (2 \times_5 2 \times_5 2 \times_5 2) +_5 4 = 1 +_5 4 = 0,$$

$$f(3) = 3^4 +_5 4 = 1 +_5 4 = 0 \text{ and } f(4) = 4^4 +_5 4 = 1 +_5 4 = 0.$$

By factor theorem, we know that if f (x) is a polynomial over the field F and $a \in F$, then x – a is a factor of f (x) iff f (a) = 0.

So the only factors of f (x) over the given field F are x – 1, x – 2, x – 3 and x – 4.

Now in the given field ({0, 1, 2, 3, 4), $+_5$, $\times_5$), we have

$$1 +_5 4 = 0 \text{ and } 2 +_5 3 = 0.$$

$\therefore$ $- 1 = 4$ and $- 4 = 1$. Also $- 2 = 3$ and $- 3 = 2$.

Thus over the given field F, we have

$$x - 1 = x + 4,\ x - 2 = x + 3,\ x - 3 = x + 2 \text{ and } x - 4 = x + 1.$$

Hence over the given field ({0, 1, 2, 3, 4), $+_5$, $\times_5$), we have

$$x^4 + 4 = (x + 1)(x + 2)(x + 3)(x + 4).$$

Example 43:

If p is a prime integer, show that it need not be prime Gaussian integer.

Solution:

We shall give an example of prime integer which is not a prime Gaussian integer.

A complex number a + ib is called a Gaussian integer if its real and imaginary parts a and b are both in integers.

Consider the integer 5. In the ring of integers the only divisors of 5 are ± 1 and ± 5. Thus 5 has no proper divisors in the ring of integers and so 5 is a prime integer.

In the ring of Gaussian integers we can factorize 5 as

$$5 = (2 + i)(2 - i).$$

The only units of the ring of Gaussian integers are ± 1 and ± i. So neither 2 + i nor 2 – i is a unit of the ring of Gaussian integers.

Consequently 2 + i and 2 – i are proper divisors of 5 in the ring of Gaussian integers. Hence 5 is not a prime element in the ring of Gaussian integers.

Thus 5 is a prime integer, but it is not a prime Gaussian integer.

Example 44:

Add and multiply the following polynomials over the ring (I_5, $+_5$, $\times_5$):

$f(x) + g(x) = 4 + 2x + x^2 + 2x^3.$

Solution:

$$f(x) + g(x) = 4 + 2x + x^2 + 2x^3,$$

$$f(x)\, g(x) = 3 + 3x + x^2 = 3x^3 + x^4 + 4x^5.$$

Example 45:

Let a be a given fixed real number and R be the ring of real numbers. Let ϕ be the mapping which associates with every polynomial p (x) – with real coefficients–the real number p (a). Show that ϕ is a homomorphism of the ring R [x] onto the ring R.

Solution:

Here R is the ring of real numbers and R [x] is the ring of polynomials over the ring R.

We have $R[x] = \{a_0 + a_1x + \ldots a_nx^n : a_0, a_1, \ldots, a_n \in R$ and n is any integer $\geq 0\}$.

If $p(x) = a_0 + a_1x + \ldots + a_nx^n \in R[x]$, then the mapping $\phi: R[x] \to R$ has been defined as

$$\phi(p(x)) = p(a) = a_0 + a_1a + \ldots + a_na^n.$$

To show that the mapping ϕ is a homomorphism of the ring R [x] onto the ring R.

The mapping ϕ is onto. Let $a \in R$. Then $f(x) = a \in R[x]$ and by the definition of the mapping ϕ, we have

$$\phi(f(x)) = f(a) = a.$$

Thus $a \in R \Rightarrow \exists\, f(x) = a \in R[x]$ such that $\phi(f(x)) = f(a) = a$.

Therefore the mapping ϕ is onto R.

Now to show that the mapping ϕ is a homomorphism.

Example 46:

In the ring Q of rational numbers, we have $3 \mid 7$ since we have $7 = 3 \times (7/3)$ and $7/3 \in Q$.

Solution:

Do yourself.

Example 47:

(a) Show that the set R of all elements of the form m + in, m, n $\in$ Z, form a ring for addition and multiplication of complex numbers. Here Z is the ring of integers.

(b) Is the map q: $R \to Z$, q (m + in) = m, a ring homomorphism?

(c) Give an example of a proper ideal in R.

Solution:

Here Z is the ring of integers and

R = {m + in,: m, n ∈ Z} *i.e.*, R is the set of Gaussian integers.

To show that R is a ring for addition and multiplication of complex numbers.

Let a + ib and c + id be any two elements of R, where a, b, c, d ∈ Z.

Then (a + ib) + (c + id) = (a + c) + i (b + d)

and (a + ib) (c + id) = (ac – bd) + i (ad + bc).

These are again elements of R because a + c, b + d, ac – bd and ad + bc are all integers. Therefore R is closed with respect to ordinary addition and multiplication of complex numbers.

Further in complex numbers both addition and multiplication are associative as well as commutative compositions. Also multiplication distributes with respect to addition. The Gaussian integer 0 + i0 is the additive identity. If a + ib ∈ R, then (– a) + i (– b) *i.e.*, – a – ib is also an element of R and it is the additive inverse of a + ib because

[(– a) + i (– b)] + (a + ib) = (– a + a) + i (–b + b) = 0 + i0
= additive identity.

Hence R is a ring for addition and multiplication of complex numbers.

(b) Let θ: R → Z be defined as

θ (m + in) = m, ∀ m + in ∈ R.

Let u = a + ib and v = c + id be any two elements of R. Then

θ (u + v) = θ [(θ + ib) + (c + id)] = θ [(a + c) + i (b + d)]
= a + c, by definition of the mapping θ
= θ (a + ib) + θ (c + id) = θ (u) + θ (v).

Again θ (uv) = θ [(a + ib) (c + id)] = θ [(ac – bd) + i (ad + bc)]
= ac – bd

while θ (u) θ (v) = θ (a + ib) θ (c + id) = ac.

We observe that θ (uv) is not necessarily equal to θ (u) θ (v).

For example if u = 1 + 2i and v = 2 + 3i, then uv = (2 – 6) + (3 + 4) i = – 4 + 7i so that θ (u v) = – 4 while θ (u) θ (v) = 1.2 = 2. Thus,

θ (uv) ≠ θ (u) θ (v).

Thus the mapping θ preserves additions in R and Z but it does not preserve multiplications in R and Z.

We have $\theta(u + v) = \theta(u) + \theta(v)$, $\forall\, u, v \in R$ but there exist elements u, v in R such that $\theta(uv) \neq \theta(u)\,\theta(v)$.

Hence the mapping θ is not a homomorphism of the ring R into the ring Z.

(c) *An example of a proper ideal i R.* To only units of the ring of Gaussian integers R are ± 1 and $\pm i$.

Take a non-zero element of R which is not a unit of R. For example take $u = 2 + 3i \in R$.

Let $S = \{zu : z \in R\}$.

If $z_1 u$, $z_2 u$ are any two elements of S, then

$z_1 u - z_2 u = (z_1 - z_2)\, u \in S$ because $z_1 - z_2 \in R$.

Also if zu is any element of S and v is any element of R, then

$v(zu) = (vz)\, u \in S$ because $vz \in R$

and also $(zu)\, v \in S$ because $v(zu) = (zu)\, v$.

$\therefore$ S is an ideal of R.

Obviously S is not the null ideal because

$1 = 1 + i0 \in R \Rightarrow 1\,(2 + 3i) = 2 + 3i \in S$ and so S contains some non-zero elements of R. Hence S is not the null ideal.

Also $S \neq R$ because $1 \in R$ but $1 \notin S$ as shown below.

If 1 is to be an element of S, then there must exist some $z \in R$ such that

$$zu = 1$$

$\Rightarrow$ $z(2 + 3i) = 1$

$\Rightarrow$ the complex number z is the multiplicative inverse of the complex number $2 + 3i$

$$\Rightarrow \quad z = \frac{1}{2+3i} = \frac{2-31}{(2+3i)(2-3i)} = \frac{2-3i}{13} = \frac{2}{13} - \frac{3}{13}\, i$$

which is not an element of R because 2/13 are not integers.

Thus there exists no $z \in R$ such that $zu = 1$.

$\therefore \quad 1 \notin S.$

Thus $1 \in R$ while $1 \notin S$.

$$S \neq R.$$

Thus S is an ideal of R and S is neither the null ideal nor $S = R$.

Hence S is a proper ideal of R.

Example 48:

Every field is a Euclidean ring.

Solution:

Let F be any field. Let the d function on the non-zero elements of F be defined as

$$d(a) = \forall\ 0 \neq a \in F.$$

Thus we have assigned the integer zero to every non-zero element in F.

If a and b are non-zero elements in F then ab is also a non-zero element in F. We have therefore

$$d(ab) = 0 = d(a).$$

Thus we have $d(ab) \geq d(a)$.

Finally if $a \in F$ and $0 \neq b \in F$, then we can write

$$a = (ab^{-1})\, b + 0$$

i.e., $a = qb + r$ where $q = q = ab^{-1}$ and $r = 0$.

Hence every field is a Euclidean ring.

Example 49(a):

If R is a ring with unit element 1 and ϕ is a homomorphism of R onto R′ prove that ϕ (1) is the unit element of R′.

Solution:

Since f is a homomorphism of R onto R', therefore R' is a homomorphic image of R. If 1 is the unity element of R, then $\phi(1) \in R'$. Let a' be any element of R'. Then $a' = \phi(a)$ for some $a \in R$ since ϕ is onto R'. We have

$$\phi(1)\, a' = \phi(1)\, \phi(a) = \phi(1a) = \phi(a) = a'$$

and $$a'\, \phi(1) = \phi(a)\, \phi(1) = \phi(a1) = \phi(a) = a'.$$

$\therefore$ $\phi(1)$ is the unity element of R'.

Example 49(b):

If R is a finite commutative ring (i.e. has only a finite number of elements) with unit element prove that every prime ideal of R is a maximal ideal of R.

Solution:

Let R be a finite commutative ring with unit element. Let S be a prime ideal of R. Then to prove that S is a maximal ideal of R.

Since S is a prime ideal of R, therefore the residue class ring R/S is an integral domain. Now

$$R/S = \{S + a: a \in R\}.$$

Since R is a finite ring, therefore R/S is a finite integral domain. But every finite integral domain is a field. Therefore R/S is a field. Since R is a commutative ring with unity and R/S is a field, therefore S is a maximal ideal of R.

Example 50:

Give an example of a ring in which some prime ideal is not a maximal ideal.

Solution:

Let us consider **I** [x] be the ring of polynomials over the ring of integers **I**. Let S be the principal ideal of **I** [x] generated by x *i.e.*, let S = (x). We shall show that (x) is prime but not maximal.

We have $S = (x) = \{x\ f(x): f(x) \in \mathbf{I}[x]\}$.

Now we shall prove that S is prime.

Let $a(x), b(x) \in \mathbf{I}[x]$ be such that $a(x)\, b(x) \in S$. Then there exists a polynomial $c(x) \in \mathbf{I}[x]$ such that

$$x\, c(x) = a(x)\, b(x). \qquad ...(1)$$

Let $a(x) = a_0 + a_1x + a_1x + a_2x^2 + ...$, $b(x) = b_0 + b_1x + b_2x^2 + ...$,

$c(x) = c_0 + c_1x + c_2x^2 + ...$. Then (1) becomes

$x(c_0 + c_1x + ...) = (a_0 + a_1x + ...)(b_0 + b_1x + ...)$.

Equation the constant term on both sides, we obtain

$a_0b_0 = 0$

$\Rightarrow a_0 = 0$ or $b_0 = 0$. [$\because$ I is without zero divisors]

Now $a_0 = 0 \Rightarrow a(x) = a_1x + a_2x^2 + ...$

$\Rightarrow a(x) = x(a_1 + a_2x + ...) \Rightarrow a(x) \in (x)$.

Similarly $b_0 = 0 \Rightarrow b(x) = b_1x + b_2x^2 + ...$

$\Rightarrow b(x) = x(b_1 + b_2x + ...) \Rightarrow b(x) \in (x)$.

Thus $a(x)\, b(x) \in (x) \Rightarrow$ either $a(x) \in (x)$ or $b(x) \in (x)$.

Hence (x) is a prime ideal.

Now to prove that (x) is not a maximal ideal of **I** [x]. For this we must show an ideal N of **I** [x] such that (x) is properly contained in N, while N itself is properly contained in **I** [x]. The ideal N = (x, 2) serves this purpose.

Obviously $(x) \subseteq (x, 2)$. In order to show that (x) is properly contained in $(x, 2)$ we must show an element of $(x, 2)$ which is not in (x). Clearly $2 \in (x, 2)$. We shall show that $2 \notin (x)$. Let $2 \in (x)$. Then we can write,

$$2 = x\,f(x) \text{ for some } f(x) \in \mathbf{I}[x].$$

Let $f(x) = a_0 + a_1x + \ldots$

Then $2 = x\,f(x) \Rightarrow\ = x(a_0 + a_1x + \ldots)$

$\Rightarrow 2 = a_0x + a_1x^2 + \ldots$

$\Rightarrow 2 = 0 + a_0x + a_1x^2 + \ldots$

$\Rightarrow 2 = 0$ [by equality of two polynomials]

But $2 \neq 0$ in the ring of integers. Hence $2 \notin (x)$. Thus (x) is properly contained in $(x, 2)$.

Now obviously $(x, 2) \subseteq I[x]$. In order to show that $(x, 2)$ is properly contained in $I[x]$ we must show an element of $I[x]$ which is not in $(x, 2)$. Clearly $1 \in I[x]$. We shall show that $1 \notin (x, 2)$. Let $1 \in (x, 2)$. Then we have a relation of the form

$$1 = x\,f(x) + 2\,g(x), \qquad \text{where } f(x), g(x) \in I[x].$$

Let $f(x) = a_0 + a_1x + \ldots\ g(x) = b_0 + b_1x + \ldots$

Then $1 = x(a_0 + a_1x + \ldots) + 2(b_0 + b_1x + \ldots)$

$\Rightarrow$ $1 = 2b_0$ [Equating constant term on both sides]

But there is no integer b_0 such that $1 = 2b_0$.

Hence $1 \notin (x, 2)$. Thus $(x, 2)$ is properly contained in $I[x]$.

Therefore (x) is not a maximal ideal of $I[x]$.

Example 51:

Let R be the field of real numbers and S the set of all those polynomials f (x) ∈ R [x] such that f (0) = 0 = f (1). Prove that S is an ideal of R [x]. Is the residue class ring R [x]/S an integral domain? Given reasons for your answer.

Solution:

Let $f(x)$, $g(x)$ be any elements of S. Then

$$f(0) = 0 = f(1) \text{ any } g(0) = 0 = g(1).$$

Let $h(x) = f(x) - g(x)$. Then

$h(0) = f(0) - g(0) = 0 - 0 = 0$ and $h(1) = f(1) - g(1) = 0 - 0 = 0$.

Thus $h(0) = 0 = h(1)$. Therefore $h(x) \in S$.

Thus f (x), g (x) $\in$ S $\Rightarrow$ h (x) = f (x) – g (x) $\in$ S.

Further let f (x) be any element of S and r (x) be any element of R [x]. Then f (0) = 0 = f (1), by definition of S.

Let t (x) = r (x) f (x) = f (x) r (x). [$\because$ R [x] is a commutative ring]

Then t (0) = r (0) f (0) = r (0).0 = 0

and t (1) = r (1) f (1) = r (1).0 = 0.

$\therefore$ t (x) $\in$ S.

Thus r (x) $\in$ R [x], f (x) $\in$ S $\Rightarrow$ r (x) f (x) $\in$ S.

Hence S is an ideal of R [x].

Now we claim that S is not a prime ideal of R [x]. Let f (x) = x (x – 1). Then f (0) = 0 (0 – 1) = 0, and f (1) = 1 (1 – 1) = 0.

Thus f (x) = x (x – 1) is an element of S.

Now let p (x) = x, q (x) = x – 1.

We have p (1) = 1 $\neq$ 0. Therefore p (x) $\notin$ S.

Also q (0) = 0 – 1 = –1 $\neq$ 0.

Therefore q (x) $\notin$ S. Thus x (x – 1) $\in$ S while neither x $\in$ S nor x – 1 $\in$ S. Hence S is not a prime ideal of R [x].

Since S is not a prime ideal of R [x], therefore the residue class ring R [x]/S is not an integral domain.

Example 52:

If r is a ring with unit element 1 and ϕ is a homomorphism of R into an integral domain R' such that kernel of ϕ i.e., I (ϕ) $\neq$ R, then prove that ϕ (1) is the unit element of R'.

Solution:

ϕ is a homomorphism of a ring R into an integral domain R'. Then kernel of ϕ.

= I (ϕ) = {x: x $\in$ R and ϕ (x) = 0 $\in$ R'}.

Since I (ϕ) $\neq$ R, therefore there exists an element a $\in$ R such that

ϕ (a) $\neq$ 0 $\in$ R'.

We have ϕ (1) ϕ (a) = ϕ (1a) = ϕ (a).

Now let b' be any element of R'. We have

ϕ (a) b' = ϕ (a) b'

$\Rightarrow$ ϕ (1) ϕ (a) b' = ϕ (a) b' [$\therefore$ ϕ (1) ϕ (a) = ϕ (a)]

$\Rightarrow$ $\phi(a)[\phi(1)b'] = \phi(a)b'$

[$\because$ $\phi(1), \phi(a) \in R'$ which, being an integral domain, is a commutative ring]

$\Rightarrow$ $\phi(a)[\phi(1)b'] - \phi(a)b' = 0$

$\Rightarrow$ $\phi(a)[\phi(1)b' - b'] = 0$

$\Rightarrow$ $\phi(1)b' - b' = 0$

[$\therefore$ $\phi(a) \neq 0$ and R' is without zero divisors]

$\Rightarrow$ $\phi(1)b' = b' = b'\phi(1)$. [$\because$ R' is a commutative ring]

Thus $\phi(1)b' = b'\phi(1) \ \forall \ b' \in R'$.

$\therefore$ $\phi(1)$ is the unit element of R'.

Example 53(a):

Show that every homomorphic image of a commutative ring is commutative.

Solution:

Let R be a commutative ring. Let f be a homomorphic mapping of R onto a ring R'. Then R' is a homomorphic image of R.

Let a', b' be any two elements of R'. Then f (a) = a', f (b) = b' for some a, b $\in$ R because f is onto R'. We have

$$a'b' = f(a)\,f(b) = f(ab)$$

$$= f(ba) \qquad [\because \text{R is commutative}]$$

$$= f(b)\,f(a) = b'a'.$$

$\therefore$ R' is a commutative ring.

Example 53(b):

Prove that any homomorphism of a field is either an isomorphism or takes each element into 0.

Show that a field has no proper homomorphic image.

Solution:

Let ϕ be a homomorphism of a field F into a ring R. Let S be the kernel of ϕ. Then S is an ideal of the field F. We know that field has no proper ideal. Therefore either S = F or S = (0).

If S = F, then by definition of kernel of ϕ, we have $\phi(x) = 0 \ \forall \ x \in F$. Thus, in this case ϕ takes each element of F into the zero element of R. In other words in this case $\phi(F)$ is the zero subring of the ring R.

If S = (0), then the kernel consists of zero element alone. So in this case ϕ is an isomorphism of F into R. Since the isomorphic image of a field is a field, therefore in this case ϕ (F) is a field isomorphic to the field F.

Example 53(c):

Show that the polynomial $x^3 - 9$ is reducible over the ring of integers modulo 11.

Solution:

Let R be the ring of integers modulo 11.

Then R is the ring

$$(\{0, 1, 2, ..., 10\}, +_{11}, \times_{11}).$$

Let $f(x) = x^3 - 9.$

In the ring of integers modulo 11, we have

$$2 +_{11} 9 = 0.$$

$\therefore \quad -9 = 2.$

So over the ring of integers modulo 11, we have

$$f(x) = x^3 - 9 = x^3 + 2.$$

In order to show that f (x) is reducible over the ring of integers modulo 11, we shall produce a proper factor of f (x) over the ring of integers modulo 11.

We have

$$f(4) = 4^3 +_{11} 2 = (4 \times_{11} 4 \times_{11} 4) +_{11} 2 = 9 +_{11} 2 = 0.$$

So by factor theorem x – 4 *i.e.*, x + 7 is a factor of f (x) over the ring of integer modulo 11. But x + 7 is not a polynomial of zero degree over the ring of integers modulo 11 and so x + 7 is not a unit of the polynomial ring R [x]. Consequently x + 7 is a proper factor of f (x) over the ring of integers modulo 11.

Hence f (x) = $x^3 - 9 = x^3 + 2$ is reducible over the ring of integers modulo 11.

Example 53(d):

If R/S is a ring of residue classes of S in R, prove that

(i) If R is commutative, so also is R/S.

(ii) if R has a unity element 1 so also has R/S, namely S + 1.

Solution:

(i) Suppose R is a commutative ring Let S + a, S + b be any two elements of R/S. Then a, b $\in$ R and ab = ba.

We have (S + a) (S + b) = S + ab = S + ba = (S + b) (S + a).

(ii) Suppose R is a ring with unit element 1. Then S + 1 $\in$ R/S. If S + a is any element of R/S, we have

$$(S + 1)\,(S + a) = S + (1a) = S + a$$

and $$(S + a)\,(S + 1) = S + (a1) = S + a.$$

$\therefore$ S + 1 is the unit element of R/S.

Example 54:

Resolve $x^2 + 1$ into factors over the field Z_5.

Solution:

The field Z_5 is $(\{0, 1, 2, 3, 4\}, +_5, \times_5)$.

Let $$f(x) = x^2 + 1.$$

We have $f(0) = 0^2 +_5 1 = 1 \neq 0,$

$$f(1) = 1^2 +_5 1 = 2 \neq 0$$

$$f(2) = 2^2 +_5 1 = 4 +_5 1 = 0,$$

$$f(3) = 3^2 +_5 1 = (3 \times_5 3) +_5 1 = 4 +_5 1 = 0,$$

$$f(4) = 4^2 +_5 1 = (4 \times_5 4) +_5 1 = 1 +_5 1 = 2 \neq 0.$$

By factor theorem, we know that if f (x) is a polynomial over the field F and a $\in$ F, then x – a divides f (x) if and only if f (a) = 0.

So the only factors of $f(x) = x^2 + 1$ over field Z_5 are x – 2 and x–3.

Here – 2 and – 3 are the additive inverse of 2 and 3 respectively in the field Z_5. In the field Z_5, we have

$$2 +_5 3 = 0.$$

$\therefore -2 = 3$ and $-3 = 2$.

So over the field Z_5, we have

$$x - 2 = x + 3 \text{ and } x - 3 = x + 2.$$

Thus over the field Z_5, we have

$$x^2 + 1 = (x - 2)(x - 3) = (x + 3)(x + 2).$$

Over the field Z_5,

$$x^2 + 1 = (x + 2)(x + 3)$$

Example 55:

Find the solution of the equation 3x = 2 in the field $(Z_7, +_7, \times_7)$.

Solution:

The field $(Z_7, +_7, \times_7)$ is the field $(\{0, 1, 2, 3, 4, 5, 6\}, +_7, \times_7)$.

If $x \in Z_7$ and $x = 3$, we have

$$3x = x +_7 x +_7 x = 3 +_7 3 +_7 3 = 2.$$

Thus $x = 3$ is a solution of the equation $3x = 2$ in the field $(Z_7, +_7, \times_7)$.

Also no other element of field Z_7 satisfies the equation $3x = 2$.

If $x = 0$, we have $3x = 0$,

if $x = 1$, we have $3x = 1$,

if $x = 2$, we have $3x = 6$,

if $x = 4$, we have $3x = 5$,

if $x = 5$, we have $3x = 1$,

and if $x = 6$ we have $3x = 4$.

Hence $x = 3$ is the only solution of the equation $3x = 2$ in the given field $(Z_7, +_7, \times_7)$.

Example 56:

If $f(x) = 3x^7 + 2x + 3$, $g(x) = 5x^3 + 2x + 6$ *be two polynomials over the field* $Z_7 = (\{0, 1, 2, 3, 4, 5, 6\}, +_7, \times_7)$, *determine*

(i) $\frac{d}{dx} f(x)$ *(ii)* $f(x) \cdot g(x)$, *and (iii)* $f(x) + g(x)$.

Solution:

(i) We have

$$= (7 +_7 7 +_7 7)\, x^6 + (1 +_7 1)$$

(ii) We have f (x) g (x)

$$= (3 + 2x + 3x^7)(6 + 2x + 5x^3)$$

$$= (3 \times_7 6) + [(3 \times_7 2) +_7 (2 \times_7 6)]\, x + (2 \times_7 2)\, x^2$$

$$+ (3 \times_7 6)\ (2 \times_7 5)\, x^4 + (3 \times_7 6)\, x^7 + (3 \times_7 2)\, x^8 + (3 \times_7 5)\, x^{10}$$

$$= 4 + 4x + 4x^4 + 3x^4 + 4x^7 + 6x^8 + x^{10}$$

(iii) We have f (x) + g (x)

$$= (3 + 2x + 3x^7) + (6 + 2x + 5x^3)$$

$$= (3 +_7 6) + (2 +_7 2)\, x + (0 +_7 5)\, x^3 + (3 +_7 0)\, x^7$$

$$= 2 + 4x + 5x^3 + 3x^7$$

$$= 3x^7 + 5x^3 + 4x + 2.$$

Example 57:

Let $f(x) = x^6 + 3x^3 + 4x^2 - 3x + 2$ *and* $g(x) = x^2 + 2x - 3$ *be in* $Z_7[x]$. *Find*

(i) Sum and product of $f(x)$ *and* $g(x)$ *in* $Z_7[x]$.

(ii) Two polynomials $q(x)$ *and* $r(x)$ *in* $Z_7[x]$ *such that*

$f(x) = q(x) + r$ *with degree* $r(x) < 2$.

Solution:

The field Z_7 is $(\{0, 1, 2, 3, 4, 5, 6), +_7, \times_7)$.

In Z_7, $3 +_7 4 = 0$ so that $4 = -3$.

$\therefore$ we can write

$$f(x) = x^6 + 3x^5 + 4x^2 - 3x + 2 = x^6 + 3x^5 + 4x^2 + 4x + 2$$

and $\quad g(x) = x^2 + 2x - 3 = x^2 + 2x + 4.$

(i) We have $f(x) + g(x)$

$$= (2 + 4x + 4x^2 + 3x^5 + x^6) + (4 + 2x + x^2)$$

$$= (2 +_7 4) + (4 +_7 2)x + (4 +_7 1)x^2 + 3x^5 + x^6$$

$$= 6 + 6x + 5x^2 + 3x^5 + x^6$$

$$= x^6 + 3x^5 + 5x^2 + 6x + 6.$$

Also $f(x)\, g(x)$

$$= (2 + 4x + 4x^2 + 3x^5 + x^6)(4 + 2x + x^2)$$

$$- (2 \times_7 4) + [(2 \times_7 2) +_7 (4 \times_7 4)]x + [(2 \times_7 1) +_7 (4 \times_7 2)$$

$$+_7 (4 \times_7 4)]x^2 + [(4 \times_7 1) +_7 (4 \times_7 2)]x^3 + (4 \times_7 1)x^4$$

$$+ (3 \times_7 4)x^5 + [(3 \times_7 2) +_7 (1 \times_7 4)]x^6$$

$$+ [(3 \times_7 1) +_7 (1 \times_7 2)]x^7 + (1 \times_7 1)x^8$$

$$= 1 + 6x + 5x^2 + 5x^3 + 4x^4 + 5x^5 + 3x^6 + 5x^7 + x^8$$

$$= x^8 + 5x^7 + 3x^6 + 5x^5 + 4x^4 + 5x^3 + 5x^2 + 6x + 1.$$

In $Z_7[x]$, let us divide $f(x)$ by $g(x)$ by long division method.

$$x^2 + 2x + 4 \;\; \overline{\begin{array}{l} x^4 + x^3 + x^2 + x + 5 \\ \hline x^6 + 3x^5 + 4x^2 + 4x + 2 \end{array}}$$

$$\begin{array}{l} x^6 + 2x^5 + 4x^4 \\ \hline x^5 + 3x^4 + 4x^2 + 4x + 2 \end{array}$$

$$\begin{array}{l} \underline{x^5+2x^4+4x^3} \\ x^4+3x^3+4x^2+4x+2 \\ \underline{x^4+2x^3+4x^2} \\ x^3+4x+2 \\ \underline{x^3+2x^2+4x} \\ 5x^2+2 \\ \underline{5x^2+3x+6} \\ 4x+3 \end{array}$$

[∵ in Z_7, – 4 = 3]

[∵ in Z_7, – 3 = 4 and $2 - 6 = 2 +_7 (-6) = 2 +_7 1 = 3$]

The degree of the remainder 4x + 3 is 1 which is less than 2 *i.e.*, the degree of the divisor g (x) = $x^2 + 2x + 4$.

Thus we have

$x^6 + 3x^5 + 4x^2 + 4x + 2 = (x^4 + x^3 + x^2 + x + 5)(x^2 + 2x + 4) + 4x + 3$

i.e., f (x) = q (x) g (x) + r (x).

Example 58:

Show that f (x) = $x^2 + 8x - 2$ is irreducible over the field of rational numbers Q. Is irreducible over real? Give reasons for your answer.

Solution:

First we solve the equation f (x) = 0 over the field C of complex numbers.

We have $x^2 + 8x - 2 = 0$

$$\Rightarrow \qquad x = \frac{-8 \pm \sqrt{(64+8)}}{2} = -4 \pm 3/2.$$

Thus, the only proper factors of f (x) = $x^2 + 8x - 2$ over the field of complex numbers are

$$x - (-4 + 3\sqrt{2}) \text{ and } x - (-4 - 3\sqrt{2}).$$

Since neither $-4 + 3\sqrt{2}$ nor $-4 - 3\sqrt{2}$ is a rational number, therefore neither x $(-4 + 3\sqrt{2})$ nor $x - (-4 - 3\sqrt{2})$ is a polynomial over the field Q. Hence f (x) has no proper factor over the field of rational numbers Q and so f (x) is irreducible over the field Q.

But both $-4 + 3\sqrt{2}$ and $-4 - 3\sqrt{2}$ are real numbers and so both $x - (-4 + 3\sqrt{2})$ and $x - (-4 - 3\sqrt{2})$ are polynomials over the field of real numbers R. Thus $x - (-4 + 3\sqrt{2})$ and x $(-4 - 3\sqrt{2})$ are proper factors of

$f(x) = x^2 + 8x - 2$ over the field of real numbers R. Hence $f(x) = x^2 + 8x - 2$ is reducible over the field R.

So $f(x) = x^2 + 8x - 2$ is not irreducible over the field of real number R.

Example 59:

Let $f(x) = 2x^4 + 3x^3 + 2$ and $g(x) = 3x^5 + 4x^3 + 2x^3 + 3$ be two polynomials over the field $Z_5 = (\{0, 1, 2, 3, 4)\}, +_5, \times_5)$.

Determine (i) $\frac{d}{dx} f(x)$, (ii) $f(x).\ g(x)$.

Solution:

(i) We have

$$\frac{d}{dx} f(x) = 4\ (2)\ x^3 + 3\ (3)\ x^2$$

$$= (2 +_5 2 +_5 2 +_5 2)\ x^3 + (3 +_5 3 +_5 3)\ x^2 = 3x^3 + 4x^2.$$

(ii) We have $f(x)\ g(x)$

$$= (2 + 3x^3 + 2x^4)\ (3 + 2x^2 + 4x^3 + 3x^5)$$

$$= (2 \times_5 3) + (2 \times_5 2)\ x + [(2 \times_5 4) +_5 (3 \times_5 3)]\ x^3 + (2 \times_5 3)\ x^4 + [(2 \times_5 3) +_5 (3 \times_5 2)]\ x^5$$

$$+ [(3 \times_5 4) +_5 (2 \times_5 2)]\ x^6 + (2 \times_5 4)\ x^7 + (3 \times_5 3)\ x^8 + (2 \times_5 3)\ x^9$$

$$= 1 + 4x^2 + 2x^3 + x^4 + x^4 + 2x^5 + x^6 + 3x^7 + 4x^8 + x^9.$$

where the quotient $q(x) = x^4 + x^3 + x^2 + x + 5$.

and the remainder $r(x) = 4x + 3$.

Example 60(a):

Show that the polynomial $x^2 + x + 4$ is irreducible over F, the field of integers modulo 11.

Solution:

The field F is $\{0, 1, ..., 10\}, +_{11}, \times_{11})$

Let $f(x) = x^2 + x + 4$.

If $a \in F$, then by a^n we shall mean $a \times_{11} a \times_{11} a \times_{11} a...$ upon n times.

Now $f(0) = 0^2 +_{11} 0 +_{11} 4 = 4$, $f(1) = 1^2 +_{11} 1 +_{11} 4 = 6$.

$f(2) = 2^2 +_{11} 2 +_{11} 4 = 10$, $f(3) = 3^2 +_{11} 3 +_{11} 4 = 5$, $f(4) = 2$,

$f(5) = 1$, $f(6) = 6^2 +_{11} 6 +_{11} 4 = 2$, $f(7) = 5$, $f(8) = 10$, $f(9) = 6$, $f(10) = 4$.

Since $f(a) \neq 0 \ \forall \ a \in F$, therefore by factor theorem $x - a$ does not divide $f(x) \ \forall \ a \in F$. Therefore $f(x)$ has no proper divisors in $F[x]$. Hence $f(x)$ is irreducible over F.

Example 60(b):

Let R be the ring of all real valued continuous functions defined on the closed interval [0, 1]. Let

$$M = \{f(x) \in R : f\left(\frac{1}{3}\right) = 0\}.$$

Show that M is a maximal ideal of R.

Solution:

First of all we observe ·'· t M is non empty because the real valued function $e(x)$ on [0, 1] defined by

$$e(x) = 0 \ \forall \ x \in [0, 1]$$

belongs to M.

Now let $f(x)$, $g(x)$ be any two elements of M. Then

$$f\left(\frac{1}{3}\right) = 0, \ g\left(\frac{1}{3}\right) = 0, \text{ by definition of M.}$$

Let $h(x) = f(x) - g(x)$.

Then $h\left(\frac{1}{3}\right) = f\left(\frac{1}{3}\right) - g\left(\frac{1}{3}\right) = 0 - 0 = 0.$

Therefore $h(x) \in M$.

Thus $f(x), g(x) \in M \Rightarrow h(x) = f(x) - g(x) \in M$.

Further let $f(x)$ be any element of M and $r(x)$ be any element of R. Then $f\left(\frac{1}{3}\right) = 0$, by definition of M.

Let $t(x) = r(x) f(x) = f(x) r(x)$. [$\because$ R is a commutative ring].

Then $t\left(\frac{1}{3}\right) = r\left(\frac{1}{3}\right) f\left(\frac{1}{3}\right) = r\left(\frac{1}{3}\right) . 0 = 0$. Therefore $t(x) \in M$.

Thus $r(x) \in R, f(x) \in M \Rightarrow r(x) f(x) \in M$.

Hence M is an ideal of R.

Clearly $M \neq R$ because $i(x) \in R$ given by $i(x) = 1 \ \forall \ x \in [0, 1]$ does not belong to M.

The ring R is with unity and the element $i(x)$ is its unity element.

Let N be an ideal of R properly containing M *i.e.*, $M \in N$ and $M \neq N$. Then M will be a maximal ideal of R if N = R, which be so if the unity i (x) of R belongs to N. Since M is a proper subset of N, therefore there exists $\lambda(x) \in N$ such that $\lambda(x) \notin M$. This means $\lambda\left(\frac{1}{3}\right) \neq 0$. Put $\lambda\left(\frac{1}{3}\right) = c$ where $c \neq 0$.

Let us define $\beta(x) \in R$, by $\beta(x) = c \ \forall \ x \in [0, 1]$. Now consider $\mu(x) \in R$ given by $\mu(x) = \lambda(x) - \beta(x)$.

$$\text{We have } \mu\left(\frac{1}{3}\right) = \lambda\left(\frac{1}{3}\right) - \beta\left(\frac{1}{3}\right) = c - c = 0.$$

Therefore $\mu(x) \in M$ and so $\mu(x)$ also belongs to N because N is a superset of M. Now N is an ideal of R and $\lambda(x)$, $\mu(x)$ are in N. Therefore $\lambda(x) - \mu(x) = \beta(x)$ is also an element of N.

Now define $\lambda(x) \in R$ by $\lambda(x) = 1/c \ \ \forall \ x \in [0, 1]$. Since N is an ideal of R, therefore $\lambda(x) \in R$ and $\beta(x) \in N \Rightarrow \lambda(x)\,\beta(x) \in N$. We shall show that $\lambda(x)\,\beta(x) = i(x)$.

For every $x \in [0, 1]$, we have

$$\lambda(x)\,\beta(x) = (1/c)c = 1.$$

Therefore $\lambda(x)\,\beta(x) = i(x)$, by definition of i(x).

Thus the unity element i(x) of R belongs to N and consequently N = R.

Hence M is a maximal ideal of R.

Example 60(c).

Consider the ring R of all 3 × 3 matrices of the type

$$\begin{bmatrix} a & b & c \\ 0 & d & e \\ 0 & 0 & f \end{bmatrix},$$

a, b, c, d, e, f are real numbers. Show that the set I of all matrices of the form

$$\begin{bmatrix} a & 0 & 0 \\ 0 & 0 & 0 \\ 0 & 0 & 0 \end{bmatrix}$$

is a left ideal of R, which is not a right ideal.

Solution:

First to show that I is a left ideal of R.

Let $A = \begin{bmatrix} a_1 & 0 \\ 0 & 0 \end{bmatrix}$ and $B = \begin{bmatrix} a_2 & 0 \\ 0 & 0 \end{bmatrix}$ be nay two elements of I.

Then $A - B = \begin{bmatrix} a_1 - a_2 & 0 \\ 0 & 0 \end{bmatrix} \in I.$

∴ I is a subgroup of the additive group of the ring R.

Now let $U = \begin{bmatrix} a & b & c \\ 0 & d & e \\ 0 & 0 & f \end{bmatrix}$ be any element of R and $A = \begin{bmatrix} p & 0 & 0 \\ 0 & 0 & 0 \\ 0 & 0 & 0 \end{bmatrix}$ be any element of I.

Then $UA = \begin{bmatrix} a & b & c \\ 0 & d & e \\ 0 & 0 & f \end{bmatrix} \begin{bmatrix} p & 0 & 0 \\ 0 & 0 & 0 \\ 0 & 0 & 0 \end{bmatrix} = \begin{bmatrix} ap & 0 & 0 \\ 0 & 0 & 0 \\ 0 & 0 & 0 \end{bmatrix} \in I$

Therefore I is a left ideal of the ring R.

But I is not a right ideal of R, since

$$\begin{bmatrix} 1 & 1 & 1 \\ 0 & 1 & 2 \\ 0 & 0 & 1 \end{bmatrix} \in R, \begin{bmatrix} 1 & 0 & 0 \\ 0 & 0 & 0 \\ 0 & 0 & 0 \end{bmatrix} \in I$$

and the product $\begin{bmatrix} 1 & 0 & 0 \\ 0 & 0 & 0 \\ 0 & 0 & 0 \end{bmatrix} \begin{bmatrix} 1 & 1 & 1 \\ 0 & 1 & 2 \\ 0 & 0 & 1 \end{bmatrix} = \begin{bmatrix} 1 & 1 & 1 \\ 0 & 0 & 0 \\ 0 & 0 & 0 \end{bmatrix} \notin I$.

Hence I is not a right ideal or R.

Example 61:

Prove that the intersection of two ideals of a ring R is an ideal of R.

Solution:

Let s and T be two ideals of a ring R. Then S, T are subgroups of R under addition. Therefore $S \cap T$ is also a subgroup of R under addition.

Now to show that $S \cap T$ is a ideal of R, we are only to show that

$$r \in R, s \in S \cap T \Rightarrow rs \in S \cap T, sr \in S \cap T.$$

We have $s \in S \cap T \Rightarrow s \in S, s \in T$.

But S and T are ideals of R. Therefore

$r \in R, s \in S \Rightarrow rs \in S, sr \in S$

and $r \in R, s \in T \Rightarrow rs \in T, sr \in T$.

Now $rs \in S, rs \in T \Rightarrow rs \in S \cap T$

and $sr \in S, sr \in T \Rightarrow sr \in S \cap T$.

$\therefore$ $S \cap T$ is also an ideal of R.

Example 62:

Show that $Z(\sqrt{-5})$, the set of complex numbers $a + b\sqrt{(-5)}$ where a, b are integers, is an integral domain.

Solution:

Let $a + b\sqrt{(-5)}$ and $c + d\sqrt{(-5)}$ be any two complex numbers belonging to the set $Z[\sqrt{-5}]$, where a, b, c, d are any integers.

Then $[a + b\sqrt{(-5)}] + [c + d\sqrt{(-5)}] = (a + c) + (b + d)\sqrt{(-5)}$

and $[a + b\sqrt{(-5)}]\,[c + d\sqrt{(-5)}] = (ac - 5bd) + (ad + bc)\sqrt{(-5)}$.

There are again members of the set $Z[\sqrt{-5}]$ because $a + c$, $b + d$, $ac - 5bd$, $ad + bc$ are all integers. Therefore $Z[\sqrt{-5}]$ is closed with respect ordinary addition and multiplication of complex numbers.

Further in complex numbers both addition and multiplication are associative as well as commutative compositions. Also multiplication distributes with respect to addition. The complex number $0 + 0\sqrt{(-5)}$ is a member of the set $Z[\sqrt{-5}]$ and is the additive identity. The additive inverse of $a + b\sqrt{(-5)} \in Z[\sqrt{-5}]$ is $(-a) + (-b)\sqrt{(-5)}$. The complex number $1 + 0\sqrt{(-5)}$ is a member of the set $Z[\sqrt{-5}]$ and is the multiplicative identity.

Therefore the set of complex numbers $a + b\sqrt{(-5)}$ where a, b, are integers is a commutative ring with unity for the addition and multiplication of complex numbers as the two ring compositions.

Also this ring is free from zero divisors since the product of two non-zero complex numbers cannot be zero. Therefore $Z[\sqrt{-5}]$ is an integral domain for ordinary addition and multiplication of complex numbers as the two ring compositions.

Example 63:

If a, b are any elements of a ring R, prove that

1. $-(-a) = a$;

2. $-(a + b) = -a - b$;
3. $-(a - b) = -a + b$.

Solution:

1. Since R is a group with respect to addition, therefore $-(-a) = a$. [Remember that in a group $(a^{-1})^{-1} = a$].
2. Since R is a group with respect to addition, therefore $-(a + b) = (-b) + (-a)$. But addition in R is commutative.

$\therefore \quad (-b) + (-a) = (-a) + (-b) = -a - b.$

$\therefore \quad -(a + b) = -a - b.$

3. $-(a - b) = -[a + (-b)] = -a + [-(-b)] = -a + b.$

Example 64:

The ring ({0, 1, 2, 3, 4, 5}. $+_6$, $\times_6$) is a ring with zero divisors. We have $2 \times_6 3 = 0$, $3 \times_6 4 = 0$ i.e., the product of two non-zero elements is equal to the zero element of the ring.

Solution:

Do yourself.

Example 65(a):

Suppose M is a ring of all 2 × 2 matrices with their elements as integers, the addition and multiplication of matrices being the two ring compositions. Then M a ring with zero divisors.

The null matrix $O = \begin{bmatrix} 0 & 0 \\ 0 & 0 \end{bmatrix}$ is the zero element of this ring.

Now $A = \begin{bmatrix} 1 & 0 \\ 0 & 0 \end{bmatrix}, B = \begin{bmatrix} 0 & 0 \\ 1 & 0 \end{bmatrix}$ are two non-zero elements of this ring *i.e.*, $A \neq O$, $B \neq O$. We have

$$AB = \begin{bmatrix} 1 & 0 \\ 0 & 0 \end{bmatrix} \begin{bmatrix} 0 & 0 \\ 1 & 0 \end{bmatrix} = \begin{bmatrix} 0 & 0 \\ 0 & 0 \end{bmatrix} = O.$$

Thus the product of two non-zero elements of the ring is equal to the zero element of the ring. Therefore M is a ring with zero divisors.

Also it is interesting to note that

$$BA = \begin{bmatrix} 0 & 0 \\ 1 & 0 \end{bmatrix} \begin{bmatrix} 1 & 0 \\ 0 & 0 \end{bmatrix} = \begin{bmatrix} 0 & 0 \\ 1 & 0 \end{bmatrix} \neq \begin{bmatrix} 0 & 0 \\ 0 & 0 \end{bmatrix}.$$

Thus, in a ring R it is possible that $ab = 0$ but $ba \neq 0$.

Example 66:

Prove that if a, b $\in$ R then $(a + b)^2 = a^2 + ab + ba + b^2$, where by x^2 we mean xx.

Solution:

We have

$(a + b)^2 = (a + b)(a + b)$

$= a(a + b) + b(a + b)$ [by right distributive law]

$= (aa = ab) + (ba + bb).$ [by left distributive law]

$= a^2 = ab = ba + b^2.$

Example 67:

If a, b, c are elements of a ring R, then evaluate $(a + b)(c + d)$.

Solution:

We have

$(a + b) + (c + d) = a(c + d) + b(c + d)$ [by right distributive law]

$= ac + ad + bc = bd.$ [by left distributive law]

Example 68(a):

The ring of integers is a ring without zero divisors. The product of two non-zero elements is equal to the zero element of the ring.

Solution:

Do yourself.

Example 68(b):

Prove that the set of residue classes modulo p is a commutative ring with respect to addition and multiplication of residue classes. Further show that the ring of residue classes modulo p is a field if and only if p is a prime.

Solution:

Let I_p be the set of residue classes modulo p. Then I_p has p distinct elements. Thus $I_p = \{[0], [1], [2], ..., [p - 1]\}$.

Let $[a], [b] \in I_p$. Then we define addition and multiplication or residue classes as follows:

$[a] + [b] = [a + b],$ (addition)

$[a]\,[b] = [ab]$ (multiplication)

Since $[a + b]$ and $[ab]$ are both residue classes modulo p, therefore I_p is closed with respect to addition and multiplication.

Now let [a], [b], [c] be any elements of I_p. Then we observe:

Commutativity of Addition:

$[a] + [b] = [a + b]$ (by def, of addition of residue classes)

$= [b + a]$ ($\because$ addition of integers is commutative)

$= [b] + [a].$

Associativity of Addition: We have

$$([a] + [b]) + [c] = [a + b] + [c] = [(a + b) + c] = [a + (b + c)]$$
$$= [a] + [b + c] = [a] + ([b] + [c]).$$

Existence of Additive Inverse: We have $[0] \in I_p$. If $[a] \in I_p$, then $[0] + [a] = [0 + a] = [a]$. Therefore [0] is the additive identity.

Existence of Additive Inverse: Let $[a] \in I_p$. Then $[-a] \in I_p$. We have $[-a] + [a] + [a] = [-a + a] = [0]$. Therefore $[-a]$ is the additive inverse of [a].

Associativity of Multiplication: We have

$$([a]\,[b])\,[c] = [ab]\,[c] = [(ab)\,c] = [a\,(bc)] = [a]\,[bc] = [a]\,([b]\,[c]).$$

Commutativity of Multiplication: We have

$$[a]\,[b] = [ab] = [ba] = [b]\,[a].$$

Distributive Laws: We have

$$[a]\,([b] + [c]) = [a]\,[b + c] = [a\,(b + c)] = [ab + ac]$$
$$= [ab] + [ac] = [a]\,[b] + [a]\,[c].$$

Similarly $([b] + [c])\,[a] = [b]\,[a] + [c]\,[a]$.

Thus, I_p is a commutative ring. Note that it is a finite ring because it has p elements. Now suppose that p is a prime number. Then to prove that I_p is a field. Let $[a], [b] \in I_p$. Then

$[a]\,[b] = [0]$

$\Rightarrow [ab] = [0]$

$\Rightarrow$ p is a divisor of ab *i.e.*, $p \mid ab$

$\Rightarrow p \mid a$ or $p \mid b$. [Note that if a and b are any two integers and p is a prime number, then $p \mid ab \Rightarrow p \mid a \Rightarrow p \mid b$]

$\Rightarrow [a] = [0] \Rightarrow [b] = [0]$.

Thus, I_p is without zero divisors. Therefore I_p is an integral domain. But every finite integral domain is a field. Hence I_p is a field. Conversely suppose that I_p is a field. Then I_p is an integral domain. Therefore I_p is without zero divisors. We are to prove that p is a prime number. Suppose p is not prime, but p is composite. Let p = mn, where $1 < m < p$, $1 < n < p$. Then

$$[mn] = [p]$$

$$\Rightarrow [m]\,[n] = [0]. \qquad (\because\ [p] = [0])$$

Also $[m] \neq [0]$, since $1 < m < p$. Similarly $[n] \neq 0$.

Thus $[m]\,[n] = [0]$ while neither $[m] = [0]$ nor $[n] = [0]$.

Therefore I_p possesses zero divisors and thus we get a contradiction. Hence p must be prime.

Example 68(c):

Do the following sets form integral domains with respect to ordinary addition and multiplication? If so state if they are fields.

1. *The set of numbers of the form b $\sqrt{2}$ with b rational.*
2. *The set of even integers.*
3. *The set of positive integers.*

Solution:

1. Let $A = \{b\sqrt{2} : b \in Q\}$.

   We have $3\sqrt{2} \in A$ and $5\sqrt{2} \in A$. Then $(3\sqrt{2})(5\sqrt{2}) = 30$.

   Now 30 cannot be put in the form b $\sqrt{2}$ where b is a rational number. Therefore $30 \notin A$. Thus A is not closed with respect to multiplication. Therefore the question of A becoming a ring does not arise.

2. Let R be the set of all even integers. Then R is a ring with respect to addition and multiplication of integers. Also the multiplication is a commutative composition. r is without zero divisors since the product of two non-zero even integers cannot be equal to zero which is the zero element of this ring. Since the integer $1 \notin R$, therefore R is a ring without unity.

   R will be an integral domain if we do not require the existence of the unit element for an integral domain.

   But R is not a field since the multiplicative identity does not exist.

3. Let N be the set of positive integers. Since the integer $0 \notin N$, therefore the additive identity does not exist. So N will not be a ring.

Example 69(a):

An element a of a ring R is said to be nilpotent if $a^n = 0$ for some positive integer n. Prove that a = 0 is the only nilpotent element of an integral domain.

Solution:

Let R be an integral domain. We have $0^n = 0$ for every positive integer n.

$\therefore$ The element 0 of R is surely nilpotent.

Now let a be any non-zero element of R. Since in an integral domain the product of two non-zero elements cannot be zero, therefore

$$a \neq 0 \Rightarrow a^2 = aa \neq 0 \Rightarrow a^3 = aa^2 \neq 0 \Rightarrow a^4 = aa^3 \neq 0, \text{ and so on.}$$

Thus, in an integral domain if $a \neq 0$, then $a^n \neq 0$ for every positive integer n and so a cannot be nilpotent.

Hence 0 is the only nilpotent element of an integral domain.

Example 69(b):

Prove that the set M of 2 × 2 matrices over the field of real numbers is a ring with respect to matrix addition and multiplication. It is a commutative ring with unity element ? Find the zero element. Does this ring possess zero divisors?

Solution:

Let A, B, $\in$ M. Then A + B $\in$ M and AB $\in$ M. Therefore M is closed with respect to addition and multiplication of matrices.

$$\therefore \quad A + (B + C) = (A + B) + C \ \forall\ A, B, C \in M$$

and $$A (BC) = (AB) C \ \forall\ A, B, C \in M.$$

Addition of matrices is a commutative composition. Therefore for all A, B $\in$ M, we have A + B = B + A.

If O be the null matrix of the type 2 × 2, then O $\in$ M and O + A + A $\forall$ A $\in$ M.

Further multiplication of matrices is distributive with respect to addition.

$$\therefore \quad A (B + C) = AB + AC$$

and $$(B + C) A - BA + CA \vee A, B, C \in M.$$

$\therefore$ M is a ring with respect to the given compositions.

Multiplication of matrices is not in general a commutative composition.

For example, if $A = \begin{bmatrix} 2 & 4 \\ 3 & 5 \end{bmatrix}, B = \begin{bmatrix} 1 & 2 \\ 0 & 1 \end{bmatrix}$, then

$$AB = \begin{bmatrix} 2 & 4 \\ 3 & 5 \end{bmatrix}\begin{bmatrix} 1 & 2 \\ 0 & 1 \end{bmatrix} = \begin{bmatrix} 2 & 8 \\ 3 & 11 \end{bmatrix} \text{ and } BA = \begin{bmatrix} 1 & 2 \\ 0 & 1 \end{bmatrix}\begin{bmatrix} 2 & 4 \\ 3 & 5 \end{bmatrix} = \begin{bmatrix} 8 & 14 \\ 3 & 5 \end{bmatrix}$$

Thus $AB \neq BA$ and so the ring is a non-commutative ring.

If I be the unite matrix of the type 2×2 *i.e.*, if $I = \begin{bmatrix} 1 & 0 \\ 0 & 1 \end{bmatrix}$, then $I \in M$. Also we have $AI = A = IA \ \forall \ A \in M$.

$\therefore$ I is the multiplicative identity.

Thus the ring possesses the unit element and we have I = 1 (the unit element of the ring).

Thus mull matrix $O = \begin{bmatrix} 0 & 0 \\ 0 & 0 \end{bmatrix}$ is the additive identity and is therefore the zero element of the ring *i.e.,* O = 0 (the zero element of the ring).

The ring possesses zero divisors. For example if

$$A = \begin{bmatrix} 0 & 1 \\ 0 & 1 \end{bmatrix}, B = \begin{bmatrix} 2 & 3 \\ 0 & 0 \end{bmatrix}, \text{ then } AB = \begin{bmatrix} 0 & 1 \\ 0 & 1 \end{bmatrix}\begin{bmatrix} 2 & 3 \\ 0 & 0 \end{bmatrix} = \begin{bmatrix} 0 & 0 \\ 0 & 0 \end{bmatrix}$$

Thus, the product of the two non-zero elements of the ring is equal to the zero element of the ring.

Example 69(c):

Prove that the set I ($\sqrt{2}$) of number of the form $a + b\sqrt{2}$, with a and b as integers is an integral domain with respect to ordinary addition and multiplication. Is it a field ?

Solution:

We can easily verify that the given system is a commutative ring with unity element $1 + 0\sqrt{2}$. Also $0 + 0\sqrt{2}$ is the zero element of this ring. Now in order to prove that its ring is an integral domain, we should prove that this ring is with out zero divisors.

Let $a + b\sqrt{2}$ and $c + d\sqrt{2}$ be any two elements of this ring. Then

$$(a + b\sqrt{2})(c + d\sqrt{2}) = 0 + 0\sqrt{2}$$

$\Rightarrow$ $ac + 2bd = 0$ and $bc + ad = 0$ and this will happen only when either

$$a = 0 \text{ and } b = 0$$

$\Rightarrow$ $c = 0$ and $d = 0$.

Thus $(a + b\sqrt{2})(c + d\sqrt{2}) = 0 + 0\sqrt{2}$

$\Rightarrow$ either $a + b\sqrt{2} = 0 \Rightarrow e + d\sqrt{2}$.

Thus, the given ring is without zero divisors. Therefore it is an integral domain. But it is not a field. Obviously $5 + 3\sqrt{2}$ is a non-zero element of this ring. Its inverse would have been

$$\frac{1}{5+3\sqrt{2}} = \frac{5-3\sqrt{2}}{\left(5+3\sqrt{2}\right)\left(5-3\sqrt{2}\right)} = \frac{5-3\sqrt{2}}{7} = \frac{5}{7} - \frac{3}{7}\sqrt{2}$$

which is not an element of this ring.

Example 70:

If R is a system satisfying all the conditions for a ring with unit element with the possible exception $a+b=b+a$, prove that the axiom $a+b = b+a$ must hold in R and that R is thus a ring.

Solution:

Since 1 is an element of R, we have

$(a + b)(1 + 1) = a(1 + 1) + b(1 + 1)$ [by right distributive law]

$= (a1 + a1) + (b1 + b1) = (a + a) + (b + b)$ ...(i)

Also $(a + b)(1 + 1) = (a + b)1 + (a + b)1$ [by left distributive law]

$= (a + b) + (a + b)$ ...(ii)

[$\because$ 1 is the unit element]

From (i) and (ii), we get

$(a + a) + (b + b) = (a + b) + (a + b)$

$\Rightarrow$ $[(a + a) + b] + b = [(a + b) + a] + b$ [by associativity of addition]

$\Rightarrow$ $(a + a) + b = (a + b) + a$ [by right cancellation law for addition in R since with the given postulates R is a group with respect to addition].

$\Rightarrow$ $a + (a + b) = a + (b + a)$ [by associativity of addition in R]

$\Rightarrow$ $(a + b) = (b + a)$ [by left cancellation law for addition in R]

Thus addition is commutative in R. Hence R is a ring.

Example 71(a):

The set M of all $n \times n$ matrices with their elements as real numbers (rational numbers, complex numbers, integers) is a non-commutative ring with unity, with respect to addition and multiplication of matrices as the two ring compositions.

Solution:

By matrix property we know that the sum and product of two matrix is also a matrix. Therefore M is closed with respect to addition and multiplication of matrices.

Further we know that

1. $A + (B + C) = (A + B) + C \ \forall \ A, B, C \in M$, since the addition of matrices is an associative the composition
2. $A + B = B + A \ \forall \ A, \in M$ since the addition of matrices is commutative.
3. If O is the null matrix of the type $n \times n$, then $O \in M$ and we have $O + A = A \ \forall \ A \in M$. Hence O is null matrix.
4. To each matrix $A \in M$ there exists a matrix $- A \in M$ such that $(-A) + A = O$ (null matrix). Hence, $-A$ is additive inverse A.
5. $(AB) C = A (BC), \ \forall \ A, B, C \in M$, since multiplication of matrices is associative.
6. $A (B + C) = AB + AC$,

   and $(B + C) A = BA + CA \ \forall \ A, B, C \in M$, since matrix multiplication is distributive with respect to matrix addition.

Hence M is a ring with respect to the two given compositions. The null matrix O of the type $n \times n$ is the zero element of this ring *i.e.*, $O = 0$.

We know that the multiplication of matrices is not in general commutative, therefore the ring is a non-commutative ring. $[n > 1]$

Hence, if I be the unit of the type $n \times n$, then $I \in M$ and we have IA = A AI " $A \in M$. Therefore the matrix 1 is the multiplicative identity. Thus, the ring is with unity and the matrix I is the unity element of the ring *i.e.* I = 1.

Example 71(b):

Show that the set of numbers of the form $a + b \sqrt{2}$, with a and b as rational numbers is a field.

Solution:

Let $R = \{a + b \sqrt{2} : a, b, \in Q\}$.

Let $a_1 + b_1\sqrt{2} \in R$ and $a_2 + b_2\sqrt{2} \in R$. Then

$$a_1, b_1, a_2, b_2 \in Q.$$

We have $(a_1 + b_1 \sqrt{2}) + (a_2 + b_2 \sqrt{2}) = (a_1 + a_2) + (b_1 + b_2) \sqrt{2}$

$\in R$ since $a_1 + a_2, b_1 + b_2 \in Q$.

Also $(a_1 + b_1 \sqrt{2}) + (a_2 + b_2 \sqrt{2}) = (a_1a_2 + 2b_1b_2) + (a_1b_1 + a_2b_1) \sqrt{2}$
$\in R$ since $a_1a_2 + 2b_1b_2$, $a_1b_2 + a_2b_1 \in Q$.

The R is closed with respect to addition and multiplication.

All the elements of R are real numbers and we know that addition and multiplication are both associative as well as commutative compositions in the set of real numbers.

Further we have $0 + 0 \sqrt{2} \in R$ since $0 \in R$

If $a + b \sqrt{2} \in R$, then

$$(0 + 0 \sqrt{2}) + (a + b \sqrt{2}) = (0 + a) + (0 + b) \sqrt{2} = a + b \sqrt{2}.$$

$\therefore$ $0 + 0 \sqrt{2}$ is the *additive identity.*

Again if $a + b \sqrt{2} \in R$, then $(-a) + (-b) \sqrt{2} \in R$ and we have

$$[(-a) + (-b) \sqrt{2}] + (a + b \sqrt{2}) = 0 + 0 \sqrt{2}.$$

$\therefore$ Each element of R possesses additive inverse.

Further in the set of real numbers multiplication is distributive with respect to addition.

Again $1 + 0 \sqrt{2} \in R$ and we have

$$(1 + 0 \sqrt{2})(a + b \sqrt{2}) = a + b \sqrt{2} = (a + b \sqrt{2})(1 + 0 \sqrt{2}).$$

$\therefore$ $1 + 0 \sqrt{2}$ is the multiplicative identity.

Thus R is a commutative ring with unity. The zero element of the ring is $0 + 0 \sqrt{2}$ and the unit element is $1 + 0 \sqrt{2}$.

Now R will be a field, if each non-zero element of R possesses multiplicative inverse.

Let $a + b \sqrt{2}$ be any non-zero element of this ring *i.e.* at least one of a and b is not zero.

$$\text{Then } \frac{1}{a + b\sqrt{2}} = \frac{a - b\sqrt{2}}{\left(a + b\sqrt{2}\right)\left(a - b\sqrt{2}\right)} = \frac{a - b\sqrt{2}}{a^2 - 2b^2}$$

$$= \left(\frac{a}{a^2 - 2b^2}\right) + \left(-\frac{b}{a^2 - 2b^2}\right)\sqrt{2}.$$

Now if a and b are rational numbers, then we can have $a^2 = 2b^2$ only if $a = 0, b = 0$. Since here at least one of the rational numbers a and b is not 0, therefore we cannot have $a^2 = 2b^2$ *i.e.*, $a^2 - 2b^2 = 0$.

$\therefore$ $\dfrac{a}{a^2 - 2b^2}$ and $-\dfrac{b}{a^2 - 2b^2}$ are both rational numbers and at least one of them is not zero.

$$\therefore \left(\frac{a}{a^2 - 2b^2}\right) + \left(-\frac{b}{a^2 - 2b^2}\right)\sqrt{2}$$ is a non-zero element of R and is the multiplicative inverse of a + b $\sqrt{2}$.

Hence the given system is a field.

Example 71(c):

Show that the set R of all real valued continuous functions defined in the closed interval [0, 1] is a commutative ring with unity with respect to the addition and multiplication of functions defined pointwise as follows:

$$(f + g)(x) = f(x) + g(x)$$

and $$(fg)(x) = f(x)\,g(x),$$

where f, g are any two members of R.

Solution:

If f is a real valued function in the closed interval [0, 1], then we mean that f (x) is a real number $\forall$ x $\in$ [0, 1]. We know that the sum and product of two real numbers are also real numbers. Also the sum and product of two continuous functions is also a continuous function. Therefore R is closed with respect to the given compositions.

Now let f, g, h b any three elements of R. Then we make the following observations:

Associativity of Addition: For every x $\in$ [0, 1], we have

$$[(f + g) + h](x) = [(f + g)(x)] + h(x)$$

$$= [(f(x) + g(x)] + h(x) = f(x) + [g(x) + h(x)]$$

[$\because$ f(x), g(x), h(x) are real numbers and addition of real numbers is associative]

$$= f(x) + (g + h)(x) = [f + (g + h)](x).$$

$\therefore$ (f + g) + h = f + (g + h), by the equality of two mappings.

Commutativity of Addition: We have

$$(f + g)(x) = f(x) + g(x) = g(x) + f(x)$$

[$\because$ addition of real numbers is commutative]

$$= (g + f)(x).$$

$$\therefore \quad f + g = g + f.$$

Existence of Additive Identity: Let us define a function e by the rule e(x) = 0 $\forall$ x $\in$ [0, 1]. Then e $\in$ R. Also if f $\in$ R. we have (e + f) (x) = e(x) + f(x) = 0 + f(x) = f(x).

Existence of Additive Inverse: Let $f \in R$. Let us define a function $-f$ by the formula $(-f)(x) = -[f(x)] \ \forall \ x \in [0, 1]$.

Then $-f \in R$ and we have

$[-f + f](x) = (-f)(x) + f(x) = -f(x) + f(x) = 0 = e(x)$.

$\therefore \quad -f + f = e$.

$\therefore$ The function $-f$ is the additive inverse of f.

Associativity of Multiplication: We have

$$[(fg)h](x) = [(fg)(x)]\,h(x)$$
$$= [f(x)\,g(x)]\,h(x) = f(x)\,[g(x)\,h(x)]$$
$$= f(x)\,[(gh)(x)] = [f(gh)](x).$$
$$\therefore \quad (fg)h = f(gh).$$

Distributive Laws: We have

$$[f(g + h)(x) = f(x)\,[(g + h)(x)] = f(x)\,[g(x) + h(x)]$$
$$= f(x)\,g(x) + f(x)\,h(x) = (fg)(x) = (fh)(x)$$
$$= [fg + fh](x).$$

$\therefore \quad f(g + h) = fg + fh$.

Similarly we can prove that $(g + h)f = gf + hf$.

$\therefore$ R is a ring with respect to the given compositions.

Commutativity of Multiplication: We have

$$(fg)(x) = f(x)\,g(x) = g(x)\,f(x) = (gf)(x).$$
$$\therefore \quad fg + gf.$$

$\therefore$ R is a commutative ring.

Existence of Multiplicative Identity: Let the define a function i by the formula $i(x) = 1 \ \forall \ x \in [0, 1]$. Then $i \in R$. If $f \in R$, we have $(if)(x) = i(x)\,f(x) = 1\,f(x) = f(x)$.

$\therefore$ if + f = fi [by commutativity of multiplication]

$\therefore$ The function i is the multiplicative identity

Thus, the ring R is with unity element.

Example 71(d):

If addition and multiplication modulo 10 is defined on the set of integers $R = \{0, 2, 4, 6, 8\}$, prove that the resulting system is a ring with unity. Is it an integral domain ?

Solution:

We have $R = \{0, 2, 4, 6, 8\}$. To prove that $(R, +_{10}, \times_{10})$ is a ring with unity.

First we shall show that $(R, +_{10})$ is an abelian group. The composition table for R for the operation $+_{10}$ is as given below.

$+_{10}$	0	2	4	6	8
0	0	2	4	6	8
2	2	4	6	8	0
4	4	6	8	0	2
6	6	8	0	2	4
8	8	0	2	4	6

We see that, all the entries in the composition table are elements of the set R. Therefore R is closed with respect to $+_{10}$

The composition $+_{10}$ on the set R is commutative. For if a, b, $\in$ R, then

$a +_{10} b$ = least non-negative remainder when a + b is divided by 10

= least non-negative remainder when b + a is divided by 10

$= b +_{10} a.$

The composition $+_{10}$ on the set R is associative. For if a, b, c $\in$ R, then

$a +_{10} (b +_{10} c) = (a +_{10} b) +_{10} c$ because $a + (b + c) = (a + b) + c.$

Existence of Additive Identity: We have $0 \in R$

and $$0 +_{10} a = a = a +_{10} 0 \ \forall \ a \in R.$$

$\therefore$ 0 is the identity for the operation $+_{10}$ and will be the zero element of the ring $(R, +_{10}, \times_{10})$.

Existence of Additive Inverse: From the composition table we see that the inverses of 0, 2, 4, 6, 8 for the operation $+_{10}$ are 0, 8, 6, 4, 2 respectively. Thus each element of R possesses inverse for $+_{10}$.

$\therefore (R, +_{10})$ is an abelian group.

Now we prepare the composition table for R for the operation $\times_{10}$.

$+_{10}$	0	2	4	6	8
0	0	0	0	0	0
2	0	4	8	2	6
4	0	8	6	4	2
6	0	2	4	6	8
8	0	6	2	8	4

We see that all, the entries in the composition table are elements of the set R. Therefore R is closed with respect to $\times_{10}$.

The operation $\times_{10}$ on the set R is associative. For if a, b, c $\in$ R, then $a \times_{10}(b \times_{10} c) = (a \times_{10} b) \times_{10} c$ because a (bc = (ab) c.

Also the operation $\times_{10}$ distributes over the operation $\times_{10}$. For if a, b, c $\in$ R, then

$$a \times_{10}(b \times_{10} c) = (a \times_{10} b) \times_{10} (a \times_{10} c)$$

and
$$(b \times_{10} c) \times_{10} a = (b \times_{10} a) +_{10} (c \times_{10} a).$$

$\therefore$ The algebraic structure $(R, +_{10}, \times_{10})$ is a ring. The zero element of this ring to 0 which is identity for the operation $+_{10}$.

This ring is a ring with unity because it possesses multiplicative identity. From the composition table for $\times_{10}$ we see that 6 $\in$ R is identity for the operation of multiplication modulo 10.

We have

$$6 \times_{10} a = a = a \times_{10} 6, \forall a \in R.$$

Thus, 6 is identity for the operation $\times_{10}$ and is therefore the unity element of the ring $(R, +_{10} \times_{10})$.

The operation $\times_{10}$ on the set R is commutative because the corresponding rows and columns in the composition table of R for the operation $\times_{10}$ are identical. Thus

$$a \times_{10} b = b \times_{10} a, \forall a, b \in R.$$

Also from the composition table of R for the operation $\times_{10}$ we observe that the product of no two non-zero elements of R is the zero element or R. Thus if a, b, $\in$ R, then

$$a \times_{10} b = 0 \Rightarrow a = 0 \text{ or } b = 0.$$

$\therefore$ The ring $(R, +_{10} \times_{10})$ does not possess zero divisors.

Since $(R, +_{10} \times_{10})$ is a commutative ring with unity and without zero divisors, therefore it is an integral domain.

Example 72:

Let C be the set of the ordered pairs (a, b) of real numbers. Define addition and multiplication in C by the equations

$$(a, b) + (c, d) = (a + c, b + d)$$

$$(a, b)(c, d) = (ac - bd, bc + ad).$$

Prove that C is a field.

Solution:

We see that C is closed with respect to the two compositions since a + c, b + d, ac – bd, bc + ad are all real numbers. Now let (a, b), (c, d), (e, f) be any elements of C. Then we make the following observations:

Associativity of Addition: We have

[(a, b) + (c, d)] + (e, f) = (a + c, b + d) + (e, f)

= ([a + c] + e, [b + d] + f) = (a + [c + e], b + [d + f])

= (a, b) + (c + e, d + f) = (a, b) + [(c, d) + (e, f)].

Commutativity of Addition: We have

(a, b) + (c, d) = (a + c, b + d) = (c + a, d + b) = (c, d) + (a, b).

Existence of Additive Identity: We have (0, 0) $\in$ C.

Also (0, 0) + (a, b) = (0 + a, 0 + b) = (a, b).

$\therefore$ (0, 0) is the additive identity.

Existence of Additive Inverse: If (a, b) $\in$ C, then

(–a, – b) $\in$ C. We have

(–a, – b) + (a, b) = (–a + a, – b + b) = (0, 0).

$\therefore$ (–a, – b) is the additive inverse of (a, b).

Associativity of Multiplication: We have

[(a, b) (c, d)] (e, f) = (ac – bd, bc + ad) (e, f)

= ([ac – bd] e – [bc + ad] f, [bc = ad] e + [ac – bd] f)

= (a [ce –df] – b [de + ef], b [ce –df] + a [de + ef])

= (a, b) (ce – df, de + cf) = (a, b) [(c, d) (e, f)].

Distributive Laws: We have

(a, b) [(c, d) + (e, f)] = (a, b) (c + e, d + f)

= (a [c + e] – b [d + f], b [c + e] + a [d + f])

= ([ac – bd] + [ae – bf], [bc + ad] + [be + af])

= (ac – bd, bc + ad) + (ae – bf, be + af) = (a, b) (c, d) + (a, b) (e, f).

Similarly we can show that the other distributive law also holds good.

Therefore C is a ring with respect to the two compositions. The ordered pair (0, 0) is the zero element of the ring.

Commutativity of Multiplication: We have

(a, b) (c, d) = (ac – bd, bc + ad) = (ca = db, da + cb) = (c, b) (a, b).

Existence of Multiplicative Identity: We have $(1, 0) \in C$. If $(a, b) \in C$, then $(a, b)(1, 0) = (a1 - b0,\ b1 + a0) = (a, b) = (1, 0)(a, b)$. Therefore $(1,0)$ is the unity element of the ring.

Existence of multiplicative inverse of each non-zero element of c. Let (a, b) be any non-zero element of C. Then a and b are not both simultaneously zero. If (c, d) is the multiplicative inverse of (a, b), then we should have

$$(a, b)(c, d) = (1, 0)$$

$$\Rightarrow \quad (ac - bd, bc + ad) = (1, 0).$$

By the definition of the equality of two ordered pairs, we have

$$ac - bd = 1 \text{ and } bc + ad = 0.$$

Solving these equations for c, d, we get

$$c = \frac{a}{a^2 + b^2},\ d = \left(-\frac{b}{a^2 + b^2}\right).$$

Now $a \neq 0$ or $b \neq 0 \Rightarrow a^2 + b^2 \neq 0$. Therefore either c or d or both are non-zero real numbers. Thus $\left(\frac{a}{a^2 + b^2}, -\frac{b}{a^2 + b^2}\right)$ is the multiplicative inverse of (a, b). Hence c is a field.

Note: In this question C is nothing but the set of complex numbers defined as ordered pairs of real numbers. Thus, we have proved that the set of complex numbers is a field with respect to addition and multiplication of complex numbers.

Example 73(a):

Show that for a field F, the set of all matrices of the form

$$\begin{bmatrix} a & b \\ 0 & 0 \end{bmatrix}$$

For a, b, $\in F$ is a right ideal but not a lift ideal of the ring of all 2×2 matrices over the field F.

Solution:

Let R be the ring of all 2×2 matrices with elements in the field F for addition and multiplication of matrices as the two ring operations. Let M be the subset of R consisting of matrices of the form

$$\begin{bmatrix} a & b \\ 0 & 0 \end{bmatrix},\ a, b, \in F.$$

First to show that M is a right ideal of the ring R.

Let $A = \begin{bmatrix} a_1 & b_1 \\ 0 & 0 \end{bmatrix}$ and $B = \begin{bmatrix} a_2 & b_2 \\ 0 & 0 \end{bmatrix}$ be any two elements of M.

Then $A - B = \begin{bmatrix} a_1 - a_2 & b_1 - b_2 \\ 0 & 0 \end{bmatrix} \in M$.

$\therefore$ M is a subgroup of the additive group of the ring R.

Now let $U = \begin{bmatrix} w & x \\ y & z \end{bmatrix}$ be any element of R and $A = \begin{bmatrix} a & b \\ 0 & 0 \end{bmatrix}$ be any element of M.

$$\text{Then } AU = \begin{bmatrix} a & b \\ 0 & 0 \end{bmatrix}\begin{bmatrix} w & x \\ y & z \end{bmatrix} = \begin{bmatrix} aw+by & ax+bz \\ 0 & 0 \end{bmatrix} \in M$$

Therefore M is a right ideal of the ring R.

But M is not a left ideal of R, since

$$\begin{bmatrix} 0 & 0 \\ 1 & 0 \end{bmatrix} \in R, \begin{bmatrix} 1 & 1 \\ 0 & 0 \end{bmatrix} \in M,$$

$$\text{and the product } \begin{bmatrix} 0 & 0 \\ 1 & 0 \end{bmatrix}\begin{bmatrix} 1 & 1 \\ 0 & 0 \end{bmatrix} = \begin{bmatrix} 0 & 0 \\ 1 & 1 \end{bmatrix} \notin M.$$

Hence M is not a left ideal of R,

Example 73(b):

If R is ring and $a \in R$ let $T = \{x \in R : ax = 0\}$. Prove that T is a right ideal of R.

Solution:

First we see that U is not empty because

$0 \in R$ is such that $a0 = 0$.

Let x_1, x_2 be any two elements of T. Then $ax_1 = 0$, $ax_2 = 0$.

We have $a(x_1 - x_2) = ax_1 - ax_2 = 0 - 0 = 0$.

$\therefore x_1 - x_2 \in T$.

$\therefore$ T is a subgroup of R under addition.

Now to show that T is a right ideal of R we are to show that $x \in T$, $y \in R \Rightarrow xy \in T$. But $x \in T \Rightarrow ax = 0$. If we show that $a(xy) = 0$, then xy will be an element of T.

We have a (xy) = (ax) y = 0y = 0.

$\therefore$ xy $\in$ T.

$\therefore$ T is a right ideal of R.

Example 73(c):

If m is a fixed integer, the set P of integers given by,

$$P = \{xm : x \text{ is an integer}\}$$

is an ideal of the ring R of all integers.

Solution:

Let $x_1m - x_2m$ be any two elements of P. Then x_1 and x_2 are some integers. We have $x_1m - x_2m = (x_1 - x_2)\, m \in P$ since $x_1 - x_2$ is also an integer.

$\therefore$ P is a subgroup of R under addition.

Now let r be any integer *i.e.*, r be any element of R and xm be any element of P. Then r (xm) = (rn) m $\in$ P since rx is also an integer. Therefore P is a left ideal of R. But R is a commutative ring. Hence P is an ideal of R.

Example 73(d):

Distinguish between subrings and ideals in a ring. Show that the 2-rowed matrices of the form

$$\begin{bmatrix} a & 0 \\ b & c \end{bmatrix}$$

where a, b, c are integers form a subring of the ring of all 2-rowed matrices with integral entries. Is this subring an integral domain?

Solution:

Do yourself.

Example 73(f):

If a, b, c, d are any elements of a ring R, prove that

1. (a – b) (c – d) = (ac + bd) – (ad + bc).
2. a – b = c – d $\Leftrightarrow$ a + d = b + c.

Solution:

1. We have

$$(a - b)(c - d) = (a - b)\,c - (c - b)\,d \qquad [\because \; a(b - c) = ab - ac]$$

$$= (ac - bc) - (ad - bd) = (ac - bc) - ad + bd$$

$= (ac + bd) - bc - ad$

[$\because$ addition is commutative and associative]

$= (ac + bd) - (bc + ad)$.

2. We have $a - b = c - d$

$\Leftrightarrow a + (-b) = c + (-d)$,

$\Leftrightarrow a + (-b) + d = c + (-d) + d$

$\Leftrightarrow a + d + (-b) = c + 0$

[$\because$ in a ring addition is commutative as well as associative and $- d = d = 0$]

$\Leftrightarrow a + d + (-b) = c$ [$\because$ $c + 0 = c$]

$\Leftrightarrow a + d + (-b) + d = c + (-d) + d$

$\Leftrightarrow a + d + 0 = c + b \Leftrightarrow a + d = c + b$.

Example 74(a):

Prove that in the list of axioms for a ring R with unity the axiom demanding commutativity under addition may be omitted.

Solution:

Suppose an algebraic structure (R, + , .) satisfies all the axioms for a ring with unity except the axiom of commutativity of addition. Then we have to show that addition must be commutative on R and thus R must be a ring.

Example 74(b):

Let p be a prime number. Prove that the set of integers I_p,

$I_p = \{0, 1, 2, 3, ..., p - 1\}$

Forms a field with respect o addition and multiplication modulo p.

Solution:

As in the chapter on groups we can prove that $(I_p, +_p)$ is an abelian group. Also I_p is closed with respect to '$\times_p$'. The composition '$\times_p$' is distributive with respect to '$+_p$' as can be proved as below:

Let $a, b, c \in I$.

Then $a \times_p (b +_p c)$

$= a \times_p (b + c)$ [$\because$ $b =_r c \equiv b + c \pmod p$]

= least non- negative remainder when a (b + c) is divided by p

= least non- negative remainder when ab + ac is divided by p

$= (ab) +_p (ac)$

$= (a \times_p b) +_p (ac)$ $\qquad [\because \ ab \equiv b \times_r b \pmod p]$

$= (a \times_p b) +_p (a \times_p c)$.

Similarly we can prove the other distributive law.

1 is the identity element for '$\times_p$'.

The zero element of the ring $(I_p +_p \times_p)$ is 0. As in groups we can prove that each non-zero of I_p has multiplicative inverse.

In short the non-zero elements of I_p from an abelian group with respect to '$\times_p$'. Therefore $(I_p, +_p, \times_p)$ is a field.

Note: If p is not prime, then this ring will have zero divisors and so it cannot be a field.

Example 74(c):

If R is a commutative ring, prove by induction that

$$(a + b)^n = a^n + {}^nC_1 a^{n-1}b + {}^nC_2 a^{n-2}b^2 + ... + b^n$$

for every positive integer n; here a and b are elements of R.

Solution:

We have $a + b = a^1 + b^1$.

Now $\quad (a + b)^2 = (a + b)(a + b)$

$= a^2 + ab + ba + b^2$, by list. Laws

$= a^2 + ab + ab + b^2$

$[\because \ ab = ba$, the ring R being commutative$]$

$= a^2 + 2ab + b^2$

$= a^2 + {}^2C_1 ab + b^2$.

Thus the theorem is true for $n = 2$.

Now assume that the theorem is true for any positive integer n *i.e.*,

$$(a + b)^n = a^n + {}^nC_1\, a^{n-1}b + ... + {}^nC_r\, a^{n-1}b^r + {}^nC_{r+1}\, a^{n+r-1}\, b^{r+1} + ... + b^n.$$

Then $\quad (a + b)^{n+1} = (a + b)(a + b)^n$

$= (a + b)(a^n + {}^nC_1\, a^{n-1}b + {}^nC_2\, a^{n-2}b^2 + ...$

$+ {}^nC_r\, a^{n-1}b^r + {}^nC_{r+1}\, a^{n-r-1}\, b^{r+1} + ... + b^n)$

$= a^{n+1} + (ba^n + {}^nC_1 a^n b) + ...$

$+ ({}^nC_r\, ba^{n-r}\, b^r + {}^nC_{r+1}\, a^{n-r}\, b^{r+1}) + ... + b^{n+1}$.

Since the ring R is commutative, therefore

$ab^n = a^n b,\ ba^{n-r} = (ba^{n-r})\, b^r = (a^{n-r}\, b)\, b^r$

$= a^{n-r}\, bb^r = a^{n-r}\, b^{r+1}.$

Also $1 = {}^nC_0,\ {}^nC_r + {}^nC_{r+1} = {}^{n+1}C_{r+1}.$

Hence $(a + b)^{n+1} = a^{n+1} + ({}^nC_0 + {}^nC_1)\, a^n b + \ldots$

$+ ({}^nC_r + {}^nC_{r+1})\, a^{n-r}\, b^{r+1} + \ldots + b^{n+1}$

$= a^{n+1} + {}^{n+1}C_1\, a^{(n+1)-1}\, b + \ldots$

$+ {}^{n+1}C_{r+1}\, a^{(n+1)-(r+1)}\, b^{r+1} + \ldots + b^{n+1}.$

Thus, the theorem is true for n + 1, if it is true for n. But it is true for n = 1 and n = 2. Hence by induction it is true for all positive integral values of n.